Service-Learning in Technical and Professional Communication

The Allyn and Bacon Series in Technical Communication

Series Editor: Sam Dragga, Texas Tech University

Thomas T. Barker
Writing Software Documentation: A Task-Oriented Approach

Carol M. Barnum
Usability Testing and Research

Deborah S. Bosley
Global Contexts: Case Studies in International Technical Communication

Melody Bowdon and J. Blake Scott
Service-Learning in Technical and Professional Communication

Paul Dombrowski
Ethics in Technical Communication

David K. Farkas and Jean B. Farkas
Principles of Web Design

Laura J. Gurak
Oral Presentations for Technical Communication

Sandra W. Harner and Tom G. Zimmerman
Technical Marketing Communications

Richard Johnson-Sheehan
Writing Proposals: Rhetoric for Managing Change

Dan Jones
Technical Writing Style

Charles Kostelnick and David D. Roberts
Designing Visual Language: Strategies for Professional Communicators

Carolyn Rude
Technical Editing, Third Edition

Gerald J. Savage and Dale L. Sullivan
Writing a Professional Life: Stories of Technical Communicators On and Off the Job

Service-Learning in Technical and Professional Communication

Melody Bowdon
University of Central Florida

J. Blake Scott
University of Central Florida

New York San Francisco Boston
London Toronto Sydney Tokyo Singapore Madrid
Mexico City Munich Paris Cape Town Hong Kong Montreal

Senior Vice President/Publisher: Joseph Opiela
Marketing Manager: Christopher Bennem
Production Manager: Denise Phillip
Project Coordination, Text Design, and Electronic Page Makeup: WestWords, Inc.
Cover Design Manager: John Callahan
Cover Designer: Teresa Ward
Manufacturing Buyer: Roy Pickering
Printer and Binder: Hamilton Printing
Cover Printer: Coral Graphics

Library of Congress Cataloging-in-Publication Data

Bowdon, Melody,
Service-learning in technical and professional communication / Melody Bowdon, J. Blake Scott.
p. cm.—(Allyn and Bacon series in technical communication)
Includes bibliographical references and index.
ISBN 0-205-33560-8
1. Communication of technical information. 2. Student service. I. Scott, J. Blake, II. Title. III. Series.

T10.5 .B67 2002
601′.4—dc21

2002016298

Please visit our website at http://www.ablongman.com

ISBN 0-205-33560-8

1 2 3 4 5 6 7 8 9 10—HT—05 04 03 02

We dedicate this book to our students and our teachers.

CONTENTS

FOREWORD by the Series Editor

The Allyn & Bacon Series in Technical Communication is designed for the growing number of students enrolled in undergraduate and graduate programs in technical communication. Such programs offer a wide variety of courses beyond the introductory technical writing course—advanced courses for which fully satisfactory and appropriately focused textbooks have often been impossible to locate. This series will also serve the continuing education needs of professional technical communicators, both those who desire to upgrade or update their own communication abilities as well as those who train or supervise writers, editors, and artists within their organization.

The chief characteristic of the books in this series is their consistent effort to integrate theory and practice. The books offer both research-based and experienced-based instruction, describing not only what to do and how to do it but explaining why. The instructors who teach advanced courses and the students who enroll in these courses are looking for more than rigid rules and ad hoc guidelines. They want books that demonstrate theoretical sophistication and a solid foundation in the research of the field as well as pragmatic advice and perceptive applications. Instructors and students will also find these books filled with activities and assignments adaptable to the classroom and to the self-guided learning processes of professional technical communicators.

To operate effectively in the field of technical communication, today's students require extensive training in the creation, analysis, and design of information for both domestic and international audiences, for both paper and electronic environments. The books in the Allyn & Bacon Series address those subjects that are most frequently taught at the undergraduate and graduate levels as a direct response to both the educational needs of students and the practical demands of business and industry. Additional books will be developed for the series in order to satisfy or anticipate changes in writing technologies, academic curricula, and the profession of technical communication.

Sam Dragga
Texas Tech University

PREFACE

Our goal in writing this book was to create a resource for helping a range of technical and professional communication students and teachers implement service-learning projects in campus and larger communities. We've addressed this book to both advanced and beginning students, to both veteran service-learning teachers and those trying it for the first time.

We begin with three chapters that define and explain our approach to service-learning and develop a rhetorical toolbox for implementing this approach. The remainder of the book is loosely organized around the process of developing, executing, and evaluating service-learning projects. These process chapters teach rhetorical strategies, ethical concerns, genre conventions, and style principles in an integrated, contextualized way. We supplement our discussion of rhetoric and ethics with heuristics for analyzing the larger cultural effects of service-learning projects. Perhaps most importantly, the book is grounded in the related practices of ethical deliberation and civic action.

Chapters Four through Ten best illustrate our integrated approach. Each of these chapters provides guidance and problem-solving tips about managing a phase of your service-learning project, presenting you with the collective wisdom that we and others have generated from years of trial and error. Each chapter also focuses on a particular genre that will help you manage your project. These common genres of technical and professional writing are discussed in terms of your project and other workplace examples to give you a broader understanding of their functions. Our discussion of each genre also includes more focused instruction on a style topic especially relevant to that genre and phase of the project. The chapters end with invention activities, usually collaborative and often requiring library or Web research. We encourage you to adapt or add to these as you see fit.

Although the basic process we present is for a class working in groups on a semester-long project, we've designed the book to be helpful for individual and smaller projects as well. Courses working with already-proposed projects, for example, might skip parts of Chapters Four and Five. We offer advice on how to adapt assignments throughout the book. We also don't want to imply that the process we explicate must be followed in a strictly linear way. You'll notice that several of the chapters take you back to Chapter Three, that Chapter Six on Managing Your Collaboration can be applied throughout the process and might be read early in the semester, and that we discuss evaluation as an ongoing activity. Finally, don't feel intimidated by the sheer number of major and supplementary assignments we discuss. We realize that most teachers will assign only some of these, although we encourage you to read all of the material to obtain a better sense of the range of technical and professional communication tasks and to learn as much as possible from others' experiences with service-learning.

Throughout the book, you'll also find sidebars called "Other Voices" and "Student Voices." These sidebars, meant to enhance our examples and advice, include short narratives and insights from just a few of the many students, teachers, and agency sponsors with whom we've worked in service-learning projects over the years. We think you'll relate to these stories, and we know you'll benefit from their wisdom.

Finally, the book's appendixes include three sample student projects that show our approach in action. The projects were completed by students in undergraduate technical and professional writing courses in regular semesters and in a condensed summer term. Each of these projects includes some of the process assignments such as the proposal and evaluation report as well as part of the final project the students produced for the agency. These samples aren't intended to be models to emulate but fuller examples that show what your larger portfolio of work might look like. We do expect you to mine them for strategies, however. In particular, you might use them to develop ideas for possible agencies and projects, review criteria for effective and ethical communication, and practice workshopping your own texts.

We've written this book in the spirit of collaboration and civic engagement, and we therefore invite you to send us your own service-learning insights and projects at the end of the semester. If you are anything like our students, you'll gain much more than rhetorical strategies and communication skills from taking on our challenge. You'll also have a positive impact on your community.

Acknowledgments

If you glance through this book you can quickly see that it draws upon the experiences and insights of a vast network of people. Of course we cannot name each of the hundreds of students with whom we have worked during the last ten years and whose hard work and creativity we have tapped into as we've written this text, but we want to begin our acknowledgments by honoring them. They have been our teachers.

We must also thank our more traditional teachers and mentors who have shaped our thinking about teaching and writing: David Mair, who introduced us to theories of technical communication and who allowed two wildly enthusiastic but inexperienced and naïve MA students to team teach a service-learning course for the first time in 1992; Kathleen Welch, who helped us lay the foundation for our rhetorical approach with her electrifying courses and intellectual guidance; Sharon Crowley, Kevin Davis, Sam Dragga, Theresa Enos, Gwendolyn Gong, Catherine Hobbs, Thomas Miller, J. Fred Reynolds, Jack Selzer, John Warnock, and Tilly Warnock, who guided our thinking and encouraged us to explore ideas in the classroom as students and as teachers.

Friends and colleagues who shaped our thinking about service-learning and teaching writing: Andy Alexander, Sid Dobrin, Debbie Hawhee, Julie Jung, Michael

Moore, Chere Peguesse, Summer Smith, Mark Thompson, Dawn Trouard, Beth Young, Kakie Urch, and Adrian Wurr.

Thanks to all of those who enriched this book by contributing their "voices" for the Other Voices and Student Voices sections; the students who allowed us to share their classwork and their stories as examples throughout the text; the teachers who class tested the book for us—Maggie Boreman, Mary Ellen Gomrad, Michelle Manning—and their students; friends and draft readers Heather Eaton and John F. Schell.

The community representatives who have worked with our students over the years and have often become friends, including Chris Hammon, Jessa Jones, Kathleen Maroney, and Arthur Padilla.

Thanks to our editor, Joe Opiela, and editorial assistant, Julie Hallett, who helped us navigate the complex experience of putting together a project as large as this. We appreciate our peer reviewers, who spent extensive time reading and responding to many drafts of this book and offered invaluable insights into its development: Karla Armbruster, Webster University; Valerie Balester, Texas A&M University; Thomas Deans, Kansas State University; Nancy DeJoy, Millikin University; Dennis Lynch, Michigan Tech University; Beverly Zimmerman, Brigham Young University.

Our families, who provided every kind of support imaginable to us during this project: Arlene, Boyce, and Gina Bowdon; Rob, Marlena, and Hannah Croll; Jim DeVine; Rodney Gentry; Jamie and Sterling Harrell; Marthalie, Rick, Phaedra, Noel, and Carly Johns; Angie Jones; Mildred Luttrell; John and Ray Scott. Angie and Marthalie provided valuable copyediting assistance with three chapters.

And finally, we want to thank our series editor, Sam Dragga, who has shaped our thinking as writing teachers for the past ten years and who believed in us enough to make this book happen.

Melody Bowdon
J. Blake Scott

What Is Service-Learning?

In 1999, six students at the University of Central Florida created documents for their local zoo as a class project for a Writing for Business Professionals course. The group of sophomores, juniors, and seniors from a range of majors and backgrounds shared interests in animal conservation and environmental education. Together, they helped an organization with similar values enhance educational programs and raise community support for conservation efforts. This involved producing educational materials, an informational web site, and promotional brochures, flyers, and signs. Creating these various texts required the students to apply practical and theoretical concepts covered in their course. They analyzed "real" audiences such as zoo employees and patrons who had vested interests in the success of the project. They adapted writing and design principles to the documents' specific purposes, audiences, and uses. They collaborated with each other and zoo personnel in managing a major project from the initial brainstorming phase through the successful implementation of a new marketing campaign.

These students were participating in service-learning, the kind of approach we'll be discussing throughout this book. They connected with a local organization, produced writing that helped meet this organization's needs, and contributed to the larger community. They acted as writing consultants on a "real-world" project that enabled them to study the functions and effects of their writing outside the classroom.

When these students first learned that they were taking a course that involved "service-learning," some had a reaction to which you may be able to relate. This term sounds a lot like community service, an experience you may associate with punishment for a speeding ticket or something you do on your own time in connection with a church or other civic group. How could such an activity be justified as part of a writing class? Since the early 1990s, service-learning has been a growing field for research and teaching across academic disciplines. Today, this kind of approach is being used on college campuses across the country to enhance students' educational experiences and improve local communities.

At its core, service-learning is a hands-on approach that uses community service as a vehicle for teaching specific course-based skills and strategies. Service-

STUDENT VOICES 1.A First Impressions

Amy: I had never heard about service-learning before, and now I have read about it in two courses in the same semester. I can understand why service-learning is considered to be a means to help students focus on their writing skills, and it seems to me that, when used properly, service learning can help to give purpose to writing assignments.

Liz: This whole "service-learning" thing is all new to me. I've never taken a writing class that approaches professional writing from this standpoint, and I'm curious to see where it takes me throughout the semester. Learning about previous students' work has reassured me that there is a whole world out there for writers to be involved in and so many areas in which we are needed. This approach to the class seems like a great way of keeping one foot in the real world instead of being sucked into the university bubble and away from the outside world.

learning combines community service and academic learning and combines real-world action with critical reflection (Rubin 307). To support funding for this kind of learning experience, Congress passed the National and Community Service Trust Act of 1993. Thomas Huckin summarizes the definition of service-learning set forth by the act in the following way: "Service-Learning is a method by which students learn through active participation in thoughtfully organized service; is conducted in, and meets the needs of the community; is integrated into and enhances the academic curriculum; includes structured time for reflection and helps foster civic responsibility" (50).

Three Models of Service-Learning in Writing Courses

Service-learning programs were first developed for higher education in the 1960s and early 1970s, largely because of student activists' efforts. The approach emerged from the field of experiential education, which included internships and cooperatives used to give students hands-on, real-world experience in negotiating concepts that can't be taught effectively only in the classroom. Most of us have tried to learn something from a book and found little success until we could actually deploy our training in a real situation. Some experiential learning students work as paid interns, and others gain relevant experience and contacts in their fields by volunteering for local businesses or nonprofit agencies. A future civil engineer may offer to help build a new community park to learn about how major projects are completed. A student planning to be a social worker may spend time at a local domestic violence shelter to help others and to learn about what her or his future career may involve.

Although it certainly makes sense for students to take on projects directly related to their majors in internships, coops, and capstone courses, many students from a wide range of majors engage in service-learning in writing courses, too.

Writing teachers tend to use three basic models for this kind of work. The first and perhaps most common approach is to have students do nonwriting work for nonprofit agencies and write *about* this work, often in journals and on listserves. Much of the writing, then, is reflective and expressive, meaning that students mostly describe their thoughts and feelings about their experiences. In a course like this, students might also produce letters to the editor or other persuasive documents that address issues they've learned about through their service.

The second approach writing teachers have developed might be thought of as a more academic extension of the first. Students study literacy and schooling, environmental conservation, or some other theme, analyzing various related readings.

OTHER VOICES 1.A Michael Moore

Michael Moore is pursuing a PhD in technical communication at Michigan Technological University. He has taught technical and professional writing courses with a service-learning emphasis for several years at MTU and at the University of Arizona.

I've never been a big fan of textbooks—neither as a student nor as a teacher, finding that they benefited me most when I was able to sell them back at the end of the term. But a textbook designed to promote and guide us through a service-learning curriculum?

This is encouraging.

My commitment to a service-learning curriculum is based on my understanding of *praxis*, a term and concept that encourages us to constantly reflect on our actions and on the effects that our actions have on others.

For example: education and community. We learn at school how to write, design, revise, analyze, persuade, revise again, and give presentations. If we invite members of the community in which we live, work, and learn to join us in that process, we can quickly and productively broaden our understanding of the effects our work can have.

Students in my composition, design, and technical communication courses that incorporate service-learning have collaborated with community members and agencies to research and write successful grants for a women's shelter, a senior-citizen agency, and a volunteer fire department; others have facilitated intergenerational writing groups between fourth graders and residents of a retirement home; others have worked in partnership with a social service agency to design and build interactive and informational websites that are accessible to a range of potential users.

"So what?" a rigorous and reflective student or teacher might ask. "Who benefits?" "What are the long-term implications of these activities?" "What kind of learning is this?"

Well, that series of questions is part our curriculum now. If we can answer some of those questions together, we'll have gone a long way toward understanding the productive effects of our communication, our responsibility and relationship with our communities, and some ideas about the ends of our education. I hope that you enjoy your exploration of these questions as you pursue your own service-learning projects.

The service that students perform (e.g., tutoring, instituting a recycling program) is also related to the theme but is not centered on writing. The writing that students produce in such a course is usually more academic and asks them to relate, in a term paper or similar assignment, the concepts in the readings to their service experiences to gain a fuller understanding of the theme. In both the first and second models, then, the writing the students do is mostly for themselves and their teachers.

The third model, and the one we advocate in this book, is sometimes called the Stanford model, as it was developed for some of Stanford University's first-year writing classes in the early 1980s. In fact, we first became interested in service-learning after hearing speakers from Stanford discuss the benefits of their approach. In the program we heard about, students mainly wrote *as* their community service rather than *about* it. The work they did for sponsoring organizations was writing. Instead of setting up campus recycling centers for a solid waste agency, for example, students might design informative pamphlets and flyers for the agency to distribute to campus groups and students. This type of writing is primarily public rather than expressive or academic. Such an approach also requires students to do some reflective writing, but this is not the primary focus of the course.

We contend that this third model is the most appropriate for a writing course, as it most clearly connects students' service to the concepts of the course. For a technical or professional writing course, these goals might include adjusting one's style for a range of technical and professional audiences, following the conventions of different genres, applying principles of visual design, and effectively collaborating with other writers. To apply these concepts, students need to be

STUDENT VOICES 1.B The Stanford Model

Kelly: The type of service-learning I find most appropriate is the Stanford model, which our class will be taking part in. In this model the emphasis is on the actual writing and creation of documents rather than completion of a work task. Basically, we are not only attempting to help the community, but also to expand our knowledge of technical and professional writing. One of the major upsides of service-learning is that all our hard work will not sit idle at the end of the semester or, worse yet, end up discarded. Hopefully someone will be able to enjoy the fruits of our labor.

Bryan: My view of service-learning is that it's yet another method by which to apply textbook logic to a "real world" situation. Service-learning is an interesting take on the standard "know your audience" approach. The fact that the class will aim its schoolwork to meet—or at least address—the needs of an organization almost guarantees that the classwork is applicable to a real-life situation. The benefits are greater when we practice the Stanford model—to write as the chosen community service rather than about it. Besides, I'm for any endeavor that allows me to network on school time. And I like the idea that this will help us to reflect critically on our work. As someone who has worked under daily deadlines for more than 10 years, I can attest to the lack of available free time needed to consider how my work has influenced me—much less the community.

engaged in technical or professional writing. We will now return to the example of the zoo group's work to explain the main features of a good service-learning project in technical or professional communication.

Attributes of a Service-Learning Project

First, *service-learning relates directly to course goals.* This is the main reason we advocate the Stanford model of writing technical or professional documents *as* service. Service-learning isn't an extra assignment tacked on to course requirements; it is a set of meaningful learning experiences that let students see academic concepts in action. The students who worked on the zoo project applied the strategies and concepts they studied in their class to each task they performed as part of their project. They analyzed their audiences and rhetorical situations (see Chapter Three), learned about genres of professional writing, applied writing and design principles, and revised their texts. In addition to implementing rhetoric and writing strategies, they learned how to identify and develop community connections and to work as a team to meet a series of deadlines.

Second, *service-learning addresses a need in the community.* Instead of being based on artificial scenarios or cases, service-learning assignments arise organically out of actual situations that call for some type of communication. As a nonprofit agency, the zoo didn't have a great deal of personpower available to produce new educational or marketing materials. The park's resources were primarily used to perform the conservation and education efforts that its public funding supported. Thus, the students' writing helped meet the organization's needs. More importantly, though, the students' work addressed larger community-level problems—lack of education about animals and their conservation as well as lack of awareness about a rich local resource and its programs.

We should explain here that we use the phrase *address a need in the community* rather than *meet a need in the community* to move away from the notion of the service-learner as a crusader who can simply march in and immediately handle an organizational or community problem. Arising out of specific historical circumstances, such problems are often quite complex and are not easily solved. This is why service-learning requires students to study the problem in its larger historical context, collaborate with both organization personnel and community members, and work within the organization's guidelines and constraints.

It's also important for students to respect the communities whose needs they are working to meet. When students are working with people in difficult situations such as homelessness, it can be tempting to succumb to what we call "the seduction of empathy." In other words, students might tell themselves that they know what it is to be homeless because they've spent some time in a shelter or they've done a great deal of research on the topic. Be sure to maintain a realistic attitude about this kind of experience, recognizing that no matter how much we are working to help another person or group of people, we can never fully understand another's life experience. We can, though, be supportive members of a community.

Third, *service-learning involves developing reciprocal relationships between the college or university and the communities in which it is embedded.* Although collaboration between colleges or universities and local community organizations makes sense in many ways, it often fails to happen simply because no one takes an initiative. Service-learning projects can provide that first contact that can lead to an ongoing, mutually beneficial relationship. Through their successful work for the zoo, the students in the Writing for Business Professionals course opened the door for future students to collaborate on zoo writing projects or even serve as writing interns there.

We have learned from experience just how crucial it is to ensure that the course–organization relationship is reciprocal. It is all too easy for students to become involved with projects that are simply too large or involve too much non-writing work (Huckin 55). It can also be easy for an organization to spend too much time supervising student groups or not obtain usable work from students. Service-learning projects should incorporate the goals and values of both the course and the sponsoring organization. This is why we will carefully guide you through the process of negotiating your service-learning project with your teacher and a contact person from the organization.

Linda Flower and her colleagues at the Community Literacy Center in Pittsburgh point to another important reciprocal relationship—that of the writers and the larger community. Student writers may not only need to negotiate their tasks and texts with their sponsoring organization; they may also need to collaborate with the community members their texts will reach. This enables those who are "served" to become more active contributors to the writing and to the larger community problem it addresses (107). The zoo project students solicited community members' feedback about the purposes, content, and design of their promotional texts.

Finally, *service-learning involves critical reflection on the student's part.* When pursuing school and workplace projects, we are often in such a rush to meet deadlines that we don't take time to reflect critically on how our work has shaped us as thinking and feeling people. As Chris Anson explains, "Theories of service-learning value reflection for helping to create the connection between academic coursework and the immediate social, political, and interpersonal experiences of community-based activities" (167). The practical nature of most technical and professional communication makes it easy to overlook the values associated with it and the effects enabled by it. Yet, good service-learning work is more than rhetorically effective—it is also ethical and beneficial to the community.

The students who worked on the zoo project spent time writing and talking about how their service-learning experiences (e.g., inventing, writing, collaborating) affected them. For example, students reflected on the process of negotiating the features of the texts with each other, zoo personnel, and prospective readers. Each student had to think about her or his development as both a writer and a group member and about her or his values as both a writing consultant and a local community citizen. Students had to work together to move from individual reflection to group deliberation to find solutions to challenges they faced.

What Service-Learning Is and Isn't

Even after reviewing the main components of a service-learning project, you may find it difficult to envision just what kinds of activities might fall into this category. Because service-learning will be a required part of a course for most of you, it is important that you understand the kinds of activities in which you will likely be participating. At this point, you may want to review some of the sample student projects in the book's appendixes to obtain more concrete ideas about possible projects.

Some of you may choose to do general volunteer work for an organization (or agency, we use these terms interchangeably) with which you've connected to get a feel for the organization's values, learn about the clientele, or better understand its services. Others may believe very strongly in the cause your agency addresses and may choose to augment your work as writers with other kinds of service. It's important that we clarify up front, however, that the model we suggest here will not *require* you to become volunteers or advocates. Rather, we invite you to become, in a sense, unpaid writing consultants for your selected organization.

To clarify what that role entails, in this section we will provide some specific examples of the kinds of duties that would and would not be appropriate as part of a service-learning project. Some examples of activities that would *not,* in and of themselves, constitute service-learning according to our model include:

- Performing original research for the organization—interviewing clients, testing water samples, and so on.
- Doing clerical work for the agency—filing, answering telephones, or stuffing envelopes.
- Providing technology training or support for agency representatives—teaching them to use computer software or backing up their data files.
- Engaging in client services—taking care of children or driving a van for an outing.
- Dealing with such custodial duties as mowing lawns or cleaning kitchens.

Although these kinds of activities are probably not related to the major goals of your writing course, they could be transformed into activities that are. More communication-oriented service-learning assignments might involve the following:

- Writing up or editing descriptions of the agency's research or service activities in annual reports, grant proposals, research summaries, and so on.
- Compiling a manual for clerical volunteers in the agency. Nonprofit agencies especially tend to have high turnover among office workers. A clearly written manual that provides staff members with easily accessible guidelines for standard procedures would make this situation easier to handle.
- Designing training materials for the use of office equipment. Even experienced agency employees may be unfamiliar with recently donated or purchased computer hardware and software, copying machines, or

audio/video equipment. Again, clear instructions for using these tools could alleviate office stresses and help to increase office productivity.
- Producing client services materials like brochures, newsletters, and websites. Like the students who worked for the zoo in the example at the beginning of this chapter, you may work with an agency whose staff members simply don't have the time to produce marketing materials.

The key to the Stanford model for service-learning that we mentioned above is *focusing on writing as the service you provide to the agency.* This is the type of project we'll encourage you to develop in the chapters that follow. This is not to say that such a project will not require you to engage in any nonwriting activities, however, just that these activities should be directly connected to the writing. Perhaps even more than most other types of writing assignments, service-learning tasks require research, collaboration, and project management.

As part of their work for the zoological park, the students mentioned above had to complete the following nonwriting activities:

- Each writer visited the park on at least one occasion. Some of the members made site visits every week or two during the period in which they were producing documents. On these visits the students explored the park and got a feel for what it had to offer and in what areas it needed to improve. This allowed the students to develop a sense of the agency's values and needs, which made writing on their behalf more possible.
- The students talked with agency representatives, learning about the organization's goals and priorities. Because they recognized the importance of cooperation on the project, they did a lot of listening. They took notes and reviewed them together.
- They studied the zoo's existing promotional and educational materials. They paid attention to the design and content of these documents and considered what such features might suggest about the zoo's attitude toward outreach. They also studied the promotional material for other similar parks in the area, gathering ideas about effective and ineffective marketing strategies.
- They researched their audience. In Central Florida, home of Mickey Mouse and Shamu, residents and tourists have many entertainment options. These students spent time examining the reasons why people choose to visit the zoo, what they feel it represents in their community, and so on. This helped them to identify the kinds of materials they would produce and the messages they wanted to convey.
- They researched logistical considerations such as cost of printing and the availability of web space.
- They collaborated on all phases of the project. Throughout the semester the students divided up the work involved—writing and nonwriting. They took turns leading and recording the results of meetings. They stayed in touch via email to keep all the group members aware of project progress.

Like many of you, the students mentioned above had no special training in working for a local organization. They drew on their common interests and worked as local intellectual activists to become more effective and reflective civic writers and to help solve a community problem. In the chapters that follow and in the appendixes of sample projects, you will learn about the experiences of many students who have produced texts for nonprofit agencies, businesses, and campus groups as part of their work for technical and professional writing courses. You will learn about the kinds of documents they produced, the challenges they faced, and the ways in which you can draw on their experiences to pursue your own successful service-learning project.

Activities

1. Visit the following websites to develop a sense of how others in writing studies and higher education more generally have defined service-learning. What do they specify as service-learning's main components and characteristics? How, in particular, do they seem to define "reflection" as a component?

American Association for Higher Education	http://www.aahe.org
National Service-Learning Clearinghouse	http://www.servicelearning.org
Campus Compact	http://www.compact.org
Bentley College	ecampus.bentley.edu
National Council of Teachers of English	http://www.ncte.org/service

2. Go to the website of the National Council of Teachers of English, and follow the links to composition programs that use service-learning in professional and technical writing courses. In addition, read Thomas Huckin's article "Technical Writing and Community Service." Compare the ways in which these courses incorporate service-learning. Do they have students write *about* their service experiences, write about the *topic* of their experiences, write *as* their service experiences, or some combination of the three? Make a list of the kinds of service-learning assignments they include.
3. Search your campus website for references to service-learning. Find out if your school is part of the Campus Compact, a national organization of university presidents who are committed to supporting service-learning and related activities. Make a list of instructors whose websites or syllabi refer to service-learning. Interview one or more of these teachers to learn about their interests in service-learning and their reasons for using it as a teaching tool.
4. Conduct an informal survey among friends and acquaintances about their first responses to the term *service-learning*. See what kinds of associations people have with this word and what they imagine a course that integrates it might involve. Compare your findings about other people's impressions with your own sense of the two separate terms (*service* and *learning*) that make up the term.

Works Cited

Anson, Chris M. "On Reflection: The Role of Logs and Journals in Service-Learning Courses." *Writing the Community: Concepts and Models for Service-Learning in Composition.* Eds. Linda Adler-Kassner, Robert Crooks, and Ann Watters. Washington, D.C.: American Association for Higher Education, 1997. 167–80.

Flower, Linda. "Partners in Inquiry: A Logic for Community Outreach." *Writing the Community: Concepts and Models for Service-Learning in Composition.* Eds. Linda Adler-Kassner, Robert Crooks, and Ann Watters. Washington, D.C.: American Association for Higher Education, 1997. 95–118.

Huckin, Thomas N. "Technical Writing and Community Service." *Journal of Business and Technical Communication* 11 (1997): 49–59.

National and Community Service—Roles for Higher Education: A Resource Guide. Corporation for National and Community Service. 1994.

National Community Service Trust Act of 1993. Pub L. 103–83. 21 Sept. 1993. Stat. 107. 785–923.

Rubin, Sharon. "Institutionalizing Service-Learning." *Service-learning in Higher Education: Concepts and Practices.* Barbara Jacoby and Associates. San Francisco: Jossey-Bass Publishers, 1996. 297–316.

Chapter 2

Service-Learning in Technical and Professional Communication

When most people hear the phrase *technical and professional communication,* they probably imagine indecipherable instructions for using a VCR or an impersonal form letter from a business. In both academia and in the workplace, people have a range of ideas about how these terms connect with each other. Some people consider professional communication to be an umbrella term and think of technical writing as one subcategory among others, like legal or medical writing. Others consider technical communication to be the primary term and see professional writing, which might include tasks such as creating memos and letters, to be one part of technical communication. Still others think of technical and professional communication as separate but closely related areas of study that emphasize somewhat different concerns. Technical writing can be thought of as more oriented toward producing texts that convey scientific or technological information. This might include instructions for using a tool or a report on the findings in a scientific experiment. Professional writing can be thought of as more related to organizational communication, with more emphasis on interpersonal correspondence than on conveying complex data. This might include something like a business's annual report or a memo evaluating an employee's performance. In any case, we believe that service-learning is an appropriate approach for teaching this range of skills. The samples and assignments you'll see throughout this book will come from both types of classes. Your instructor may focus on one or the other approach more narrowly, but the two have many characteristics in common, and these will be the focus of the book.

Mary Lay and her coauthors define technical communication as "... applied communication, communication designed to perform specific tasks or help the audience solve specific problems" (10). Technical and professional writers act as audience advocates; they are often liaisons between engineers or physicians or CEOs or other technical specialists and lay audiences. Technical and professional writing, therefore, are *audience centered,* or designed to address audience needs. They are also *subject oriented;* that is, their primary purpose is to convey information.

Throughout this book, you'll read about the projects of many writing students who have designed a wide range of technical and professional texts including letters, résumés, reports, proposals, instructions, brochures, and informational websites.

In this chapter, we'll discuss ways in which service-learning can help you to develop as a technical or professional writer. As we present some of the advantages of this approach for the students, collaborating agencies, and colleges involved, we'll invite you into a professional conversation about some of the most complex questions and challenges facing the field today. We'll explain some of the benefits of the service-learning approach and provide an overview of the three kinds of sites where service-learning can take place within our model. To begin, we'll introduce you to one student project that will allow us to ground our discussion in a concrete scenario.

When police lieutenant Bill Wood was assigned to produce a service-learning project for a graduate class in professional writing, it didn't take him long to choose a focus. Several years before, when he was in school and didn't have children of his own, Bill had served as a Big Brother to a boy in his area. The experience meant a lot to him and his Little Brother, and in subsequent years he had often thought about ways in which he might be able to contribute once again to the agency's worthwhile program. He contacted the group and learned that although most of the young girls needing an adult friend were already assigned to someone, there was a waiting list of more than 90 boys in the area who wanted and needed a Big Brother; there just weren't enough volunteers available to work with them. Clearly, the agency was facing a recruitment crisis. Remembering how gratifying his own experience was for him as a young man, Bill set out to acquire funding for a recruitment campaign targeted at undergraduate students at his university. He researched costs for publicity venues like billboards, movie theater slide shows, and newspaper ads. He negotiated discounted costs for each of these services and designed promotional materials for each medium based on the agency's existing recruitment approach. Then he wrote a grant proposal for submission to a local civic foundation to fund the campaign.

The project Bill worked on captures many of the advantages of service-learning for technical and professional writing. Because it emerged out of his own commitments and provided a service to his community, the project gave Bill a chance to develop both as a citizen and as a professional. As a professional writer, Bill learned about the document production process by researching local publicity services and their costs. The project also enabled him to learn style, arrangement, and design conventions of several professional genres. He wrote a compelling narrative and put together a budget for the proposal. He designed and manipulated graphics for inclusion in the promotional materials. In addition, the community connections he made helped him to be a better police officer, and of course his work will benefit future Big and Little Brothers in the area.

Service and Shared Benefits

In the case of Bill Wood and Big Brothers/Big Sisters, it is clear that everyone involved benefited. Robert Coles, a Harvard University professor of child psychiatry, suggests that this shared benefit is critical to meaningful service activities.

OTHER VOICES 2.A Bill Wood

Bill Wood is a lieutenant in the Orlando, Florida, Police Department. He is pursuing a master's degree in technical communication at the University of Central Florida.

One of the advantages of undertaking a service-learning project is that it is a win-win situation for both students and community alike. The student becomes immersed in a real-world scenario that usually contains unexpected real-world twists and turns. And the community benefits because the student makes a civic commitment to better his or her environs. The student also learns to write in a style the real world requires as opposed to what the instructor mandates in the classroom. In my opinion this instills a world of confidence in the student looking to enhance his or her writing skills. Hopefully, the service-learning project will see its fruition, but in any case the student and the community organization both learn and grow from the process.

I think one of the most critical elements to consider in choosing a service-learning project is to find one that intrigues you. You will dive into the endeavor with much more enthusiasm if it has special meaning, as opposed to having one assigned to you by the instructor. Several years ago, before I had kids of my own, I was a Big Brother in Big Brothers/Big Sisters, Inc., so my choice was easy. This was an organization I could identify with and knew something about. I knew they depended a great deal on outside funding for their survival. Anything I could do to enhance the buoyancy of the organization through creating fundraising documents was very gratifying for me.

Coles is well known for his work with student community service programs and has guided many interns through the experience of working for the first time with people in need. In *The Call of Service,* Coles explains that the key to service is a mutually beneficial relationship between the server and the served. To engage in effective and meaningful service, Coles maintains, the person who offers her or his time and energy and effort to a cause must recognize the benefits received from that involvement.

In other words, it is critical that you approach whatever service-learning project you develop expecting to gain valuable experience and training as well as offering useful services to an agency that needs the help. This perspective is particularly useful for you as technical and professional writing students because most of you hope to gain immediately applicable practical knowledge that will benefit you in current or future jobs.

An effective service-learning program is valuable for the students, the sponsoring agency or organization, and the students' college or university. Here are just some of the benefits to students, agencies, and colleges.

Benefits to Students

***Students** have an opportunity to apply professional writing principles and subject-area knowledge in nonacademic situations related to their professional and civic interests.* We have already discussed how well-designed service-learning projects can help students learn and apply writing, design, collaboration, and project-management

STUDENT VOICES 2.A What Do We Get from Service-Learning?

Anne: I was excited when I found out that our class would be working for nonprofit agencies through our writing. I have been a volunteer at a children's charity for a long time, and this assignment gave me a new way to work for a cause I believe in.

Jamie: The very first day of my writing class, I left feeling very confused. I went home thinking—what in the world does professional writing have to do with service-learning (which I had never heard of previously)?!? Actually the combining of the two annoyed me. I thought that the organization was just using the class to do their work. A day or two later, the subject sort of intrigued me. Then, after I learned more, I changed my mind—I discovered that the class benefits as much as or more than the organization. This agency representative was taking out time to help us learn, and she was actually trusting us with major projects.

Cathy: When I first heard about this project, I was afraid that local private businesses wanted to reap the benefits of students working for free. The important part that I had been overlooking was that we would be receiving benefits as well. These benefits take the nontangible form of experience and the tangible form of contacts we may be able to rely on in the future.

Holly: On the very first day of this class, all I could think of was "what have I gotten myself into?" I was only supposed to be here for a general requirement. The whole concept of service-learning seemed confusing to me. I could not understand what a professional writing class had to do with community service work. I had expected that this class would be like my English Comp One and Two classes where everyone did their own research papers and that was it. I definitely received the shock of my life with this course. But I also learned about designing documents, working with a team, and making presentations. I made good connections with a nonprofit agency and improved my writing.

strategies. Technical writing teacher Thomas Huckin, in "Technical Writing and Community Service," has also pointed to these benefits to students.

But Bill's experience suggests that such projects can also help students develop as professionals more generally. As an advanced student in a technical or professional subject, you have access to the latest knowledge and trends in your field of study. If designed with your interests in mind, service-learning projects can enable you to test and further internalize that knowledge by applying it outside the classroom.

Gregory Wickliff, a technical communication professor from the University of North Carolina, has surveyed college graduates who worked on commissioned projects with local businesses and government agencies as students in technical writing courses. The respondents expressed appreciation for this opportunity, suggesting that although they might not always use the full range of their professional training in such projects, they did get a chance to see how their training did or did not apply to projects in the field (189). This opportunity to explore how your academic background will interface with work tasks is one of the most important benefits of service-learning.

***Students** are able to interact with real-world audiences, getting feedback on their work from agency representatives and community members.* It's one thing to practice writing letters, reports, proposals, surveys, instructions, and other professional and technical documents as a part of canned assignments or hypothetical scenarios. It's an entirely different and more meaningful experience to produce documents that will be read by actual audiences beyond the classroom. Though receiving and using feedback from teachers and classmates is certainly crucial to your training as a writer, it can't replace the opportunity to work with readers who will actually *use* the texts you're producing. For some projects, you may even be able to see the final, published version in use by the end of the semester.

A group of students working on a sex education campaign geared toward at-risk teenage girls learned about this benefit through their experience. They carefully applied basic principles of effective document design when drafting early versions of their brochure and posters. When they presented the materials to members of their target audience, however, their young readers set them straight about a number of problems, including inappropriate vocabulary and false assumptions about the audience's knowledge.

Many technical and professional writers gather feedback through more systematic usability tests, tests that assess how members of the target audience actually use the document in a typical setting. Such tests can entail interviews, observations, and recorded transcripts of users articulating their thoughts as they go along. In designing an online tutorial for a web text editor, a group of advanced professional writing students at the University of Florida conducted two rounds of such tests; in the process they found places where novice users needed more detailed explanations and examples. Chapter Seven, "Executing Your Project," and Chapter Nine, "Evaluating Your Project," will elaborate on strategies for assessing your work's relevance to readers.

***Students** learn to manage major projects, balancing varied responsibilities and roles.* Whether you work on your project individually or as part of a team, service-learning will require you to take on a number of roles. You will be a student, a writer, a document designer, a consultant, and a collaborator. At times, your collaboration will require you to take the reins as a leader. At other times, you will need to negotiate with people who have more power than yourself. Service-learning projects typically involve seeing a project through from the invention stage to the final production stage. The collaboration, leadership, and project management skills you apply will be crucial to your future success in school and the workplace. Indeed, many university programs in engineering, business, and other areas include entire courses dedicated to teaching such skills, allowing students to understand processes as well as facts.

Connections among the parts of a learning process are explored in a book called *Knowing and Being.* Here, philosopher Michael Polanyi offers a model of thinking that distinguishes two ways people come to understand ideas or processes—**focal** and **subsidiary.** When we look at a process from a focal perspective, we zoom in on one particular step or piece of the picture. When we take a subsidiary perspective, we consider how the pieces fit together as a whole. For example, one group of Melody's students decided to take on the challenge of creating a web page promoting a family film series sponsored by a business networking group. None of the students had created a web page before, so they had to learn the related concepts from a range of angles.

This group of students knew that they wanted to use a WYSIWYG (What You See Is What You Get) or HTML editing program to design their page. They knew that they could apply previous experience with other document design software to this process to do things like cutting and pasting images into a document and putting the pieces together to create a page that was aesthetically pleasing and informative. They learned how to use an uploading feature to transfer the information from a document file to a website, but when they viewed the document online, it looked quite different from what they saw on their editing screen.

Though they were able to accomplish every part of the process to this point by following step-by-step instructions, they were unable to correct problems they faced in the upload phase until they developed a sense of the larger process of web publishing. Until they understood that the image they saw on the WYSIWIG screen was communicated through a coding language that could be manipulated by the software they were using, for example, they weren't able to go into the source of the file and make minor corrections to its appearance. On the other hand, if they had not gone through the individual steps of designing a page, the code would have probably meant very little to them. They had to see this process in two ways—as a series of steps to be completed one-by-one *and* as a big picture, a process that would not work unless all of its parts worked together.

This model of focal and subsidiary knowledge applies to service-learning in technical and professional communication more broadly. Most of us can complete a class assignment that is presented to us in the form of a series of steps and individual pieces, but we may find it difficult to see how each of the assignments we do for a class fits into the big picture of a work context. We may understand what our textbooks, professor, and peers say makes for an effective letter of inquiry, for example, but unless we have an opportunity to follow the life cycle of the text, we will very likely miss some important considerations in the process of creating it.

As we mentioned in Chapter One, a service-learning project will require you to perform a range of duties and engage in a variety of activities. You may sometimes do research, write first drafts, revise other students' texts, or take notes during meetings. While engaging in all of these smaller activities, you will also need to keep an eye on the grand scheme.

***Students** are faced with "real" ethical dilemmas similar to those they'll confront in their careers.* Expanding technologies make it possible for humans to do new things every day—we are discovering new medical treatments, creating new computer applications, developing new agricultural techniques. Ethical problems are a predictable byproduct of these kinds of developments. New information technologies may threaten someone's right to privacy. Advancements in medical research like genetic engineering might allow future generations to do away with individuality as we understand it. Technological knowledge brings with it significant responsibility.

In an effort to develop new pedagogies for teaching the ethical implications of this responsibility, technical writing scholars such as Sam Dragga, Cezar Ornatowski, and Gregory Clark have studied ways in which the ethical systems of student writers are similar to and/or divergent from those held by practitioners in the field. Though each of these researchers asks a different set of questions to reach his conclusions, they agree that writing classes should address the kinds of concrete ethical concerns that students will face on the job. With its twin emphases on workplace writing and critical reflection, service-learning invites deliberation about such concerns.

As we will discuss in Chapter Eight, service-learning projects can also lead students to face ethically complex situations that require them to negotiate sets of competing values. Most students will encounter some kind of ethical dilemma during the course of this project, whether it has to do with dealing with a problem group member or struggling with a questionable or confusing policy or practice at your cooperating agency. The greatest benefit of tackling such issues in this context is that you have the support and counsel of your instructor, group members, and other classmates. Also, because reflection is such an important component of this kind of service-learning work, you will have many opportunities to stop and think through your decision-making processes; this will enable you to make explicit the kinds of implicit values that guide your behavior.

***Students** make professional connections with community leaders and develop writing portfolios suitable for use in job searches.* A service-learning project often requires that a student work with the managers, officers, or board members of the organizations they are serving. These leaders are sometimes potential employers who will look favorably on students who demonstrate a consistent commitment to their organization and the community it serves. This kind of work suggests that you can see yourself as a team member and citizen.

Many college students in engineering, business, pre-medicine, and other technical fields assume that their jobs will not entail much writing, but testimony from advanced professionals consistently suggests that this is not the case. Respondents to a survey of engineering alumni at the University of California, Berkeley listed technical writing as the second most important subject, after management practices, that students could take in school (in Olsen and Huckin 5). Another survey of 1,400 members of professional organizations in various disciplines (e.g., chemistry, engineering, psychology, and business management) found that the average respondent spends almost 50 percent of her/his time doing some kind of writing (in Lay et al. 4). Service-learning projects typically involve several genres of writing, including at least one complicated genre, such as a proposal, annual report, or newsletter. Documents that are published by the organization and used by actual readers make especially impressive additions to a writing portfolio. Being able to illustrate your collaboration and project management skills through concrete examples will make your job application materials and interviews more memorable to prospective employers. Thus, a service-learning project can help you generate a set of texts that showcase your skills as a writer/document designer as well as your commitment to civic action.

Benefits to Cooperating Agencies

Cooperating agencies *receive assistance with writing projects critical to their operations.* As we've mentioned before, many of the nonprofit agencies, businesses, and campus organizations you'll be working with in this class simply don't have the workforce or rhetorical expertise to create the documents they need. At the very least, sponsoring organizations get solid drafts of documents that they can modify to meet their changing needs. We have found that sponsoring agencies almost always appreciate the fresh perspectives students bring to projects. At best, agencies receive texts that they can immediately use to accomplish their missions. Beyond getting help with urgently needed written or online products, agency personnel have a chance to shift from the writer to the supervisor role.

Agencies also have an unusual opportunity to see themselves from a different perspective through the writing that students perform. Students may identify strengths, weaknesses, and opportunities that organizations have not considered. If students misinterpret missions or activities based on information and materials the agency provides, agency workers may reconsider and revise those documents. Revision of these materials may be a useful task for service-learning students.

STUDENT VOICES 2.B What Makes Service-Learning Work?

Laura: What I found especially important as I started to explore service-learning is the emphasis on this type of learning/work as symbiotic. We (the students) get great experience and an opportunity to build our portfolios, and they (the client or organization) get some much-needed help. What could be better? And oh yeah, we get the opportunity to support a cause we believe in but may not always have the time or energy to support. Doing work like this, helping out someone (or thing) that can't always help itself (due to time, money, resource, etc., restrictions) makes me feel good about myself too.

Geri: I think that the most useful part of this project will be making valuable contacts in our fields of interest. After undergraduate school, I realized (too late, unfortunately) that I missed numerous opportunities to volunteer for projects that would have put me in a position to get to know people who would be able to help me find a job once I graduated. I thought that it was unimportant to devote time to internships and the like while I was earning my degree, because money was such a big issue in my day-to-day life at the time—BIG MISTAKE! Sure, bills had to be paid, and I had to eat, buy gasoline, and so on. But once I had finished my course of study, I was completely unprepared for the task of finding my first "real" job. I didn't know anyone (or much of anyTHING) about what I had been preparing for four long years to do. Now that I'm older and somewhat wiser, I see the value in this project, and I really do appreciate the opportunity!

Two University of Central Florida Organizational Communication majors who were writing documents for a youth mentoring program in their area knew from their previous involvement that the all-volunteer staff had barely enough resources to provide a proposed tutoring and counseling service for children in the low-income, high-crime area. Though superhuman effort on the part of several volunteers had made it possible for the agency to begin the program, they needed more materials and supplies to meet the growing need. The two students wrote letters to local businesses requesting donations of craft supplies, educational materials, snacks, and other items critical to the group's work. They wrote and designed fact sheets and brochures to send along with the letters, highlighting the organization's previous accomplishments and goals for the future. Their efforts paid off in material ways—a struggling group was able to expand its services and to reach out to more children in the community.

Even when they do not result in such success, service-learning projects can still benefit the sponsoring agency. In a Penn State technical writing class students worked in groups to write grant proposals for the local AIDS Project. Although only one out of five proposals was actually revised and used successfully by the agency, the executive director was thankful that the project forced her to clarify the agency's goals and needs.

Cooperating agencies *make contacts with college representatives, increasing their access to resources and their profiles in the community.* Even though they may have offices geographically close to local universities, many of the agencies you target for your projects won't have active cooperative relationships with anyone at your school. Developing these relationships can help the agencies with outreach and raising community support for their work.

When five Environmental Science students at the University of Arizona started to work with a conservation group to develop educational materials for school-age children, they expected to use their cutting-edge insights on water conservation and ecology to help the Friends of the Santa Cruz River group produce an accurate and engaging website and brochure for use with young students. Once they became involved with the project, they realized how valuable it was for their community, and they began to promote it among their professors and classmates. Soon the program was attracting volunteers from the university and garnering praise, which allowed the agency to draw on the expertise of the larger school community.

Cooperating agencies *have the opportunity to connect with students who might someday offer other kinds of support for their work.* In our experience, many of the students who begin to work with local organizations as part of a service-learning project return after the project is over to continue offering their services and support. A service-learning project gives the agency an opportunity to win your loyalty and commitment for the future. If you become involved with an agency now, you may return in the future to volunteer your time or to donate your money.

A group of students in a summer course designed fundraising materials such as brochures and posters for a nonprofit organization that provides teachers at low-income schools with such supplies as paper, pencils, craft items, and so forth. The students selected this organization primarily because their professor had already established a relationship with the marketing director and the students needed a ready-made project to get their work done in the short summer term. When she learned about the important services the agency offers by spending time in its 5,000-square-foot warehouse and talking with teachers who came to get materials, one student realized that this was a process she wanted to be part of on a long-term basis. When the summer ended, she became a regular volunteer at the store and found ways to connect her work there with her goals in later coursework.

Benefits to Colleges

***Colleges** develop community ties and reputations of commitment to service.* Changes in community values affect everyone, including universities. Most contemporary American universities are in a position to reimagine their roles in their towns and cities, repositioning themselves with greater connection to the businesses, public schools, and nonprofit agencies in their areas. Service-learning provides an excellent forum for this kind of outreach, giving a school well-trained and supervised ambassadors to the community who provide meaningful service.

In "Partners in Inquiry: A Logic for Community Outreach," Linda Flower briefly describes the history of the relationship between what she calls "town and gown"—the community and the university. Flower points out that many universities have traditionally held elitist attitudes toward the communities around them, reinforcing the image of institutions of higher education as "ivory towers." She underscores the idea that colleges must approach their community work with a spirit of inquiry; that is, they must view the well-being of the community as inextricably tied to that of the institution, and must strive to connect the two agendas.

When their student representatives engage in service-learning, universities are often more likely to be viewed as connected with and concerned about the community. A university that is perceived this way is more likely to be in a position to have a positive impact on the community. This can create a long-term, mutually beneficial relationship.

***Colleges** develop stronger writing faculty through increased field experience for teachers.* Your writing teacher may have a solid background in your field, whether it's engineering or business or social work, but many of us have focused our training on learning strategies and theories for teaching writing and on developing our knowledge of rhetorical studies. By working with students and agency representatives, teachers, too, can learn about writing conventions and strategies in a wide range of subjects and fields.

One of the main complaints raised by some of the ethics scholars referred to earlier is the gap between the values and knowledge of teachers and those of workplace professionals. Gerald Savage has argued that if teachers want to be actively involved in ethics education, they need to be working not only with their students in classrooms, presenting theories and ideas about how one might proceed, but they also need to reach beyond this traditional boundary to learn from and potentially influence the cultures of these off-campus sites.

Certainly, one of our own strongest motivations for continuing to teach our technical and professional writing courses with a service-learning approach is the great opportunity it provides us to learn about workplace writing and about a wide range of academic fields and social concerns. When our students design websites for our state Wildlife, Game, and Fish agency or create brochures designed to teach people how to avoid skin cancer, we don't simply teach document design and audience analysis. We also learn about the content areas the documents address, and we are exposed to new organizational cultures. Such opportunities help to make us better teachers and writers, which benefits our universities.

***Colleges** enrich their curricula through increased interdisciplinary ties.* Although you may work on a service-learning project to fulfill requirements for just one class, most of you will draw upon knowledge from more than one discipline to complete your work. You may even choose to involve your professors and classmates from your major with a project. At many universities, service-learning projects are bringing faculty and students from a range of fields together to accomplish shared goals. One example of this would include an interdisciplinary community history project in process in Orlando, Florida. Students and teachers from several fields are working with local industry representatives to collect oral histories of the lives of residents of historic districts in the area. Computer science and digital media students and professors are designing the interface while writing and history teachers are collaborating to generate material. Together, these pieces will form a virtual tour of the area. This project will help to preserve the history of the community and provide training to students and faculty in a range of areas.

Challenges of Service-Learning

While we believe that a service-learning approach is an excellent tool for helping professional and technical communication students to develop their skills and portfolios, we also want to acknowledge some of the difficulties that can emerge with such projects. Throughout the chapters that follow, we'll offer a balanced perspective on the many challenges that come from working with real-world audiences and tasks by describing our students' experiences and solutions in difficult circumstances. We recommend that you start to think about these challenges before you begin your project so that you can make smart planning decisions that may help you to negotiate them.

One concern shared by many writing students is the possibility of a difference between the expectations of your agency contact person and those of your instructor. What will happen if your instructor has given you one set of criteria for evaluating an effective progress report, for example, and your contact person wants to see you do something entirely different? To prepare in advance for this kind of problem, start to keep a reflective journal early in your project. Be prepared to discuss the expectation gaps with your teacher and to demonstrate your strategies for bridging those gaps to her or him. If you find ways to explain why you make certain writing decisions, your instructor will likely be willing to work with you to solve your difficulties.

Another concern to consider is the period of adjustment you may have to go through when you begin to work with your agency. New employees and interns

OTHER VOICES 2.B Thomas Miller

Thomas P. Miller, Ph.D., is associate professor of English at the University of Arizona. His award-winning scholarship addresses technical and professional writing, histories of rhetoric, and the tradition of civic humanism.

Service-learning? The phrase seems obvious enough, rather commonplace, even innocuous to many of us in higher education. What's not to like? Everyone has heard about the need to "learn how to learn" and become a "life-long learner." "Service" has been invoked by everyone from professional politicians to venture capitalists to describe running the government and making a profit.

"Public servants" and the "service economy" may evoke the sort of cynicism that associates doing good with being a do-gooder, but broader civic and religious traditions can help us to renew our sense of the value of service as a means to learn from experiences more diverse than our own. Ancient doctrines of public duties and divine missions often value learning by doing in the assumption that we can come to know what to do by putting what we know to good use. Contemporary discussions of service-learning stress the reciprocal dynamics of learning from others by sharing what we know how to do.

Service-learning may be little more than just another trendy catchphrase unless we understand it to mean that service is a means to learn from others and not simply a means to give back to those less fortunate than ourselves. When approaching any service commitment, we need to begin by thinking of it as a learning opportunity if we are to discover what the situation and the people involved have to teach us. This stance has spiritual value as well as practical utility, for it defers judgment, invigorates our abilities to observe experience, and focuses our capacities on the possibilities of what is to be learned from the collaboration at hand. This stance empowers us to learn how to learn from others. Assuming it can help us to understand the most basic sense and highest possibility of "service-learning."

often have to go through a process of learning an organization's rules and conventions to develop a good workplace comfort level. You may have to make several preliminary visits to your site to develop a rapport with your contact person and a strong sense of the agency's objectives. Build in time for this process as you plan your work for the semester. In the section below, we'll discuss types of sites where you may choose to do your service-learning. We'll consider possible benefits and challenges specific to each option.

Three Types of Sites Appropriate for Service-Learning Work

Throughout this book, we will present examples of service-learning projects at three different types of sites—nonprofit and government agencies, local businesses, and campus organizations. Each of these types, we will explain, has distinct advantages and disadvantages. Your teacher may recommend or assign certain sites, but if you choose one yourself, you will want to begin by considering these advantages and disadvantages along with your civic and professional interests.

Nonprofit Agencies

A number of the sample projects you've read about in these first two chapters took place in nonprofit agencies, and many of you may choose such sites for your work. A nonprofit agency might be a local organization sponsored by a private group or a government organization. You might work with a group such as Forever Wild, the wildlife rescue and recuperation organization in Tucson, Arizona, for which a group of students designed promotional materials and procedural manuals. This group was created by local citizens who were concerned about the welfare of the many injured animals in the desert.

A nonprofit agency might also be the local chapter of a national organization. You might work for the local chapter of the United Way, YMCA, American Cancer Society, or, like Bill Wood, Big Brothers/Big Sisters. Your responsibilities for a national organization may differ from those for a local one. Many national groups already have standardized promotional materials and other texts such as brochures, newsletters, and annual reports. For these agencies, you might produce materials related to a particular fundraising event or, as in Bill's case, an urgent local need for recruiting volunteers. One sample student project in the appendix of this book includes a grant proposal written for a regional chapter of Habitat for Humanity.

Finally, you might work for a government agency or program, such as a county commission or a solid waste authority. A group of University of Arizona students collaborated with a public school district to develop documents for a program that encouraged homeless teenagers to stay in school. In addition to recruitment materials, they produced evaluation instruments such as surveys and questionnaires to assess how well the district was meeting student needs. Another sample student project in the appendix is a volunteer training manual for the Alachua County, Florida, Humane Society.

One of the advantages of choosing a nonprofit agency, as these examples have begun to show, is the range of options. Whatever your values and civic interests, you should have little problem finding a nonprofit agency that shares them. Another reason projects with nonprofit agencies can be especially easy to find is that most nonprofits urgently need writing help. Most nonprofit agencies are underfunded, understaffed, and overworked; as a result, they usually have several writing projects sitting on the back burner. Staff members are usually eager to gain the assistance of advanced writing students. Indeed, students are often given important research and project coordination responsibilities. Nonprofit agencies also produce a broad range of professional texts, from promotional materials to internal organizational materials to service-related materials.

The eagerness of most nonprofit agencies can also be a disadvantage if you're not careful. Staff members might have a tendency to view you as a volunteer rather than as a writing consultant and, consequently, might try to give you responsibilities beyond the purview of the assignment. In addition, because staff members are so busy with multiple tasks, they may not always provide close supervision or substantive feedback. Indeed, they may be hard to reach at times. Finally, the writing-intensive projects of nonprofit agencies must also conform to strict deadlines, costs, and other constraints, which will give you valuable project management experience but may also not allow for optimum invention and revision time.

Businesses

Doing service-learning for a business can be a tricky undertaking because, as you know from the criteria presented in Chapter One, a service-learning activity must address some kind of need in the community. In most cases, internships and other workplace-related programs provide students with professional experience and connections and provide businesses with inexpensive or even free labor. But there is a way to make working for a local business into a legitimate service-learning project. The key is a focus on outreach. Essentially, your role in such a project should be to serve as a liaison between the business and the community. You might do work for a business that offers a service to people in your area. To illustrate, a group of students worked with a private women's health clinic to promote a series of prenatal health workshops for economically disadvantaged women in Tucson, Arizona. The students produced a prenatal health manual for the workshops, helping the clinic channel its resources to community members in need.

Other examples of business-related projects include developing instructional materials to accompany science and engineering kits donated by companies to schools, designing brochures and entry forms for a company's charity golf tournament, and creating a web page where employees can find information about worthwhile local causes to which they can make tax-deductible donations.

Perhaps the most obvious advantage of working for a business is the professional experience you can gain. This type of project might be the closest to the work you will do when you obtain a job, and it will give you practice in learning a corporate culture. On an even more practical note, such a project might provide you with job contacts.

Like nonprofit agencies, businesses work under strict deadlines and standards. Although you might end up working on an important project, you might not have as much artistic freedom or responsibility as you would in another kind of project. Instead, your texts will need to conform tightly to the guidelines and ethos of the company. Many larger companies have detailed style manuals, for example. Your work will likely be supervised closely and edited thoroughly. This does not mean you will receive a lot of feedback or hand-holding, however. On the contrary, many business sponsors are not likely to provide feedback or encouragement. Because of their more complex priorities, business representatives may not be as invested in providing students with a positive learning experience as individuals from other kinds of organizations. This is not always the case, but it is something to keep in mind when you are choosing a project site.

Campus Organizations

The final type of site we'll highlight through examples in the following chapters is your own campus. Although projects with campus organizations may not seem particularly glamorous, they are probably more interesting and complex than you think, and they are everywhere. Our students have written and designed documents for the health center, the center for academic computing, university libraries, an applied research laboratory, and a host of student-run organizations. For example, a group of civil engineering students took on the task of proposing a new University of Arizona campus recycling program. They studied the school's current policies, researched other universities' approaches, and created a model designed to save the university money and lead to more recycling. In another technical writing class, groups of students produced print and online documentation for CourseTalk, interactive web discussion software available through Penn State's Center for Academic Computing. This project enabled students to learn about the technical communicator's role in new product development from performing a task analysis to conducting usability tests to aiding in final production decisions. Other possibilities for campus-based projects include working with the Early Childhood Education program to propose a campus daycare center or collaborating with campus police to design brochures and posters advertising sexual assault prevention workshops. The sample campus project in the appendixes includes promotional materials for a student-run dance marathon for children's charities.

This kind of project has several practical advantages. First, it's convenient. Perhaps you'd love to work with a nonprofit agency or business in your area but would have trouble getting there to collaborate with the representatives because of transportation or scheduling issues. Another advantage of a campus project is that it allows you to help a community of which you're a part and draw on expertise you already have. In some cases, you might design documents for use by students; as you are already a part of this audience, you would have an easy time analyzing it and getting feedback from it. Perhaps most importantly, we've found that sponsors at campus organizations are more likely to be concerned about your learning experience and therefore might more readily instruct and guide you over the course of the project. Although some campus projects may require you to follow strict deadlines, others may not, giving you more flexibility.

Campus projects have disadvantages as well. They often don't simulate a work situation as closely as an agency or business project does; therefore they might not allow you to develop as wide a range of experiences as those venues might. Because they take place on campus, these projects may not seem as impressive in your résumé or writing portfolio to future employees as others. When you work on campus with professors and other school employees, it may be more difficult to feel that you are breaking out of your standard student role. You may feel that you are not broadening your horizons as much as you might with another kind of project.

No matter what kind of organization you choose for your service-learning project, you will need to apply the basic principles of rhetoric and technical communication we will present in the next chapter. These ideas and definitions will apply to each writing task you face in your service-learning project. Above all, you will be reminded that as a writer, you have power and responsibilities. You will learn new ways to use these principles to create audience-oriented, subject-centered, and ethically sound documents.

Activities

1. Collect examples of technical and professional communication such as instructions, proposals, manuals, and correspondence. Find them in your mail, accompanying equipment or tools, on the World Wide Web, or in other places. Working with a small group of classmates, analyze the documents in terms of their audiences, their usefulness, and the ethical principles that underlie them.
2. Interview a professor or a practitioner in your academic field, focusing on questions related to writing. Ask her or him about conventions of writing in the field or about how much writing might be required during a typical workday or week. Collect general advice about writing as a professional, and share it with your classmates.
3. Write a journal entry in response to this chapter, highlighting ways in which you expect to benefit from working on a service-learning project, hesitations or concerns you have about the idea, and any leads you might have on possible projects or sponsoring agencies.
4. Explore some of the websites for national technical and professional communication organizations such as those below. Learn about publications, fields of research, and job opportunities:

Association for Business Communication	http://www.cohums.ohio-state.edu/english/organizations/abc
Association of Teachers of Technical Writing	http://www.attw.org
Society of Technical Communicators	http://www.stc.org
TECHWR/L	http://www.raycomm.com/techwhirl

Works Cited

Clark, Gregory. "Ethics in Technical Communication: A Rhetorical Perspective." *IEEE Transactions on Professional Communication* 30.3 (Sept. 1987): 190–195.

Coles, Robert. *The Call of Service.* Boston: Houghton Mifflin, 1993.

Dragga, Sam. "'Is This Ethical?' A Survey of Opinion on Principles and Practices of Document Design." *Technical Communication* 43 (1996): 255–265.

Flower, Linda. "Partners in Inquiry: A Logic for Community Outreach." *Writing the Community: Concepts and Models for Service-Learning in Composition.* Eds. Linda Adler-Kassner,

Robert Crooks, and Ann Watters. Washington, D.C.: American Association for Higher Education, 1997. 95–118.

Huckin, Thomas N. "Technical Writing and Community Service." *Journal of Business and Technical Communication* 11 (1997): 49–59.

Lay, Mary M., Billie J. Wahlstrom, Carolyn D. Rude, Cynthia L. Selfe, and Jack Selzer. *Technical Communication.* 2nd ed. Boston: McGraw-Hill, 2000.

Olson, Leslie A., and Thomas N. Huckin. *Technical Writing and Professional Communication.* 2nd ed. New York: McGraw-Hill, 1991.

Ornatowski, Cezar M. "Between Efficiency and Politics: Rhetoric and Ethics in Technical Writing." *Technical Communication Quarterly* 1 (1992): 91–103.

Polanyi, Michael. *Knowing and Being.* Chicago: University of Chicago Press, 1969.

Savage, Gerald. "Doing Unto Others Through Technical Communication Internship Programs." *Journal of Technical Writing and Communication* 27.4 (1997): 401–415.

Wickliff, Gregory A. "Assessing the Value of Client-Based Group Projects in an Introductory Technical Communication Course." *Journal of Business and Technical Communication* 11.2 (1997): 170–191.

Chapter 3

A Rhetorical Toolbox for Technical and Professional Communication

When you hear the word **"rhetoric,"** you may think of the empty promises of a politician or the slanted words of a lawyer. But rhetoric has a history and set of meanings that expand well beyond such negative associations. This chapter will offer some definitions of rhetoric and then explain how technical and professional communication—especially when grounded in a service-learning approach—are inherently rhetorical. The chapter's main purposes, though, are to outline the rhetorical theory that undergirds this book and to provide you with a rhetorical toolbox for critiquing and producing technical documents. We introduce you to rhetorical terms not to bombard you with theoretical jargon, but to begin to establish a common set of concepts to guide our inventing, discussing, workshopping, drafting, and revising as communicators.

Until relatively recently in Western history, the most important part of higher education was rhetorical study. In ancient Greece and Rome, for example, the art of speaking well in civic forums about issues of communal importance was widely regarded as the most important and difficult area of study. Rhetoric involved not only learning how to speak eloquently and persuasively, but also how to speak ethically. For Isocrates, one of the first rhetoricians to establish a permanent school in ancient Greece, the goal of rhetoric was not only to persuade, but also to deliberate about and arrive at the best course of action. He taught his students to deliberate with themselves and others about the values and effects of their rhetoric.

> ". . . when anyone elects to speak or write discourses which are worthy of praise and honor, it is not conceivable that he will support causes which are unjust or petty or devoted to private quarrels, and not rather those which are great and honorable, devoted to the welfare of humanity and our common good."
>
> Isocrates, *Antidosis* 276–77

Later, Roman rhetoricians Cicero and Quintilian extended Isocrates's model, similarly describing the ideal rhetor as the good person speaking well. The Isocratean tradition of rhetorical training was, at its heart, training in active, ethical citizenship, a goal shared by proponents of service-learning.

Aristotle, roughly Isocrates's contemporary, developed what is now the most famous definition of rhetoric—*the faculty of discovering the available means of persuasion in a given situation.* This definition points to several important aspects of rhetoric (in addition to its connection to ethics).

- Rhetoric is a faculty that can be taught and developed.
- Rhetoric involves discovering and deploying persuasive discourse.
- Rhetoric is situational.

In this definition and throughout his writings on rhetoric, Aristotle emphasizes that rhetoric or persuasive discourse does not take place in a vacuum but is *context dependent.* Later in this chapter we will elaborate on the elements of the rhetorical situation or context and categorize some of the different "means of persuasion" that a rhetor can employ in response to a situation.

Technical and Professional Communication as Rhetoric

Technical and professional communication can be considered rhetorical in several ways. Some types of texts, such as proposals, recommendation reports, and résumés, clearly have persuasion as a primary aim; the writers of such texts attempt to persuade their audiences to accept and implement their requests, solutions, or recommendations. The letter of inquiry that you will write to a local organization, for example, will attempt to interest that organization in sponsoring a project with you. Other types of texts, such as progress reports and instructions, may primarily aim to inform or help users complete tasks. Such texts must also be persuasive, however, if only to convince their audiences that they are credible or should be followed carefully. Without misrepresenting their progress, writers of progress reports often emphasize their accomplishments to instill confidence in their readers. In addition to informing, describing, explaining, or guiding, then, technical and professional communication argues or persuades, implicitly if not explicitly.

Technical and professional communication can also be considered rhetorical because of their emphasis on audience. As Mary Lay and her coauthors of *Technical Communication* explain, technical texts combine a focus on particular subjects with a strong *audience-based orientation,* meaning such texts are "designed from the point of view of the audience and what it needs" (16). Even beyond audience-based orientation, technical and professional communication follows the principle of audience advocacy. Not only must the technical communicator fully account for the audience and its values and needs, she or he must serve as the audience's advocate, as the word *accommodate* in the definition above implies. It is the communicator's job to make information accessible and user-friendly while conforming to ethical and legal standards. Indeed, writers of documentation, instructions, and other technical and professional texts can be held legally responsible for how well they accommodate users; the written or online documentation for new products, for example, is subject to product liability laws.

Dan Jones, in *Technical Writing Style,* recommends evaluating technical communication according to ethicist Vincent Ruggiero's principle of respect for persons. From this principle, Ruggiero develops three types of ethical criteria—obligations, ideals, and consequences (Jones 241). As we have been emphasizing in this chapter and the last, technical and professional communication is often defined in terms of the obligation or duty to accommodate information to the user and help the user apply it effectively and safely. Such an obligation especially applies to writers of instructional documents, which have legal as well as ethical requirements for anticipating and accommodating users' needs. Ideals can be thought of as communally defined notions of excellence. Technical and professional writers should consider their own ideals as well as those of their organization, field, audience, and larger culture. Later in this chapter, we overview such writing-related ideals as accessibility and concision. Larger cultural ideals include honesty and efficiency. In some cases writers can positively impact the goals of the organization through sharing their values. A group of students writing for the Gainesville, Florida, chapter of Surfrider drew on their environmental knowledge and values in prompting the organization to consider expansion of its beach preservation efforts to include North Florida springs and lakes. The third ethical criterion is consequences, the harmful or beneficial effects of the communication, especially for the audience. In addition to evaluating carefully what our texts enable, this criterion requires us as communicators to consider what our texts might disable. How might the design of a website, for example, prevent users with low-end browsers from easily accessing important information? How might a community service project detract attention from another set of even more pressing needs?

Like rhetoric more generally, technical and professional communication is a situated activity, addressed to an audience and produced in response to a set of specific circumstances. For example, a local women's resource center might write a proposal to a government funding agency to obtain resources to develop a job training program, or an employee might produce a set of procedures to help coworkers streamline and standardize a data entry process. Unlike much creative and academic writing, technical and professional texts are produced for actual, concrete audiences who will use them to make decisions or perform actions. This is not to say that technical and professional communicators can always predict their audiences or that all of their texts are written to a precise, easily discernable set of readers, but often such communicators have clear ideas about who will use their texts and frequently even test their texts out on these users ahead of time (in a later chapter, we take you through the process of usability testing).

As we discussed in Chapter Two, technical and professional communication's rhetoricity—that is, its situatedness and connection to specific audiences—makes service-learning particularly well suited to it. Service-learning projects are clearly situated forms of social, rhetorical action intended to help particular audiences solve particular problems. By throwing into relief the rhetoricity of technical or professional communication, service-learning projects move past the all-too-common view of such communication as a simple set of formats and skills. Further, service-learning, with its dual emphases on critical reflection and beneficial action, can also enhance technical communication's ethical goal of deliberating about and arriving at the best course.

The rest of this chapter will introduce you to rhetorical concepts for analyzing and inventing technical and professional texts or any type of discourse. Subsequent chapters will revisit and apply the ideas below to various phases of your service-learning project. For now, we just want to give you an overview of the concepts.

> "When teachers of philosophia [e.g., rhetorical theory] make their pupils conversant with the lessons of discourse, they require them to combine in practice the particular things which they have learned, in order that they may grasp them more firmly and bring their theories into closer touch with occasions for applying them."
>
> Isocrates, *Antidosis* 184–85

The Rhetorical Situation and Cultural Circuit

The most fundamental way to approach technical communication rhetorically is to account for the social and rhetorical contexts out of which it operates. To use Polanyi's terms, communicators need to step back and look at writing tasks from both focal and subsidiary perspectives as pieces and as parts of a big picture; you must pay attention to the specific design elements of your résumé, but these only make sense in relation to the qualifications of the job, your audience's expectations, and other larger-picture concerns. You may remember the concept of the rhetorical situation from earlier writing classes. Rhetorician Lloyd Bitzer has defined the *rhetorical situation* as the "context of persons, events, objects, relationships, and an exigence which strongly invites" discursive intervention (5). More simply, the rhetorical situation is the set of circumstances that calls discourse into action (2). This "calling" is what Bitzer means by the term **exigence.** Other rhetoricians have modified this concept to account for the ways communicators can *create* the need for their rhetorical intervention as well as respond to existing exigencies.

Gorgias, one of the earliest rhetoricians, defined his theory of rhetoric around exigence and the rhetor's opportune or kairotic response to it. As Sharon Crowley and Debra Hawhee explain, Gorgias's notion of **kairos**—the opportune time and/or place for deploying rhetoric—was inherently tied to the rhetorical situation (31–33). To be persuasive, a rhetor must be adept at reading the dynamics of a situation and tailoring her or his rhetoric to seize the advantage of that situation. In addition to thinking about rhetorical contexts as shaping rhetoric, we can also think of rhetoric as shaping rhetorical contexts, by, say, altering relationships or changing the stakes or influencing events. Sometimes rhetors have to make their audiences aware of a situation's exigence, and, sometimes, rhetorics must create as well as discover kairos. This might be the case for an employee writing an unsolicited proposal to solve a problem of which her coworkers and supervisors are not fully aware; she must spend more time establishing the need for her solution.

In addition to exigence, a rhetorical situation involves a **rhetor,** such as a technical writer, who responds to the situation, the rhetor's **purpose(s),** and an **audience** to whom the text is addressed. Other elements of the rhetorical situation are the **text** itself, the **subject matter,** and the larger **sociocultural contexts** that help shape the text's **production, distribution, and reception.** As Figure 3.1 shows, the rhetorical situation is often visually represented by a triangle, thus the term rhetorical triangle.

Notice how the lines of the triangle connect the elements of the rhetorical situation—writer, audience, text, and subject—forming sets of relationships.

Such relationships are also influenced by the larger contexts of the text's production, distribution, and use. Analyzing textual forms (in this case technical and professional writing) according to the recursive circuit of their production, distribution, and reception is a method developed by cultural critic Richard Johnson. The text's production is not solely determined by the writer but is also conditioned by such elements as economic constraints (e.g., the budget for reproducing and distributing a newsletter) and institutional pressures (e.g., a deadline imposed by a supervisor). Accounting for the text's distribution means accounting for the avenues through which it reaches and circulates among its readers and not just its form. If you produce a website for an organization, for example, you'll want to register it with search engines and link it to other sites visited by your targeted viewers. Analyzing the reception of the text involves analyzing the various factors that help shape its use, from the material environment to the readers' attitudes toward it.

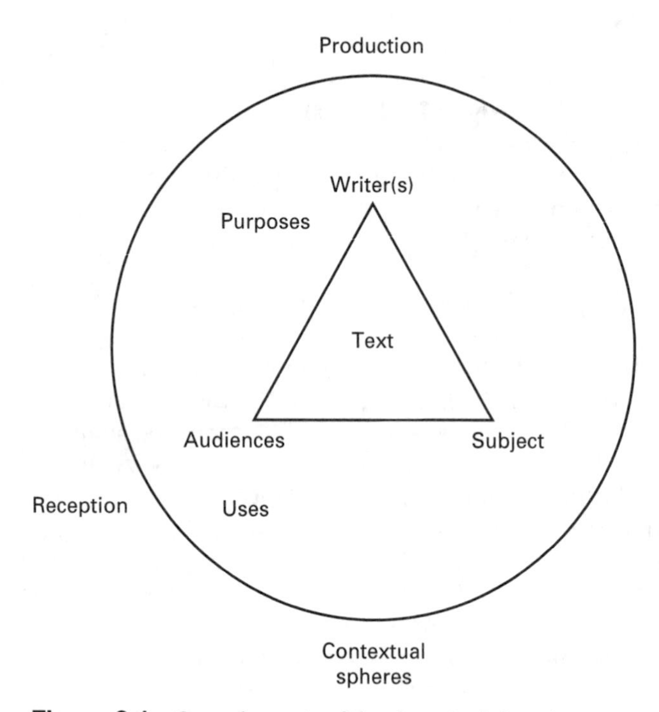

Figure 3.1 Core elements of the rhetorical situation

In the rest of this section we will expand on our explanation of the elem the rhetorical situation and cultural circuit, ending with a set of questi guide your invention and analysis of technical texts. We will illustrate our explanation with an extended example of a service-learning project by a class of professional writing students at Penn State. In this class the students worked with a staff member from the University Health Center to produce a set of brochures informing students about sexually transmitted diseases (STDs) and the center's prevention and treatment services.

Writers and Their Roles

Most workplace communication is not the product of only one writer, but a group of writers and other collaborators. Online documentation, for instance, might be produced by a team of writers, graphic specialists, and software developers. It follows, then, that authors of workplace texts represent not only themselves, but also the larger organizations for which they work. Indeed, many documents are identified only by the organization's name, and others are written for the signature of a supervisor. The corporate-based representation of technical communication has been described as one of its defining characteristics (Lay et al.). Writers must develop collaborative skills, clarify whom they represent, and understand the ethos or projected identity of the organization.

One of the most important steps in assessing a rhetorical situation is defining your role as a professional and writer. This role can be determined by a number of factors, including your job responsibilities, relationships with coworkers and supervisors, and, perhaps most importantly, relationships with the audience (see Couture and Rymer). The concept of a **discourse community** can help us think about some of these communicative relationships. A discourse community can be defined as a group of people bound by a common interest who share and regulate specialized kinds of knowledge and ways of communicating (see Porter; Anson and Forsberg 202). Discourse communities can range in scope from, say, people working on a particular project to supervisors and workers in an organization to an entire field or interest group. Discourse communities are not made up of people with equal power, but are networks of relationships among regulators or supervisors, communicators, and consumers.

> "What came to mind as I read about discourse communities is the comparison between the students in my English classes and those in my education classes. While I sometimes feel like a fish out of water in my English class, I feel confident and able to discuss all the topics in the education classes. Instead of having to reread passages to make sense of them, I can usually skim my education readings because I am so used to the conventions and phraseology."
>
> Lynn, Professional Writing Student

It is not always easy to juggle your multiple roles; your individual perspective might be different from that of the rest of the group, for example, or you may need to mediate between the needs of your audience and the expectations of your supervisors. In such instances you should consider your ethical obligations as well as the power dynamics involved.

In the case of the STD brochures, the team of writers was led by the staff member but also included several students who were not part of the center's health care discourse community. The roles of the students were mainly to assist in the research, writing, and design of the brochures and to test the brochures on a sample user group of fellow students. The staff member guided and oversaw the students' work, and a group of other health care professionals evaluated the final versions. Because the students were closer to members of the target audience than their agency collaborators, their perspectives on language and design were particularly useful. They functioned, in part, as representatives of the audience's discourse community on the writing team.

Writers' and Organization's Purposes

In technical or professional writing, you'll need to consider the purposes of your supervisor and the larger organization as well as your own. Purposes are sometimes conceptualized as objectives or as specific, measurable outcomes that you want your text to achieve. How do you want your readers to respond to your text? What do you want them to do with it or based on it? What kind of relationship with your readers do you want to foster? What kind of impression do you want to make of yourself and your organization? As students, you'll probably have the academic objective of earning a good grade. In our experience with service-learning projects, however, this purpose often takes a back seat to more community-oriented ones.

In some situations, like the one the STD brochure writers faced, your aims and purposes are mostly determined for you, and in others you may start with general aims (e.g., persuasive, informative, instructional) but need to develop more specific purposes tied to readers' responses and actions. It can be helpful to designate which purposes are primary and which ones are secondary. The brochure writers' aims were primarily informative but also persuasive; they sought to provide at-risk students with basic information about risk assessment, prevention, testing, and treatment and also to persuade these students to take action based on this information, by, say, assessing their health, making an appointment at the center or implementing preventive measures.

Audience: Background, Needs, Values, and Uses

One of the most important elements of any rhetorical situation is the relationship between the communicator(s) and **audience.** To discover the available means of persuasion and to make your text accommodating, you will need to carefully consider your relationship with your audience. If you have an established relationship with your audience, for example, you might be able to write in a less formal tone. You will likely take greater care in persuading a potential client than a coworker, a hostile audience than a friendly one. Just as you consider your roles as a writer in a rhetorical situation, you should also consider your audience's roles. You can start by determining your audience's roles in their organizations and their likely uses of the text. In many professional communication situations, such as those involving readers of empirical reports or users of instructions, the readers are primarily learners and/or implementers. In others they may be decision makers who want the bottom line or advisors who want supplementary information as well. You will also want to consider the audience's familiarity with the subject (the audience-subject relation), needs, values, expectations, and motivation to use the text.

The first, most powerful audience the brochure writers faced was the health center staff who supervised and decided whether or not to approve the brochures. The writers had to meet this audience's specifications for the text and timeline and budget for the project. The ultimate audience, though, was of course the clients who would receive the brochures if they were approved and used. To achieve their purposes, the writers needed to ask how much the target audience of at-risk students knew about STDs and the center's resources regarding them, and how motivated the audience would likely be to read and use the brochures. Asking such questions helped the writers determine what kind of information to present and how to present it in the most accessible and persuasive way. As students themselves, some of the writers were in a good position to begin to analyze the target audience, even though the audience was probably more heterogeneous and less easy to predict than we have implied from our discussion so far. In such a case it might be useful to consider Andrea Lunsford and Lisa Ede's distinction between *audience invoked*—the audience the writers call to and, in a sense, help shape through the language and design of the text—and *audience addressed*—the actual readers who encounter and use the text. Although technical writers can invoke an audience based on a thorough analysis of, and even input from, potential readers, the audience addressed is almost always more complex. It would be difficult to determine to whom clients of the health center might pass the brochures after they leave the clinic.

Communicators often have secondary audiences along with a primary one. In hypothesizing who might comprise these secondary groups of readers, it is sometimes helpful to think through the life cycle of the text, its avenues of distribution and circulation. In what sites will the text be made available, and who is likely to encounter the text in these sites? Returning to the multiple purposes of the text can also help you brainstorm who the secondary audiences might be. Because the brochures are needed, in part, to aid health care providers at the center in explaining STDs and providing clients with supplementary information, these same health care providers constitute a secondary audience. Sexually active undergraduate students are not the only patrons of the health center; other audiences might include undergraduate students who are not at risk for STDs, graduate students, and even faculty members, although these audiences are less crucial.

"I am in full agreement with the line, 'Although technical writers can invoke an audience based on a thorough analysis of and even input from potential readers, the audience addressed is almost always more complex.' When I write a proposal or other project at work, it is very hard to pinpoint exactly who will be looking at my work once it is sent out. The possibilities range from government officials, clinicians, and judges to a diverse group of people on a proposal evaluation committee.

When writing these proposals I try my hardest (and sometimes fail) to think of all the possible routes my words will take once they leave the office. Often, wondering who will be part of my secondary audience can be frustrating."

Gloria, Technical Writing Student

Text

The purposes, audiences, and uses of a document will help determine another component of the rhetorical situation—the **text** itself. As the users' advocate, you will need to tailor the text's medium and its form to the expectations of those users. The medium of the communication can be an oral presentation, print document, website or other online text, or even a multimedia text. Different people have different amounts of experience and levels of comfort with different media; users of an advanced software program, for example, may not need extensive printed documentation; new users may need several sets of instructions to complete a complex task.

Technical and professional texts typically follow general conventions of particular genres, such as empirical reports, instructions, and fact sheets. The conventions of instructions, for example, include chronological, numbered steps written in imperative voice, action-view drawings or photographs showing the steps, and troubleshooting advice. Readers of a proposal will expect the text to include a discussion of the problem, description of methods and solution, and outline of schedule and costs. In subsequent chapters we will introduce you to several common professional genres, including the letter of inquiry, proposal, progress report, and evaluation report. Following rhetorician Carolyn Miller, we find it useful to think of genres as forms of social action rather than rigid formats. Communicators follow the conventions of particular genres not simply to conform to rules or standards, but to fulfill their audiences' expectations and to enable their readers to better follow and act on the information in their texts.

Beyond generic conventions, which span across discourse communities, you'll need to adapt your text to the more specific conventions of the discourse communities you're representing and addressing. What kinds of graphics and what style does your organization use in its brochures? What form (e.g., poster, booklet, notecards, etc.) will your readers expect your instructions to take? We'll give you more heuristics for analyzing and formulating such discourse conventions in Chapter Seven.

The writers of the STD documents decided to use a fact-sheet-type of brochure—a common genre for conveying general information to patients—to motivate readers through powerful and concise statements and to make it easier for them to take the text with them. They needed to consider, then, existing brochures' language and design conventions, such as basic drawings instead of photographs, small chunks of texts and bulleted lists, a simple color scheme, and a straightforward but user-friendly style. The writers' design of the brochures was constrained by other institutional factors, too, such as the cost of reproducing the brochures and the need to include standard university information (the center used a template for the front and back covers). One student group also learned the importance of considering the values of their sponsors and discourse community (in this case health care workers) when some of their visuals were rejected for portraying clients as cartoon characters; the sponsors explained that such images can be disrespectful by trivializing clients and their problems.

Subject

As the figure of the rhetorical triangle shows, the **subject** is another core element of the rhetorical situation. Technical communicators must determine what they know about the subject as well as what the audience knows or needs to know about it. The writers of a progress report, for example, should consider how much background information the audience will need to remind them of the project. Technical communicators often try to determine the questions their audience will likely have about the subject. Many manuals and other technical texts are even arranged around reader-centered headings in the form of questions. The audience's knowledge, expectations, and needs regarding the subject are factors that help shape the text. Writers have to make decisions about considerations such as how much information to present, what information to give the most presence or emphasis, and how to arrange the information for maximum accessibility.

Larger Sociocultural Contexts of Production, Distribution, and Reception

All of the above elements are embedded in a larger set of circumstances—we might call them spheres—that both create exigence for and put constraints on the cycle of communication. Writers can begin to analyze contexts by thinking through the problems that invite their intervention. In the case of the brochures, the general problem was the increasing incidence of STDs among college-age students. The center's local problem was its lack of print materials to inform students about STDs and how to prevent them; the health care providers did not have any print materials about STDs to give students when they visited. You can also think of contexts as the larger networks of interpretation and material conditions that influence how your texts can be produced, distributed, and received.

In addition to more immediate situational factors, such as the health center's institutional requirements and the settings in which clients will read the brochures, larger cultural influences form a contextual sphere that influences the rhetorical act. By culture, we mean socially constructed behaviors, values, and conditions. Examples of larger cultural concerns include business customs, gender roles, ethical norms, and economic conditions. An audience's values are certainly conditioned by broader cultural norms. Awareness of cultural elements is becoming increasingly important as communication becomes more global. Service-learning technical and professional writing projects associated with local service organizations must also account for the cultural elements of the communities they serve, however. The brochure writers will need to consider norms and patterns relating to sex and dating when they discuss prevention strategies. They will even need to consider the various associations of specific terms and concepts (e.g., *safer sex*), as their audience will interpret the brochures within a broader network of cultural messages about STDs.

Invention Questions

Our overall message has been that the technical or professional communicator must consider the interconnected elements and relationships of the situation at hand. The following questions synthesize the concerns we've been discussing, providing a compact heuristic for rhetorically analyzing your assigned texts. We suggest that you mark this page as you will likely return to it to guide your invention for each assignment. As you apply these questions more and more, thinking rhetorically should gradually become second nature to you.

Writer

- What are your roles and responsibilities?
- Who are the other members of the writing team, and what are their roles?
- Whom will the text represent? Within what discourse community are you writing?
- What is your relationship to the audience? How well do you know the audience? What is your audience's attitude toward you?
- What are your primary and secondary aims and purposes? What do you want to accomplish with the text? What do you want readers to do in response?

Audience

- Who is your primary target audience? How complex and heterogeneous is this audience? Who else is likely to encounter and use the text in its life cycle?
- What are the roles of your readers? Are they primarily learners, implementers, decision makers, transmitters, advisors, or some combination of these?
- Of what discourse communities are your readers members?
- How much knowledge does your audience have about the subject and text? How familiar is the audience with the problem?
- What are your audience's needs regarding the problem?
- What expectations will your audience have for the text?
- What values of your audience might help shape their attitude toward, and interpretation of, the text? How can you appeal to these values?

- How motivated will your audience be to read and use the text? How can you increase this motivation?
- How is your audience likely to use the text?

Text and Subject

- What medium and genre are best suited to the purposes, audiences, and uses of the text? What form best enables you to reach your audience and make the information accessible to them?
- How much experience and comfort will the audience have with the medium and the genre?
- To what generic and discourse community conventions should you conform? How familiar will the readers be with these conventions?
- How is the text related to other texts?
- What do you know about the subject?
- What does your audience know? What questions will your audience likely have? What will their attitude likely be?

Contextual Spheres of Production, Distribution, and Reception

- What is the exigence of the situation? What aspects of the situation invite your intervention?
- What technical and institutional problems will the text help solve?
- What are the institutional constraints to which you and the text must conform?
- What would a diagram of the text's life cycle look like?
- In what environments will your audience interpret and use the text?
- What cultural norms, patterns of behavior, and other conditions will affect the audience's interpretation of the text? How can you accommodate these?

We now turn from the fundamental rhetorical concept of the rhetorical situation to a discussion of the main "means of persuasion" that technical and professional communicators can deploy in particular situations.

Rhetorical Appeals

In *On Rhetoric,* Aristotle describes three general types of persuasive appeals or means of persuasion—**logos, pathos, and ethos.** Ethos appeals relate to the writer's credibility and character; pathos appeals address the audience's values and emotions; logos appeals present the reasoning or the logic of the argument. These appeals sometimes overlap; for example, a well-reasoned argument usually increases the writer's credibility, or an ethical appeal to goodwill might play on an audience's emotions.

> "Of the *pisteis* [rhetorical proofs or appeals] provided through speech there are three species: for some are in the character [ethos] of the speaker, and some in disposing the listener in some way [pathos], and some in the argument [logos] itself, by showing or seeming to show something."
>
> Aristotle, *On Rhetoric* 37

Ethos

For Isocrates, Aristotle, and other ancient rhetoricians, ethos was the most important of the three appeals. They recognized that the communicator's character and credibility go a long way toward persuading an audience. Ethos is particularly important for technical and professional communicators because they often represent their superiors and the company or organization as well as themselves. Such communicators must carefully attend to the organization's ethos, the character the organization adopts in its communications with people outside and inside the organization (Lay et al. 117). Aristotle described ethos as something like persona, the character a writer projects in the text. We might call this the *invented ethos*—the writer invents it through the act of communication. Isocrates, Cicero, and other ancients conceptualized ethos more broadly, as both the invented ethos and the character or reputation the writer brings with her or him. We'll call this latter aspect the *situated ethos.*

In a past service-learning project, a group of students worked with United Way personnel to revise and redesign a fundraising packet that targeted corporations. These students had to discern the organizational ethos of United Way by studying other texts directed toward corporate donors. Because United Way is a well-known, well-respected social service agency and had already established connections with many corporations, the writers came into the situation with a familiar and positive situated ethos. For some readers, however, they could not rely on this reputation alone. The writers also had to establish credibility and goodwill in the packet by explaining how donations would be used, by explaining how a donation would benefit the corporation, and by other means.

As the Roman rhetorician Cicero explained, one way for rhetors to establish a positive, persuasive ethos is by demonstrating that they are knowledgeable and have done their homework about the issue. This is especially important when they don't come into a situation with already-established credentials. Communicators can establish credibility for themselves in several ways:

- by showing that they understand the audience's problem and have carefully arrived at a solution
- by demonstrating their expertise on the subject
- by citing other authorities on the subject
- by presenting the text in a polished manner, as this implies that the writers are careful and professional
- by avoiding grammatical and mechanical mistakes that might distract from the text's persuasiveness (not to mention comprehension).

A second, related ethos-building strategy Cicero discussed was showing goodwill toward the audience. Of course technical and professional communicators already have an obligation to accommodate readers' needs. Accessibility can serve the persuasive aim as well as the informative one. If readers of instructions have to flip back and forth to find needed visuals, for example, they might become less motivated or even irritated. Although the United Way corporate packet contained a substantial amount of information, it was arranged so that readers could easily find and retrieve small pieces of that information, such as a list of the former year's contributors. Goodwill can also be created through tone and voice. A professional yet friendly tone will generally be persuasive in workplace situations, although a more formal tone might be in order for supervisory or client audiences the writer doesn't know well. First and second person closes the rhetorical distance between the writers and audience, which can be reassuring to users in a text such as an online tutorial. The United Way writers used first and second person to help convey a sense of partnership between the agency and contributing corporations.

Pathos

Pathos—the appeal to the audience's values or emotions—is a second general type of appeal. Although you may think of emotional appeals as being inappropriate for technical and professional communication, this is often not the case. Audiences often are emotionally invested in the problems technical and professional documents address. Emotional appeals can be used to establish a bond with an audience, create goodwill toward an audience (a connection to ethos), or help ensure the safety of an audience. Humor might be used to reassure and motivate the users of an online tutorial, for example. Appeals to fear, in the form of warning or danger callouts, might be appropriate for users of certain instructions. Because they are often read first and are read holistically, visual aids can carry enhanced pathos appeals.

In addition, audiences interpret and evaluate texts according to a network of values. Evaluators of proposals, for example, might assess how well the proposed solutions embody the values of the funding organization. A personnel manager might look for references to the company's values in a job applicant's materials. In our student example, the writers of the corporate fundraising packet emphasized United Way's careful management of donated funds as well as the benefits for the company and the community of donating, thereby appealing to the value of positive local public relationships. Another group of students who were producing a brochure for a student-run charity dance-a-thon appealed to their audience's sense of fun and community by including photographs of past volunteers dancing in one large group and also celebrating their fundraising achievement with children who benefited from it.

Because they can be so powerful, pathos appeals should be used with care. If not suited to the rhetorical situation and not crafted carefully, appeals to emotions and values can alienate or anger audiences. The excessive use of pathos appeals may make some audiences distrustful and thus may detract from your ethos. You can prevent such rhetorical mishaps by using pathos appeals to strengthen logos and ethos appeals and by basing your appeals on careful audience analysis (and even testing) rather than on hasty assumptions about an audience. Determine the kinds of appeals that are most valued by your audiences. The writers of the corporate fundraising packet could not rely solely on emotionally touching photographs and quotations from community members who had benefited from United Way services. They instead used these means of persuasion to supplement a credible ethos and a set of well-supported reasons.

Logos

The third means of persuasion discussed by Aristotle and other early rhetoricians is logos or the appeal to reasoning. We discussed earlier that all technical and professional communication has at least a secondary persuasive aim; it follows, then, that all such texts make some kind of argument, whether or not the argument is overt. The basic elements of an argument are a claim, reasons, premises, and evidence. Some of these elements, such as a premise or even the claim itself, may be implicit if they are already clear or already accepted by the audience.

The key to an effective argument is persuasive, well-supported reasons. You can think of reasons as the because clauses that follow your claim (Lay et al. 120). For example, the company should hire you *because* you have leadership skills and extensive experience, or the agency should fund your proposal *because* your team has the cheapest solution and the most feasible management plan. Sometimes the reasons are related to pathos and ethos appeals: Hire me because I share the values of your company, or fund our proposal because we are the most capable, reputable engineering firm. It is important to remember that it is the audience, and not the writer(s), who assesses the persuasiveness of the reasons.

If an audience readily agrees with a reason, it may not need further support. Most reasons, however, need to be backed up with some type of evidence. Just as reasons support a claim, evidence supports or backs up reasons. Evidence can take the form of examples, data, and testimony from others. Rhetors should assess the types of evidence that their audiences will find most credible and make sure that the evidence they provide is both relevant and sufficient. The proposal team that argues that its solution is the best because it is the most feasible will undoubtedly need to provide evidence of this feasibility, such as an explanation of methods, a projection of the schedule, and a detailed description of costs.

Reasons are linked to a claim by what we are calling premises, which are basically the assumptions the audience must accept to move from the claim to the reasons, to follow the logical connection between the two elements. A résumé writer who argues that the company should hire her based on her formal training is hoping that her audience will accept the assumption that this training taught her skills relevant to the job. Like reasons, premises sometimes need support and sometimes do not. Depending on the audience, the job applicant might need to articulate reasons *why* her training would be relevant to the job and beneficial to the company.

Understanding the parts of an argument and how they work together to form a logos appeal can enable communicators to critique and shore up their own arguments better, especially when they are directed to skeptical audiences. The students writing for United Way conceptualized their corporate fundraising packet as an argument that needed good reasons and evidence, and their packet was the more persuasive for it. Their major claim that the corporation should contribute to United Way was supported by pathos-inflected reasons about the benefits of contributing to the corporation and the larger community. These reasons were, in turn, supported by an array of evidence, including quantitative data about the agency's budget and expenditures, descriptions of the other agencies and specific projects that United Way supports, and testimony from the recipients of community programs. A less obvious reason included in the packet (and fortified by a list) was that many other important corporations donated funds to United Way. For the most part, the writers did not need to provide backing for their premises or assumptions, as they were careful to include reasons in line with the audiences' values.

In the next section, we move from discussing means of persuasion to explaining a rhetorical system for analyzing one part of the rhetorical situation—the text itself.

The Five Canons — for composition, action (heuristic)

for research & writing

Classical rhetoricians identified five related canons or constituent parts of rhetoric—invention, arrangement, style, memory, and delivery (see Figure 3.2).

Taken together, the five canons compose a heuristic for analyzing the production and reception of any text. Although early rhetoricians used the canons to analyze oral discourse, we can adapt them for written and electronic discourse as well. As the following sections will explain and the figure below shows, the five canons

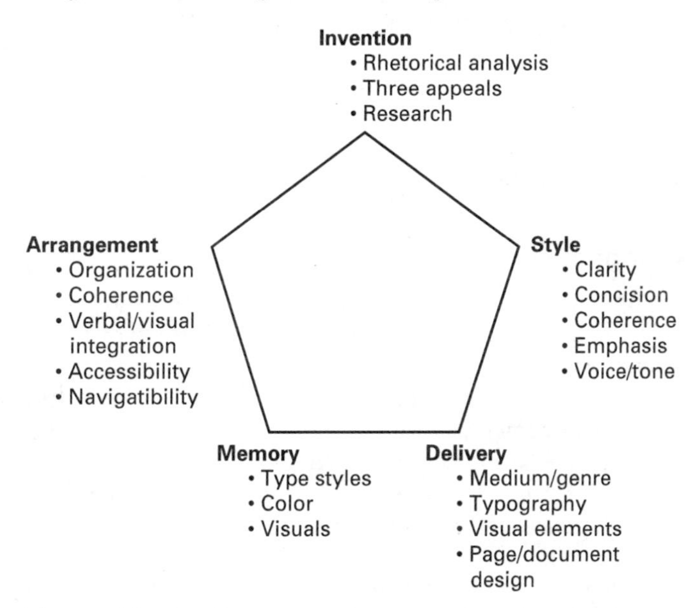

Figure 3.2 The five canons of classical rhetoric and their textual components

are interconnected and sometimes overlapping. A text's arrangement, for example, can affect its memorableness. We separate them in our discussion, however, to enable more systematic analysis of the rhetorical dimensions of a technical text.

> "Since all the business and art of orators is divided into five parts, they ought first to find out what they should say [invention]; next, to dispose and arrange their matter, not only in a certain order, but with a sort of power and judgment; then to clothe and deck their thoughts with language [style]; then to secure them in their [and their audience's] memory; and, lastly, to deliver them with dignity and grace."
>
> Cicero, *On the Character of the Orator* 40–41

Invention

The first canon of rhetoric, **invention** is often defined similarly to rhetoric itself as the discovery of the available and appropriate means of persuasion in a situation. You may associate invention with the prewriting step in the writing process; yet, this latter term falsely implies that the writing process is linear and that invention takes place only at its beginning. Rhetors invent throughout the act of communication, of course, from the first consideration of the rhetorical situation to the final round of editing.

We will not go into much detail about invention here, as the previous sections on the rhetorical situation and three main types of appeals have covered this canon. The most helpful invention strategy, we maintain, is analyzing your exigence, purposes, audiences, uses, and contextual constraints. In relating the canon of invention to technical and professional communication, some rhetoricians have emphasized not only the appropriateness of one's arguments, but also the sufficiency and relevance of the information one presents (Dragga and Gong 31). But these considerations also depend on one's audience and other elements of the rhetorical situation. Sufficiency and relevance are determined largely by one's readers. How much information will readers need to be persuaded? How well does the information meet their expectations? How well does the information match their values?

Arrangement

Arrangement refers to the selection and organization of a text's parts, both verbal and visual. You might also associate it with the words *structure* and *form*. We can think of arrangement on several different levels: the arrangement of words and sentences, the arrangement of paragraphs or sections, and the overall arrangement of the entire document. Within this latter aspect, we can consider page and document design as well as the verbal–visual relationship in a text. Our discussion of arrangement will focus on several crucial aspects of the canon: organization, document-level coherence, verbal–visual integration, and accessibility/navigatibility.

Organization The organization of most technical texts is deductive and top-down; that is, it leads with the most important information (such as conclusions) and previews the structure of the text. This straightforward type of arrangement makes sense for busy readers who want the bottom line, need to locate a particular piece of information, or even need to determine whether the document is of interest. In what is sometimes called a managerial arrangement, many reports preview the conclusion(s) and recommendation(s) up front. The introductory summaries of proposals frequently preview the scope, major parts, and overall cost of the proposed solution. Letters of inquiry typically indicate what the writer is requesting within the first couple of sentences. Not all situations call for a deductive, top-down organization, however. When delivering bad news or making a controversial argument, writers may want to, respectively, buffer the bad news with good news or begin by building the audience's trust. Writers should have good reasons for violating the common expectation of a deductive organization. If a writer has setbacks to discuss in a progress report, for example, he may not want to preview these in the beginning.

The overall organization of the text can be sequential, categorical, or a combination of these. **Sequential** patterns of organization include spatial, chronological, and hierarchical, and are usually easier for the audience to discern and predict (Dragga and Gong 49). Most readers of technical product descriptions will likely expect the descriptions to progress spatially in a single direction. Instructions, procedures, and progress reports are among genres that typically follow a chronological organization. Recommendation reports are typically written hierarchically, meaning that the recommendations are listed in either ascending or descending order (Dragga and Gong 50). Numbered (as opposed to bulleted) lists can indicate such a hierarchy.

Compared to these sequential patterns, **categorical** patterns of organization make it more difficult for readers to predict information, although some genres of technical texts are often organized around standard categories. Empirical reports, for example, typically follow the structure of introduction, methods, results, discussion, and conclusions. Organizational categories can be formed out of the subject matter of the text as well as its generic conventions. The links on the homepage of a company or university website are examples of subject-based categories.

Finally, some technical texts combine sequential and categorical principles of organization. A progress report might be organized primarily around the sequence of past work, present work, and future work, while the items under each of these headings may be organized according to specific parts of the project. Because they are often subject oriented and harder to follow, categorical arrangements should be used with the audience in mind. What categories will readers expect and most readily recognize? How can you make the pattern of categories clearer to readers?

Coherence **Coherence** might be defined as the logical flow and consistency of a text's arrangement. The arrangement of a coherent text not only flows smoothly and logically from one part to the next, but also "hangs together," to borrow a phrase from *Style: Ten Lessons in Clarity and Grace* by Joseph Williams. Readers must be able to identify the logical thread of ideas in a text. We can connect coherence to rhetorician Kenneth Burke's audience-based definition of *form* as the creation and fulfillment of desire. A text has form, Burke explains, insofar as it causes readers to expect and desire a sequence and then gratifies them with that sequence (124). Creating this form is especially important for proposals and other texts that depend on a particular argumentative sequence. Proposal readers need to see the logical progression from and connection among the problem, objectives, and solution. Tight coherence is less important for nonlinear texts such as organizational websites, although even these sites direct the flow of information in a limited number of directions.

Writers can create document-level coherence in several ways, most importantly by forecasting the structure of a text, arranging the blocks of information in a logical progression (sequential or categorical), and including transitional sentences between blocks of information. Coherence can also be an important criterion for deciding where to include visuals in a text and whether to include certain information in the main text or in an appendix. You don't want to make users flip back and forth between the verbal and visual components of a text, and you don't want supplementary information to detract from the text's flow.

Coherence is also a document design issue, especially for nonlinear professional and technical texts such as newsletters or organizational websites. Readers of such texts need to see how the multiple parts of the text work together, even if they are exploring or looking for particular pieces of information, and design features can clarify this. Well-designed websites, for example, create coherence by clearly indicating whether the user has left the site and by providing the reader with a consistent set of navigational aids.

Verbal–Visual Integration Integrating visual and verbal elements is an important step in creating cohesion or a sense of unity. Visuals that do not cohere with the rest of the text can confuse rather than help users. In general, visuals should be placed near the blocks of text that discuss them, as with instructions that place action-view drawings directly beside or below the steps they illustrate. In Chapter Seven we take you through the process of creating a visual blueprint of your document's layout, which can help ensure such coordination. Visuals can also be smoothly integrated through a three-step process of introducing them, explaining them, and helping readers draw conclusions from them (Anderson 295–96). Instead of simply placing tables or figures in a text and letting them speak for themselves, experienced technical communicators usually refer to them (e.g., "As Figure 1.3 illustrates . . .") and then help readers interpret them, pointing readers to specific parts of the visuals and emphasizing their main messages. This process may seem somewhat redundant, but it can clarify visuals and give users multiple ways to access the information.

Accessibility One of the most important concepts in technical and professional communication is **accessibility,** as this concept goes to the heart of the technical communicator's job to accommodate users. Unlike the readers of a novel, most readers of a workplace text do not read the text in its entirety from beginning to end. Instead, they read selectively and strategically to solve a problem or to find information or to perform a task. As Janice Redish and her colleagues at the American Institute for Research explain, most technical texts are written for "busy people who want to get in, get what they need, and get out of the document as quickly as possible" (131). Thus **retrievability** is also important. Think about the last time you looked in a manual or visited a website or read a memo or brochure: You probably skimmed the text and looked only for the information you wanted or needed at that particular moment. Therefore, you wanted that information to be easy to access and retrieve.

To make matters even more complicated, workplace texts often have multiple audiences who have different needs and interests. Therefore, professional communicators need to make different pathways of information accessible to different audiences. This is one way in which redundancy can be functional. A report may repeat some of the same information in the executive summary and introduction, but this is because some readers will only read one or the other.

Making a text accessible starts with arranging it from the readers' point of view, which you can begin to determine from your analysis of their questions, needs, uses, and so on. More specifically, making a text accessible involves previewing its contents, setting up signposts with headings, and making the organization of parts visually explicit (Redish et al. 145). Learning how to preview a text is one of the most important skills a communicator can learn. In correspondence, a good subject line can preview what the text is about. In line with their deductive organizations, the introductions of most technical and professional texts (from reports to proposals to instructions) typically include a **purpose statement** and a **forecasting element.** A purpose statement tells readers what the text is intended to do for them and often begins something like, "The purpose of this report is. . . ." Some purpose statements also tell readers how to approach and use the text. A forecasting element, which can be in the form of a sentence or a list, tells readers what the text will cover and usually previews the major parts of the text, which are also indicated by the major headings. Some technical texts combine the purpose statement and the forecasting element, as in the following example:

> "This tutorial is intended to guide the novice user through the process of designing a basic web page, from naming the page to adding text, links, and visuals, and color."

A longer document can benefit from such forecasting not only in the overall introduction, but also in the introduction to each major section. You'll see such sectional forecasting elements in most of the proposals and longer reports in the appendixes.

Arranging the text around headings and subheadings is another way to make it accessible. Although headings should not substitute for transitional statements in creating coherence, they can help guide readers through a text or locate a needed piece of information. Headings should be carefully crafted, however, to be both reader based and informative. Some headings are even written in the form of readers' questions. Vague headings can be worse than no headings at all, frustrating readers or causing them to overlook important information. Even headings that reflect the standard sections of a genre, such as "results" or "management plan" can be crafted to inform readers about the specific information covered. Another danger with headings is having too many levels of them; usually three levels is sufficient even for a longer document.

One way to make the arrangement of the text visually apparent is by making the headings stand out through such techniques as bolding, underlining, or capitalizing them. Blank space can also be used to visually separate sections. When using visual markers such as headings or blank space, be consistent and follow conventions (sometimes prescribed by style guides). All headings of the same level should look the same, and major headings should stand out more than minor ones.

Beyond previewing and signposting the sections of a document, technical and professional communicators deploy a variety of strategies to make the information in these sections more accessible. These strategies include:

- formatting text in bulleted or numbered lists
- breaking up text into short, manageable paragraphs
- using blank space to create blocks of information and guide the reader through them
- placing the most important information in the beginnings or ends of paragraphs.

You will utilize all of these strategies when you design your résumé for the next chapter.

In web-based documents, accessibility might be thought of in terms of **navigability.** In addition to finding and retrieving information, navigation involves moving through it in a strategic way to make connections. In general, hypertext documents are more dynamic than print ones; that is to say, users help shape the arrangement through the hyperlinks and pathways they follow. Although some hypertext documents, such as web tutorials, guide readers along a mostly linear pathway, others enable more varied reading movement. This potential for variation makes hypertext a good medium for texts with multiple audiences. Online documentation, for example, often provides differing pathways of help for experienced users and novice users.

As a creator of hypertext, it is important to remember that your arrangement of each screen and of the overall network of links makes certain pathways possible and others not possible. Because readers' navigation of hypertext is harder to predict than their navigation of a traditional print document, user testing may prove to be especially helpful. Students in one service-learning professional writing course performed user testing of the university computing center's website to determine how easy it was to navigate the site and to locate specific information within it. They then sent the center short reports describing their tests, outlining their findings, and recommending specific changes to the site.

To aid navigation across web pages, hypertext designers often include a frame or a table with the site's major headings and links on each page. This can also help users determine whether they have left the site. Other navigation-aiding techniques include arranging information to minimize the audience's horizontal and vertical scrolling, placing anchors to help readers quickly move from one part of a page to another, and creating multiple links to the same information. Headings, blank space, lists, and other techniques can increase the accessibility of hypertext as well as print documents.

Style

Style, the third canon of rhetoric, can be generally defined as the expression of a text shaped by word choice and sentence composition. It includes such aspects of the text as its clarity or readability, coherence, register, and tone. You may think of a technical style as fairly formal, objective, and full of jargon. But technical and professional style is as varied as the documents it is used to create. Appropriate style depends on a number of rhetorical elements, including the purpose, subject, type of communication (e.g., email or formal report), institutional conventions, and, most importantly, writer–reader relationship. The closer and more established the relationship, for example, the less formal the style probably needs to be. We can point to certain characteristics of style that most workplace audiences value and expect, however, based on their positions as busy, task-oriented users; these include clarity, active style, concision, coherence, and emphasis.

As a reader of instructions, reports, and correspondence yourself, you have probably noticed that these types of texts are more straightforward and concise than, say, novels, academic journal articles, or even some textbooks. **Clarity** refers to the readability and understandability of a text. A clear style enables readers to follow, understand, and use the information more easily. In *Style: Ten Lessons in Clarity and Grace,* Joseph Williams discusses clarity in terms of the actors and action in a sentence. Clarity is closely related to **active style,** our style focus in Chapter Eight. Williams advises writers to, whenever possible, make a sentence's subject an agent of action and a sentence's verb an expression of that action. Consider the following example:

> "The proposal meticulously followed the guidelines of the request for proposals (RFP)."

as opposed to,

> "The guidelines of the RFP were meticulously followed."

Action-oriented sentences are easier to comprehend and can also be more energetic and efficient. This is why technical and professional writers should consider focusing on the verb fairly quickly in the sentence, avoiding weak verbs such as *is* and *has,* and avoiding expletives or sentences that begin with "There are . . ." and "It is . . ." In addition, unless they have a good reason to use the passive voice (e.g., the actor in the sentence is not important), writers should use active voice, as this type of construction follows the abovementioned actor–action pattern and is therefore clearer.

We have found from our teaching experience that most students err on the side of wordiness rather than **concision,** often because they are trying to sound authoritative or are paying more attention to the subject matter than the audience. Ironically, a wordy text is more likely to detract from the writer's ethos than boost it. In addition to using active voice, writers can work toward concision by identifying and deleting meaningless or unnecessarily redundant words, replacing phrases with synonymous words, and combining short, related sentences. Although workplace texts should usually be concise, they should not be so concise that they seem overly simplistic (and thus insulting) or dull. Using various sentence structures from simple to compound–complex can make the style of a text more engaging and interesting. We elaborate on concision in the next chapter.

Sentence-level coherence is an important stylistic goal. Choppy, confusing, or otherwise incoherent prose can irritate readers or cause them to stop reading. An executive evaluating a proposal may dismiss it if she has to work too hard to follow the argument. A novice user may not be able to complete a task if the instructions are confusing or require constant rereading. In *Rhetorical Grammar,* Martha Kolln describes stylistic coherence in terms of a contract between the writer and reader(s). The writer has an obligation or contract, Kolln explains, to fulfill the reader's expectation that each sentence be connected to what has gone before (44). As we discussed in the section on arrangement, coherence requires that the parts of a text hang together as well as flow. This does not necessarily mean that all the sentences in a text or passage should create a completely unified whole, but that the sentences should work together to create larger topical patterns that readers can easily discern.

Coherence, or logical flow of thought, can be improved on the sentence level by using several techniques, which we will suggest that you apply to your project proposal in Chapter Five:

- creating given–new information chains
- using transitional words and phrases
- repeating key words or phrases
- ensuring that the progression of sentences makes logical sense
- maintaining a sense of focus among a group of connected sentences.

In a given–new information chain, a sentence will begin with a word or idea that ended the previous sentence; this enables the reader to enter the new sentence on more familiar ground and to assimilate more easily the new idea into her frame of reference. Transitional words and phrases can show relationship between sentences, though they should not substitute for linkages in meaning. Writers can rely too heavily on transitional elements and other cues to create coherence that should also be apparent from the progression of ideas. You may have read passages that include transitions but nevertheless required you to go back and reread sentences to discern where the argument was going.

Like coherence, stylistic **emphasis** is important but can be difficult to plan. In addition to emphasizing key information in a text visually, through placement, font variation, or other strategies, communicators can emphasize this information through its placement in the sentence. Generally speaking, readers pay the most attention to the information at the end of the sentence, particularly if the end is part of the sentence's main clause. Thus, the ending is a place of emphasis. We revisit this style topic in Chapter Nine's discussion of the evaluation report.

Tone and Voice Through their linguistic expression, communicators project a particular **voice** and **tone.** Tone, you may remember from previous writing classes, can be defined as the writer's attitude toward the subject and the audience. Although the most appropriate tone is determined by the writer's relationships to readers, the readers' needs, and other rhetorical considerations, a respectful yet friendly tone usually works well; a flippant or sarcastic tone rarely does. Whatever the tone, it should project goodwill toward the audience. Novice users of a tutorial may require a supportive tone, readers of reports may expect a straightforward tone, and evaluators of a proposal may call for a confident tone. Job application letters are an ideal type of text in which to experiment with tone, as they require you to draw the line between confidence and cockiness.

Voice is a related stylistic concept that can refer to the level of formality and rhetorical distance between writer and reader. The use of jargon, formal phrasing, and longer sentences can enhance formality, as can the use of third person. Although some of these characteristics are still common in some professional areas, such as those of science and law, the trend in workplace communication is toward informality and the use of first and second person. In some types of texts, such as instructions, first and second person are essentially requirements.

Delivery and Memory

Cicero described **delivery** as the physical presentation of a text. Although delivery has historically been viewed as the icing on the text's cake, a less substantive part of the text to be applied at its completion, this element is increasingly being recognized as crucial to a text's effectiveness. Effective delivery is both functional and aesthetically polished. The former quality creates goodwill, the latter credibility. The functionality of a text's delivery is, of course, situationally defined by its purposes, audiences, uses, and so on. In deciding the type of delivery that would be most appropriate for a situation, communicators should first consider medium, such as face-to-face discussion, print, email, or the web. Genre and format should be considered next, as they can influence the document's visual design. Brochures and résumés are but two genres that follow distinct design conventions.

The main components of delivery or physical presentation on which we will focus here are typography, nonlinear components, and document design. **Typography** may seem inconsequential, but it can significantly affect the readability and tone of a text. Typefaces can be classified as serif or sans serif, the former having finishing strokes on the ends of letters.

> Can you see the horizontal line strokes on the letters of this Times New Roman typeface?
>
> As you can see, this Helvetica font contains no such line strokes and is therefore a sans serif typeface.

Serif typefaces, such as Palatino, Garamond, and Times New Roman, are generally considered more readable for blocks of verbal text, as the serifs help guide the reader's eyes across the page. Some designers consider serif typefaces to be more elegant as well. In contrast, sans serif typefaces are often used in workplace texts for headings and verbal components of visuals such as labels and captions, as they allow for more holistic viewing. Because of their simplicity and, thus, readability, sans serif types such as Arial and Helvetica are also preferred for screens, where resolution can vary. Other elements of typefaces that can affect their readability are line thickness, width (i.e., the space between letters), size, and x-height (i.e., the difference in height between the lower and upper-case letters, which affects how bubbly the font looks).

Type styles, such as boldface, italics, and underlining, are examples of what John Frederick Reynolds has termed the "extratextual features" of delivery. Along with all capital letters, these styles can be used in headings to make the arrangement of a text visually apparent and in blocks of verbal text to emphasize key ideas or to make them more memorable. Although type styles are useful for emphasis, they can, in some situations, detract from readability. Italicized text is less readable on the screen, and all caps should rarely be used. As with typefaces in general, you should be highly selective in how you vary type styles because too many variations can detract from the impact and create an unprofessional tone not unlike that of junk mail.

In relation to emphasis, type styles can serve the fifth canon, **memory.** In ancient rhetoric, this canon referred to the successful memorization of speeches. Many contemporary composition and rhetoric scholars associate it with the memorability of a text, or how well the audience will be able to recall its features and content.

Nonlinear components, a term used by Charles Kostelnick and David Roberts, include visuals or illustrations as well as spacing and shading or color. One of the defining characteristics of technical communication is its heavy use of tables and figures. Proposals typically include timelines, organizational charts, and tables of costs. Annual reports often utilize pie and bar graphs. Photographs or action-view drawings are essential elements of most sets of instructions. Visuals can serve a variety of functions in technical texts, such as:

- emphasizing important information (e.g., callout of warning)
- showing what something looks like (e.g., photograph, drawing)
- showing how to perform an action (e.g., action-view drawing)
- showing relationships or trends among data (e.g., pie, bar, or line graph)
- rendering a large amount of data more accessible (e.g., table).

Visuals can also work with the verbal text in a variety of ways; they can complement it, supplement it, elaborate on it, emphasize it, or illustrate it, for example. For most of us, it may be easier to connect visuals to genres and purposes than to audiences. But workplace communicators should nevertheless consider with which types of visuals their audiences will be more familiar. Although novice audiences will likely understand simple tables, action-view drawings, and pie graphs, they may have a difficult time interpreting a schematic diagram, exploded-view drawing, or 100 percent bar graph.

Color can be a tricky element to manage. Although it certainly can emphasize information and add polish to the document, it is often used in a way that detracts from rather than enhances readability. Like type styles, color should be used selectively so that it maintains its impact. In addition to using color for emphasis, communicators can use it for reinforcing associations in meaning and for unifying the overall communication (Anderson 291). Instructions, for example, sometimes put cautions in red because many readers will immediately read this color as a cue to be alert. Many organizations, such as the American Cancer Society, use specific colors to represent themselves.

Spacing, or the distribution of blank space, is an often-underrecognized element of design. Blank space can serve many important functions. As Dragga and Gong explain, it "organizes and unifies the page, divides blocks of discourse, implies elegance, guides the eye through the page, and links neighboring pages" (171). Most readers prefer a generous amount of blank space, probably for both practical and aesthetic reasons. You'll need to consider your spacing of margins, of indentions, before and after headings and sections, around visuals, and between columns. Spacing also includes the **leading,** or spacing of the lines, as in single or double spaced. In addition to using blank space strategically, communicators should take care to use it consistently. Indeed, **consistency** is one of the most important criteria for good document design; because readers use spacing and other visual cues to interpret information, inconsistent cues will likely confuse them or inadvertently send them the wrong messages.

Summary

Technical and professional communication can be considered rhetorical in several respects: it is socially constructed in specific situations, it is audience oriented, and it is written for persuasive as well as informative purposes. Rhetoric, or the art of discovering the available and appropriate means of persuasion in a given situation, offers communicators a loaded toolbox of concepts for effectively critiquing, assessing, and composing technical texts. The most fundamental way communicators can approach writing rhetorically is by assessing their rhetorical situations, which include purposes, audiences, the subject, the text, and larger institutional and cultural influences. The means of persuasion that communicators can employ can be classified as ethos, pathos, or logos appeals, depending on whether they emphasize the writer's character, the emotions and values of the audience, or the reasoning of the argument, respectively. Communicators can conceptualize the texts they critique and construct as being comprised of five interconnected and overlapping components or canons: invention, arrangement, style, memory, and delivery. You will apply all of these conceptual tools in your service-learning project.

Activities

1. Apply the questions in the heuristic to the last major text you wrote for a course and then to the last nonacademic text you created (electronic texts count). What differences can you note between the two rhetorical situations? How can you account for these differences?

2. Discuss the term *rhetoric* in class. What negative and positive associations do you have with this term? Compare your understandings of the term with the description we've offered here.
3. With Ruggerio's three types of ethical criteria in mind (obligations, ideals, and consequences), list the qualities that you think make rhetoric ethical. Which of these might apply even more strongly to technical and professional texts? Both the Society for Technical Communication (STC) and the Association of Teachers of Technical Writing (ATTW) have established a code of ethics. Keep in mind that the latter organization's code applies to technical writing teachers as well as nonacademic professionals. Visit the following websites and relate the statements in their codes to Ruggerio's three types of ethical criteria.

   STC: http://www.stc.org/ethical.html
   ATTW: http://www.attw.org/ATTWcode.asp
4. Find a brochure or website of two student organizations on campus. Identify pathos, ethos, and logos appeals in the verbal and visual components of these documents. How do the different types of appeals reinforce each other? Which type of appeal does each organization primarily use to represent itself? Why do you think this is the case?
5. Go to one of the progress reports in the appendixes and make an outline of its arrangement (don't just depend on the headings and subheadings to do this). What patterns of organization (e.g., chronological, categorical, or both) does the progress report follow? In what ways might the report's arrangement meet or fail to meet the expectations of the audience (in this case, the instructor and agency contact people)?

Works Cited

Anderson, Paul V. *Technical Communication: A Reader-Centered Approach.* 4th ed. Fort Worth: Harcourt Brace, 1999.

Anson, Chris M., and L. Lee Forsberg. "Moving Beyond the Academic Community: Transitional Stages in Professional Writing." *Written Communication* 7.2 (April 1990): 200–231.

Aristotle. *On Rhetoric.* Trans. George A. Kennedy. New York: Oxford, 1991.

Bitzer, Lloyd F. "The Rhetorical Situation." *Philosophy and Rhetoric* 1 (1968): 1–14.

Burke, Kenneth. *Counter-Statement.* Berkeley: University of California Press, 1968.

Cicero. *On Oratory and Orators.* Trans. J. S. Watson. Carbondale: Southern Illinois University Press, 1970.

Couture, Barbara, and Jone Rymer Goldstein. *Cases for Technical and Professional Writing.* New York: Little, Brown & Co., 1985.

Crowley, Sharon, and Debra Hawhee. *Ancient Rhetorics for Contemporary Students.* 2nd ed. Boston: Allyn & Bacon, 1999.

Dobrin, David N. "What's Technical about Technical Writing?" *New Essays in Technical and Scientific Communication: Research, Theory, Practice.* Eds. Paul V. Anderson, R. John Brockmann, and Carolyn R. Miller. Farmingdale, NY: Baywood, 1983. 227–50.

Dragga, Sam, and Gwendolyn Gong. *Editing: The Design of Rhetoric.* Amityville, NY: Baywood, 1989.

Isocrates. *Isocrates.* Trans. George Norlin. 3 vols. Loeb Classical Library. Cambridge: Harvard University Press, 1992.

Jones, Dan. *Technical Writing Style.* Boston: Allyn & Bacon, 1998.

Kolln, Martha. *Rhetorical Grammar: Grammatical Choices, Rhetorical Effects.* 3rd ed. New York: Longman, 1999.

Kostelnick, Charles, and David D. Roberts. *Designing Visual Language: Strategies for Professional Communicators.* Boston: Allyn & Bacon, 1998.

Lay, Mary M., Billie J. Wahlstrom, Carolyn D. Rude, Cynthia L. Selfe, and Jack Selzer. *Technical Communication.* 2nd ed. Boston: McGraw-Hill, 2000.

Lunsford, Andrea, and Lisa Ede. "Audience Addressed/Audience Invoked: The Role of Audience in Composition Theory and Pedagogy." *College Composition and Communication* 35 (1984): 155–71.

Miller, Carolyn R. "Genre as Social Action." *Quarterly Journal of Speech* 70 (1984): 151–67.

Porter, James E. "Intertextuality and the Discourse Community." *Rhetoric Review* 5 (1986): 34–47.

Redish, Janice C., Robbin M. Battison, and Edward S. Gold. "Making Information Accessible to Readers." *Writing in Nonacademic Settings.* Eds. Lee Odell and Dixie Goswami. New York: Guilford Press, 129–153.

Reynolds, John Frederick. "Classical Rhetoric and Computer-Assisted Composition: Extra-Textual Features as Delivery." *Computer-Assisted Composition Journal* 3.3 (1989): 101–107.

Williams, Joseph M. *Style: Ten Lessons in Clarity and Grace.* 6th ed. New York: Longman, 2000.

Chapter 4

Choosing Your Project

As you'll recall, in the first three chapters of this book we presented some basic ideas about service-learning and technical and professional communication. We offered you a sense of what service-learning is and how it might be useful to the parties affected by it. We considered some basic rhetorical principles that can help you to be a more purposeful and persuasive writer. In the remaining chapters, we will address the process of completing a service-learning project from the earliest stage—choosing an agency—through the final stages of presenting your work to others.

Throughout this book we emphasize flexibility. We offer a range of approaches and possibilities for dealing with tasks and challenges and invite you to decide what will work best in your particular situation. The remaining chapters present a wide range of possible assignments and activities. As you explore sample student documents throughout these chapters as well as complete projects in the appendixes, you'll see that many students just like you have completed a great deal of impressive work over the course of a semester. Refer to your class syllabus and your instructor's comments to find out which assignments you'll be focusing on in your class.

In this chapter we'll address the first step of a service-learning project—finding a sponsoring organization. Your first reaction to this task might be concern; you may not have a lot of free time to volunteer in the community, or you may not immediately feel deep commitments to any social causes or community issues. As we discussed earlier, however, service-learning is not social activism or volunteer work, but clearly contextualized coursework. And you surely have some connection to community problems or issues, even if they are not obviously political.

In the pages that follow we'll help you to identify causes or concerns that interest you, to locate and investigate local agencies that address those concerns, and to narrow your focus to two or three sites that might work for you. We'll show you how to write a letter or inquiry and a résumé to send to an organization. This will allow you to introduce yourself and your project and to solicit information about possible collaboration. Finally, we'll give you and your classmates some suggestions for processing the information you collect through your correspondence. We'll give you ideas for presenting your project to your classmates and for pulling together collaborative groups. This section will describe a pitch day on which you and your classmates can present what you've learned about your agencies and then divide yourselves into writing groups with shared interests and goals.

Identifying Your Interests and Concerns

Most people have significant beliefs, concerns, and contacts that can lead them to a good service-learning project. You can use the heuristics below to help identify some of yours in a kind of self-profile. First, write down any community-related problems you see or encounter in your daily life—whether at work, at home, or with your friends—but haven't had a chance to do anything about. Maybe you are concerned about the homeless people you see walking around the outskirts of campus, or maybe you are worried about the quality of your city's drinking water.

To further generate possibilities, complete the following survey. For each of the community concerns below, rate your level of interest from one (low) to five (high). Keep in mind that you may need to consider several areas of interest before identifying a well-suited agency. The following list is not all inclusive but will offer you a starting point.

1. Children
   a. Early childhood education
   b. Foster care/adoption services
   c. Mentoring programs
   d. Parent education
   e. Drug and alcohol education

2. Civic/Community Concerns
   a. Art and music festivals
   b. Immigrant education programs
   c. Preservation of historical sites
   d. Sports and recreation
   e. Transportation
   f. Volunteer centers or community foundations
   g. Voter education

3. Civil Rights
   a. Capital punishment
   b. Disability concerns
   c. Race relations
   d. Legal assistance
   e. Discrimination
   f. Reproductive rights

4. Education
    a. Adult vocational education
    b. Art and music education
    c. Charter schools
    d. Literacy
    e. School funding
    f. Special education
5. Environment
    a. Domestic animal welfare
    b. Hazardous waste concerns
    c. Land preservation
    d. Recycling
    e. Solid waste disposal
    f. Wild animal conservation
    g. Water safety
6. Medical Issues
    a. AIDS
    b. Cancer
    c. Health education
    d. Immunization drives
    e. Medical research
    f. Mental health
    g. Reproductive health
7. Social and Family Services
    a. Housing/homelessness
    b. Hunger
    c. Services for elderly
    d. Domestic violence
    e. Child abuse
8. Your College Campus
    a. Beautification
    b. Career and placement issues
    c. Community outreach programs
    d. Crime victim services
    e. Health services
    f. Parking availability and transportation
    g. Recycling
    h. Safety/security
9. Your Workplace
    a. Community support programs
    b. Customer education
    c. Drug-testing policies

d. Employee crisis assistance
e. Fairness in hiring and promotion policies
f. Safety concerns
g. Smoking policies

After completing the survey and adding the highest-ranking areas to your growing list of concerns, take an extended break. Then come back, and this time reverse directions and begin to narrow your list of interests. Mark out ideas that seem relatively less interesting; highlight those for which you know you can find a collaborating agency. As you select four or five issues on which to focus your ensuing search, keep in mind the following:

- You may have considered earning potential when you chose your field of study. As you think about your service-learning project, imagine the kind of job you'd love to do if money and job security were not a concern. Maybe you've always wanted to work with children or be an artist. Maybe you'd like to spend all of your time exercising or reading books or fishing. As you identify the activities that give you the most satisfaction, imagine how they might connect to community efforts.
- If you choose to pursue a project that connects with a deep conviction or an interest you have, be sure that you're not selecting something that will be too emotionally painful for you or difficult to work on with others. To illustrate, one student initially thought she'd work with her local organ donor liaison office; she had this interest because her brother died while waiting for an organ transplant. Despite this deep connection, however, she ultimately decided that she wasn't emotionally ready to spend an extended period of time confronting and addressing the issue with others.
- As you think about your values and beliefs, keep in mind that some organizations might participate in activities or promote values that are in conflict with your university's policies. For instance, don't choose a project that promotes hatred for a particular type of people based on race, gender, sexual orientation, and the like. It's also generally best not to work for a particular political candidate's campaign.
- Most of you will be working on your projects in collaborative groups, so you'll want to choose something that others in your class will also find interesting. This doesn't mean that you can't bring in creative and unusual ideas—in our experience, fresh and unexpected kinds of projects are among the most popular and successful. Just don't choose something that would require all members and/or peer reviewers to have extensive knowledge about a field or to have certain strong religious or political viewpoints.

Identifying Possible Sponsors

After you've narrowed the field to several community concerns, it's time to start looking for agencies that address them. Use some or all of the techniques below to find organizations whose interests intersect with yours. You can apply these lists to several interest areas.

First, Inventory Your Existing Contacts

- organizations you're familiar with that deal directly with your top interests from the survey above
- organizations you worked with (even indirectly) through high school or college projects
- organizations with which your parents, siblings, spouse, or other family members have worked
- organizations from whom you've received assistance
- campus organizations of which you or someone you know is a member
- community organizations with which your church or other place of worship is affiliated
- community organizations affiliated with your employer
- organizations supported by businesses for whom you might like to work someday
- organizations led by community members whom you respect.

Next, Do Some Research to Expand Your List

When comparing your list of interests with the organizations that you have identified so far, you will probably find interests for which you still don't have matches. Even if you have found some promising candidates, you may not be aware of other suitable organizations on your campus or in the surrounding community. We often have our students do some field and web research to expand the list of sponsors they might target. Here are some types of resources to utilize:

- Talk with your *professors and classmates.* If you're looking for a project that relates directly to your field of study, ask your teachers, mentors, and peers in that field for suggestions of an organization that has a good reputation and that might be doing the kind of work that interests you. It isn't necessary to reinvent the wheel to find a good project. Take advantage of the local knowledge base.
- Go to your *city phone book.* If you are interested in a particular social problem, chances are there will be a list of local resources that address it in the directory. Look at community resource sections, government pages, and the yellow pages to find leads.
- Search your *local newspaper's website.* Most major newspapers have websites. Using a variety of keywords, search the site to find stories related to your area of concern.
- Check out your *newspaper's local or community section* for human-interest stories featuring projects by local agencies. These might include profiles of individuals or groups working to address local problems. Most newspapers contain such a section at least on a weekly basis.
- Contact your *area's volunteer center.* This might also be called a community alliance or community foundation. Most areas have such an agency, which provides information about local agencies and offers services that coordinate

potential volunteers and organizations with needs. Community foundations also help manage and distribute money for service agencies. If your area doesn't have one, ask a librarian to direct you to a database of regional, state, and national volunteer centers. Consider contacting the closest agency in your general area that addresses your area of concern and asking them where they would refer clients with such an interest in your town.

- Check out your *county's United Way website or publications.* This nationwide organization serves as a clearinghouse for a large number of community service agencies, distributing funds to them and aiding them with projects. Most United Way websites have links to, or at least information about, local affiliated agencies, from well-known organizations such as Big Brother/Big Sisters and Planned Parenthood to small, lesser-known ones such as a local consumer credit counseling service or community food bank. If you visit the local United Way office, you can collect not only its print publications describing affiliated agencies, but also brochures produced by the affiliated agencies themselves.
- Contact your area *Chamber of Commerce.* Most local chambers have websites with the contact information of, and even links to, member businesses and other organizations in the area. Most chambers also publish directories with information about their members; you can usually purchase such a book for a few dollars. Finally, your Chamber of Commerce may provide other publications such as a newsletter or brochures that tell about community services provided by local groups.
- Contact a *volunteer center on your campus.* More and more colleges and universities are integrating service into their general missions and curricula. Your campus might have a volunteer or service office connected with the office of student services or affairs. At the University of Florida, an organization called TreeHouse links students to local agencies looking for volunteers and helps instructors develop service-learning projects.
- Search for *campus organizations from the website of the office of student services, the student activities center, or the student union.* At the University of Florida, for example, all three have websites with information about or links to the web pages of numerous student organizations. If student organizations have offices in the student union or another place on campus, visit them to collect brochures and other print publications.
- Find out if your school is part of the *Campus Compact program,* a nationwide network of college and university presidents who are committed to emphasizing service on their campuses. If it is, speak with someone in your president's office. They can likely provide you with a list of cooperating businesses and agencies.
- Do *other online research* using a good search engine like Google. Enter keywords related to your issue and your geographic area. Explore the results; even those that don't point you to an organization might give you some hints about where to look or what kinds of projects you might pursue once you do find the right site.

OTHER VOICES 4.A Andrea Pacini

Andrea Pacini is a recent graduate of the University of Central Florida where she studied creative writing. She is a marketing and public relations associate for Vive *magazine.*

For my service-learning project I chose to work with the yearbook class at a local high school (Trinity Preparatory). I chose this agency because it was a fusion of my interests—I enjoy working with children and am going into the journalism field.

The most important thing I encountered when choosing a service-learning project was finding an agency whose cause I identified with and was interested in promoting. I strongly believe that being committed to and excited about your agency's cause will help waylay any frustrations or obstacles you might encounter along the way.

Another important factor to consider when choosing a project is the availability of your contact person. After many attempts to contact other agencies, our group decided to work with Trinity Prep because of the enthusiasm and prompt reply of its contact person. Chances are that if the contact person at an agency is too busy to accept an offer for free help, she or he will also be too busy to assist you over the course of the project.

Targeting a Few Organizations

Just as you narrowed your list of interests, you must narrow your list of possible sponsoring organizations. After all, you won't have time to further research all of them. At this point, try to narrow the list down to two or three that you will continue researching. These organizations will become the subjects of the agency profile that we discuss in the next section. Because you may subsequently discover that these organizations are not suitable, you should also have a couple of back-up choices.

When narrowing your choices, consider as many of the following questions as you can:

- How much do you already know about the organization?
- How well do the organization's values seem to intersect with yours?
- How physically accessible is the organization? How easily would you be able to get there?
- How well do the types of texts produced by the organization mesh with the texts that you've produced or want to produce?
- What professional benefits might your work with the agency provide you?

Gathering Information about Your Target Organizations: The Agency Profile

After you've identified the two or three organizations that hold the most promise for your service-learning project, you'll need to gather more information about them to determine the one to target in your letter of inquiry. We sometimes have students write a short memo in which they profile the two organizations that seem most promising and then present this to their classmates and instructor for feedback. The **agency profile** will not only be a selection and reflection tool for you, but will provide your instructor with helpful information about the promising agencies. Below we provide a list of questions your profile might answer. You can probably answer most of these from studying the organization's website and print informational materials. You should also be able to collect the information from a phone conversation with any staff member. We suggest starting with the websites and going from there.

General Information

- phone number, address, URL
- name of agency or department director and email address
- if different from above, name of contact person and email address
- location
- hours of operation
- number of staff members.

Services and Values

- What are the organization's mission and main objectives?
- What community problems does the organization address with its services?
- To what extent do the agency's expressed or enacted values intersect or clash with yours?

Texts

- What kinds of texts does the organization produce? What kinds of texts do they need revised or produced relatively soon?
- Who writes or produces these texts? Who supervises this production?
- Which of these texts seem most interesting? Which do you have experience writing? Which do you need experience writing?
- How would you describe the ethos projected in the agency's texts?

Collecting More Details: The Preliminary Visit and Trip Report

Some organizations do not have websites. Others don't describe the kinds of texts they produce on their sites. As a result, you may need to make a preliminary visit to one or more of your target agencies to gather the information you need. It's best to call ahead of time to make sure that you visit at a good time and that someone will be there to answer your questions. You should explain why you are visiting and why you are interested in the agency.

If you get a chance to have an extended conversation with an agency staff person, you might also get answers to the questions below, answers that you could add to your agency profile. Remember that you will also have an opportunity to request more information at your first major meeting with a contact person.

- Do they have student interns? Have they sponsored student projects from your school or other schools in the past?
- Who is typically the contact person for such projects?
- How equipped is the organization to work with groups of students? Does it have enough staff and space? Does it have regular hours?
- What is the agency's level of need? Are they drowning in work and therefore unlikely to have much time to work with you? Are they in an inactive period and therefore unlikely to need much assistance at this time?
- What kinds of projects interest them? Are they asking for projects that would require expertise or resources you don't have?

If you choose to make a preliminary visit, your instructor may ask you to write a **trip report,** a type of document that many businesspeople write on a regular basis. Even if you don't write one at this point, you will probably write one about your first major meeting with the agency's contact person. A trip report is a short, relatively informal account of a business trip or meeting. Written to an internal audience such as a manager (in this case your instructor and classmates), this report is usually presented in memo form. Many trip reports contain the following parts:

1. A brief overview of the trip's purpose, including the objectives you hoped the trip would accomplish.
2. A summary of the trip. Instead of a play-by-play narrative of the trip, this section should highlight its most important parts. With whom did you meet and what did you discuss? What did you observe? What materials did you see? Which questions were answered?
3. An analysis section. In this section explain what you think about your observations and activities on the trip. You might include reflections about the contact person's apparent attitude toward the collaboration or about the tone and atmosphere in the office. You might discuss previous projects that students have completed with the agency, or you might note specific ideas you have for a major project. You might also elaborate on your list of advantages and disadvantages of working with this agency.
4. A conclusion. In this final section, present a short and concise statement of your bottom-line opinion about working with the organization.

As you move closer to making a decision about which organization you'll address in the letter of inquiry, take some time to review all of the information you've compiled. Consider your lists of concerns alongside your list of promising agencies and your agency profile to ensure that they connect. Talk to your classmates about the agencies that seem the most suitable and interesting. After you decide on one, you are ready to make a more serious inquiry.

OTHER VOICES 4.B Barbara Heifferon

Barbara Heifferon, PhD., is assistant professor of technical writing in the English department of Clemson University in South Carolina.

Because I use problem-based learning, I present my students with a problem within the first week or two of the semester. During a recent semester, for example, I collaborated in writing and received a large USDA grant to help increase the number of applications for Food Stamps in the upstate of South Carolina to address the increasing malnutrition rates being reported among K–12 children and youth. For the first couple of weeks, the students in my Honors Technical Writing course did intensive research and wrote reports and graphed statistics based on the problem itself. We met as a class with the directors of the Department of Social Services (DSS) for the state of South Carolina and received further information. The students and I worked with the resources available in the Multimedia Authoring and Training Lab in which I teach. We have Adobe design products such as PageMaker, PhotoShop and Illustrator, Macromedia products, CD burners, plus video and audio digitizing capability. In addition, we have access to recording and videotaping studios. Part of the project-choosing process, therefore, involves familiarizing the students with the lab and the tools available to them.

Following presentations and discussions of our preliminary research, our class discussed and debated various ways to approach the problem and get the word out to people who needed to consider their possible eligibility for food stamps. We brainstormed via dry-erase boards and overhead computer projectors until we had five projects that addressed the problem using different media and conceptual approaches. Five teams then each chose one of the projects and prepared an eleven-page proposal to argue for their project's viability. To be a viable project, it would need to be completed in one semester and effectively address at least one aspect of the problem. Students completed two very different types of brochures, wrote newspaper stories, prepared advertisements, and designed, produced, and delivered radio and television public service announcements. Because all of our data is tracked by DSS, even before the class ended, we already received confirmation that student efforts had increased applications in the communities we targeted.

What seems to work best for students, communities, and me as facilitator in choosing projects is beginning with a problem that has many possible approaches, conducting shared preliminary research, introducing the tools available, brainstorming projects collaboratively, and having teams argue for and explore through collaboratively written proposals the viability of the projects they choose.

Letter of Inquiry—Rhetorical Situation

After selecting the most promising organization and doing some preliminary research about that organization and the kinds of documents they produce, the next step is formally contacting the organization regarding your interest. This brings us to the **letter of inquiry,** a common type of correspondence in technical and professional communication. As we noted in a previous section, you'll generally write print correspondence to readers outside of your organization as letters and write correspondence to readers within your organization as memos; that is, letters are for external audiences, and memos are for internal ones. Although your instructor may allow you to send the initial correspondence as email (which is already set up to resemble memo format), you will more likely contact the prospective organization through the more formal print genre of a letter.

If you are hoping to work with a nonprofit agency, your primary audience for this letter of inquiry will likely be the volunteer coordinator. In most agencies this person handles initial contacts with people who are interested in providing services. If you're not sure whom to address, you could write the letter to the head of the organization or the head of the department that produces the texts with which you want to help. In any case, be sure you direct the letter to a particular person, preferably someone who is a decision maker. Secondary audiences include other staff or board members who might read the letter, your classmates, and your instructor. Because your readers may not know you, and because they are likely busy and overworked, assume that they will give your letter a quick read rather than a careful perusal. Concision and accessibility will therefore be crucial.

The primary purpose of the letter is persuasive—to persuade the organization to respond to your inquiry and, ultimately, to sponsor a project for you. To accomplish this you will need to persuade them that you are qualified to help them produce texts and that they, like you, will benefit from the project. (You may enclose your résumé to help accomplish the former.) Because this letter may be the first piece of communication the readers receive from you, it may be used to assess your strengths as a writer. Your letter should persuade a staff person to respond, to answer your questions, to help you develop a project, and to function as your contact person and supervisor for the project. Your letter also has an informative purpose: to explain how the project fits into your course and what its parameters are.

Conventions of Document Design

Most of you are probably already familiar with conventional letter design and parts. As a reminder, Figure 4.1 illustrates the following parts, from top to bottom.

- the return address and date, the former usually designed as letterhead
- the inside address, or the address of the recipient
- the subject line, which is optional and used mainly in letters that function as short reports
- the salutation or greeting
- the body of the letter, which can include headings or lists
- The complimentary close and signature block
- Notations regarding enclosures or copies sent to others.

This letter is written in block format, which means that all items are flush left. Notice in Figure 4.1 that the writer takes care to use the reader's professional title in

Professor Melody A. Bowdon
Department of English
University of Central Florida
Orlando, FL 32816

December 1, 2000

Professor Blake Scott
Department of English
University of Florida
PO Box 117310
Gainesville, FL 32611

RE: Current Textbook Project

Dear Professor Scott,

Paragraph one starts here ____
____
____
____
____

Paragraph two starts here ____
____
____
____
____

Paragraph three starts here ____
____
____
____
____

Sincerely,

Melody A. Bowdon

Enclosure (1)

Figure 4.1 Letter format, block style

the inside address and salutation. Never use sexist titles such as Miss or Mrs., and avoid antiquated ones such as Sir or Madam. Don't begin a letter like this with the phrase "To Whom It May Concern." The tone should be friendly but professional.

Keep the letter to one page if possible. At the same time, remember that most readers prefer a document with a generous amount of blank space to one that looks crowded. Although Figure 4.1 shows a certain number of skipped lines between each element of the letter, you can adjust these, to some extent, to create a balanced page design; for example, you can skip extra lines between the first few elements to move the main text of the letter down further. Ideally, a one-page letter should fit like a picture in a frame.

Letter Parts

Introduction

The introduction of the letter should be short but must accomplish several things. First, it should tell the reader who you are and why you're writing. Your explanation of the latter should mention the course assignment as well as your specific interest in the organization. Perhaps you have a personal connection to the organization, share some of the organization's goals and values, and/or have skills that would be particularly useful to the organization. If you have already met the reader or someone else from the organization, mention this up front as well. If you've had a telephone conversation with the person, mention that contact, including the date.

Figure 4.2 shows a sample letter written by a student at the University of Central Florida. The student, Kristianna, is inquiring about the possibility of pursuing a project for the local Habitat for Humanity chapter. Notice how Kristianna establishes an early connection with her reader in the third sentence. Notice, too, how she ends the opening paragraph by emphasizing what she could do for the organization, couching her major qualifications in a reader-centered way.

This approach is a crucial part of a successful letter of inquiry. Many job application letter and résumé writers make the mistake of emphasizing themselves and their needs rather than the interests and expectations of the reader. One indication of such a problem is that numerous sentences start with *I*. Although the subject of the letter is partially you, it is also your readers, their needs, and their potential benefits from the project. Note Kristianna's primary emphasis on what she and her classmates have to offer the agency rather than on what they hope to gain from it.

Explanation

The body of the letter should explain more about the assignment and its parameters, explain the role of the organization in the project, and suggest a possible project idea. Your readers will need to know the scope and type of work you hope to do, the number of students (e.g., a group of three), and the time frame for your assignment. These parameters will depend on your course; the assignment may call for a group project or an individual one, a unit-long project or a semester-long one.

Kristianna Hope Fallows
888 Chipley Ct.
Winter Park, FL 32792

407-677-8888
khf2@pegasus.ucf.edu

September 7, 1999

Mr. Chris Jepson, Executive Director
Habitat for Humanity of Greater Orlando Area, Inc.
808 West Central Boulevard
Orlando, FL 32805

Dear Director Jepson,

My name is Kristianna Fallows and I am a Communications major at the University of Central Florida. As part of a Business Communication course, my classmates and I have been directed to assist a local community service organization with a major writing project. I am especially interested in working with your organization, as I very much share your goal or mission of helping struggling families in need. It is this shared commitment as well as my writing, design, and project management skills that would enable me to help you meet your communication needs (the enclose resume will give you a betters sense of these skills).

There are lots of options for our writing project, including the writing, document design, and/or editing of a number of genres, such as a newsletter, annual report, fundraising materials, website, or a set of office procedures. Most projects include one major text and perhaps a couple of related shorter ones as well. I noticed from your website that you were seeking a volunteer to produce your quarterly newsletter; this job along with editing your website would be ideal in scope. Another possible course project could involve producing promotional materials for your upcoming Hometown Heroes fundraising campaign.

There are two or three other students who would also assist you with the project. Not only would we work with you weekly on site, but we would implement strategies for producing and refining the project in our Business Communication course. While part of the project can involve research and other preparatory work, most of it should be dedicated to writing and designing the document(s). We can begin working in about a week. Our assignment must be completed by November 15. We would need a contact person at the agency who would serve along with our instructor as a co-supervisor.

Although each student in the course is sending out a letter like this one, we will only be able to work with a few organizations. If you think you could benefit from such a project, please contact me at your earliest convenience at the phone number or email address above. The next step in the project would be to meet with the contact person to develop a more specific proposal about what the project would entail. I will contact you in a few days to follow up on this letter. My classmates and I are quite excited and eager to help

(continued)

Figure 4.2 Letter of inquiry

Figure 4.2 *(continued)*

facilitate your efforts to improve substandard housing for working families. I look forward to talking with you about possible communication projects. Thank you for your consideration.

Sincerely,

Kristianna Fallows
Enclosure (1)

Perhaps the best way to explain scope and type of work is through examples of possible projects. In Kristianna's letter, for example, she not only mentions several examples of possible texts, but also impressively suggests two possible projects based on specific texts the organization produces. This move strengthens her ethos in two ways: It shows that she has done her homework and has given some thought to the project, and it displays her goodwill toward her audience and its needs.

After giving examples of projects, Kristianna describes some of the assignment's other parameters, including its collaborative nature and time frame. Her purpose here is to give the reader a general sense of how the project would work, not to overwhelm her with details that can be worked out later.

Call to Action

The final paragraph should call for some kind of response and perhaps inform the reader of when you plan to make contact again. Because Kristianna and her classmates are writing to more organizations than the class will actually work with, she suggests that the reader, if interested, should contact her soon. Kristianna wisely explains what the next step will be and tells the readers that she will contact them again as a follow up. The trick to writing a persuasive call to action in letters of inquiry, job application letters, and other similar correspondence is to sound energetic and confident but not conceited or overconfident. The clause "I look forward to talking with you, . . ." confidently assumes that another contact will be made but doesn't assume anything about the reader's reaction.

Once again, the point is to elicit the reader's response, not to overwhelm her or him with requests. You can work out the date and time of the next meeting and ask the reader about other possible projects during the next contact.

Style Focus: Concision

A concise letter will not only garner the appreciation of your busy readers, but it will also showcase your writing skills. If you exhaust your readers with a wordy letter or come across as a windy, ineffective communicator, they may not respond or even read the entire letter. In this subsection we will go over six techniques for making your technical or professional communication more concise.

1. Delete empty modifiers.
2. Replace redundant pairs with single, more precise words.
3. Replace phrases with single words.
4. Delete words that the reader can easily infer or doesn't otherwise need.
5. Combine two closely related simple sentences into a complex or a compound sentence (this may also help you create a more varied rhythm in your prose).
6. Replace expletive constructions with more direct, active constructions (this will also make your prose more active).

The first three techniques are adapted from Joseph Williams' *Style: Ten Lessons in Clarity and Grace,* one of our favorite style handbooks. If your course requires this book, you might study the lesson on concision before reading the examples below.

Figure 4.3 shows an earlier draft of Kristianna's letter with revisions for concision noted. The changes are documented with the "track changes" function in Microsoft Word; additions are underlined and deletions are crossed through.

Delete Empty Modifiers

Probably the most common empty modifier that plagues student writing is *very.* Look for the following types of words to determine whether or not the emphasis or qualification is needed (most of the time it is not or it could be conveyed less amorphously):

- very
- definitely
- certain
- quite
- basically
- various
- really
- generally
- actually
- practically

Williams aptly describes these words as "verbal tics that we use as unconsciously as we clear our throats" (141). As you can see, Kristianna's earlier draft contains two such unnecessary "tics," a *very much* in the first paragraph and a *quite* in the last one.

Replace Redundant Pairs

Redundant pairs of modifiers or other words, common in novice and expert writing alike, are sometimes created when writers haven't thought through the precise ideas they want to express. When such pairs are pointlessly redundant (i.e., the redundancy isn't an intentional strategy to create emphasis), they should either be replaced with a more appropriate and precise word or deleted altogether. Here are some examples of redundant pairs:

- each and every
- any and all
- first and foremost
- happy and excited
- certain and particular
- future plans

The pairs in the first row are also mentioned in Williams (140–142). In the first paragraph of her letter, Kristianna uses the words *goal* and *mission* to convey the

Kristianna Hope Fallows
888 Chipley Ct. 407-677-8888
Winter Park, FL 32792 khf2@pegasus.ucf.edu

September 7, 1999

Mr. Chris Jepson, Executive Director
Habitat for Humanity of Greater Orlando Area, Inc.
808 West Central Boulevard
Orlando, FL 32805

Dear Director Jepson,

My name is Kristianna Fallows and I am a Communications major at the University of Central Florida. As part of a Business Communication course, my classmates and I have been directed to assist a local community service organization with a major writing project. I am especially interested in working with your organization, as I ~~very much~~ share your goal ~~or mission~~ of helping struggling families in need. ~~It is t~~This shared commitment as well as my writing, design, and project management skills ~~that~~ would enable me to help you meet your communication needs (the enclose resume will give you a betters sense of these skills).

~~There are lots of options for our writing project, including~~ The course project can involve the writing, document design, and/or editing of a number of genres, such as a newsletter, annual report, fundraising materials, website, or a set of office procedures. Most projects include one major text and perhaps a couple of related shorter ones~~ as well.~~ I noticed from your website that you were seeking a volunteer to produce your quarterly newsletter; this job along with editing your website would be ideal in scope. Another possible course project could involve producing ~~promotional~~ materials for your upcoming Hometown Heroes fundraising campaign.

There are two or three other students who would also assist you with the project. Not only would we work with you weekly on site, we would implement strategies for producing and refining the project in our Business Communication course. While part of the project can involve research and other preparatory work, most of it should be dedicated to writing and designing the document(s). ~~We can begin working in about a week. Our assignment must be completed by~~Our time frame is from next week to November 15. We would need a contact person at the agency who would serve along with our instructor as a co-supervisor.

Although each student in the course is sending out a letter like this one, we will only be able to work with a few organizations. If you think you could benefit from such a project, please contact me at your earliest convenience at the phone number or email address above. The next step in the project would be to meet with the contact person to develop a more specific proposal ~~about what~~ for the project ~~would entail.~~ I will contact you in a

(continued)

Figure 4.3 Revised letter of inquiry

Figure 4.3 *(continued)*

few days to follow up on this letter. My classmates and I are ~~quite excited and~~ eager to help facilitate your efforts to improve substandard housing for working families. I look forward to talking with you, ~~about possible communication projects. T~~and I thank you for your consideration.

Sincerely,

Kristianna Fallows
Enclosure (1)

same meaning. Although either word would work, *goal* is probably closer to what she means, as mission statements are somewhat general. We see another redundant pair in the last paragraph when Kristianna conveys her excitement and eagerness. *Excited* is the less precise and persuasive word of the pair and could easily be deleted.

Replace Phrases with Words

Writers sometimes use wordy phrases, such as *due to the fact that,* to sound formal. Regardless of whether this formality is appropriate, however, readers of technical and professional writing will appreciate concision more. Therefore, you should replace the phrase with a single word whenever possible. Some such phrases function as transitional elements and others as modifiers. Here is a list of some common wordy phrases and the words they might be replaced with:

- due to the fact that → because
- in order to → to
- the reason for → why
- in regard to → regarding
- despite the fact that → although
- it is possible that → may
- in the event that → if
- take into consideration → consider

The two phrases in the last row contain nominalizations or nouns that have been formed from verbs; notice how the single-word replacement is simply the nominalization turned back into a verb.

In her earlier draft, Kristianna uses the phrase *it is important that* in the third paragraph, partially to emphasize the information that follows. Yet she can achieve this same emphasis by replacing the phrase with the word *must,* a move that also enables her to more easily combine the sentence with the previous one.

Delete Inferable Words

As part of perusing your prose for unnecessary words, look for words that other words in the sentence already imply. For example, the word *promotional* in the last sentence of Kristianna's second paragraph is already implied by the words *fundraising campaign*—fundraising campaigns are necessarily promotional. The deleted *as well* in the second paragraph and the deleted words in the third paragraph are also implied or otherwise unnecessary.

Combine Simple Sentences

Although combining sentences won't automatically trim your prose, it can have that effect by eliminating the need to repeat subjects or other words. Note how Kristianna combined the two closely related sentences in the third paragraph. Although she could have combined them with a comma and an *and,* she further condensed them into one simple sentence. Kristianna also combined the last two sentences of the letter, this time leaving the second sentence as an independent clause. This seems appropriate because both sentences are closing lines that point to future action in a way that creates goodwill.

Combining sentences can offer the bonus of making relationships among ideas in sentences clearer. If the function of one sentence is to modify another, more important one, for example, this relationship can be better conveyed by making the former sentence a dependent clause of the latter one. Consider the following two sentences: "My classmates and I have been directed to assist a community agency with a writing project. This is part of our Business Comm-unication course." The second sentence essentially supplements the first and is not important enough to be a separate sentence. We could better capture its function and significance by making it an opening dependent clause as in the following sentence: "As part of our Business Communication course, my classmates and I have been directed to assist a community agency with a writing project."

Replace Expletives with More Direct Constructions

Unless used to announce a topic or otherwise convey emphasis, expletive constructions are best avoided as they rely on weak verbs and create unnecessarily wordy sentences. Expletives—the most common of which are *There are/is . . .* and *It is . . .* —are sometimes used by writers when they aren't sure how to begin sentences or introduce ideas. To revise an expletive, determine what the sentence's main subject is, and then begin the sentence with this subject and the action it is taking.

Kristianna, in her earlier draft, begins the second paragraph with the expletive *There are . . .* After deleting this construction, she refigures the sentence into a basic subject–verb–object construction in which the subject and action are slightly more specific. This pattern enables Kristianna to simplify and shorten the sentence while leaving the important information at the end where it will receive more attention from the reader. We find another expletive—an *It is . . .* —at the end of the first paragraph. In this case the expletive can simply be deleted along with the word *that.* This eliminates the need for the "to be" verb.

Writing Workshops

We now turn to the first of our writing workshop guides. These guides not only help you critique your own and classmates' texts, but also highlight the criteria your instructor will use to evaluate your work.

As you'll see when you look at the guide in this section, we like to give our students a lot of specific direction for revision. Your instructor may not expect you to follow all of our suggestions explicitly, but we encourage you to at least consider each of the issues we raise. Feel free to address concerns not in the guides and to spend more time on questions that address the main weaknesses of your classmates' texts. More importantly, read skeptically, and don't hold back negative comments. Generic positive comments such as *looks good* do not help writers improve their drafts, although you do want to indicate the texts' strengths so that they aren't diluted. Whenever possible, comments should include specific suggestions for improvement. Provide both line-specific and summary comments on your copy of the writer's draft. You might list summary comments and suggestions from the most to the least important at the end of the text. If you have time after critiquing the drafts, discuss your reactions and notes with the writer, giving her or him a chance to ask questions.

One last note: In our experience, the more complete and polished the draft a student brings to a writing workshop, the more helpful the comments she or he will receive (and the better grade she or he will ultimately earn). Drafts for writing workshops should not be "rough," but should be as finished as possible.

Writing Workshop Guide for Letter of Inquiry

1. Glance over the letter's delivery or physical presentation; make suggestions for creating a more balanced page design, for improving the clarity of the font, and for creating a more professional impression. Check the letter against the conventional format displayed in Figure 4.1.
2. Taking the role of the writer's busy and overworked audience, quickly read the letter, underlining places that seem persuasive and writing question marks where you become confused or have questions.
3. Go to the places beside which you wrote question marks, and give specific suggestions for clarifying or further explaining those sections. Where could the writer better anticipate and answer your questions as a potential service-learning sponsor? Then identify places where the writer begins to overwhelm you with too many details for an initial inquiry.
4. Put stars by sentences in which the writer points to a connection with the organization, shows knowledge about the organization, shows concern for the organization's needs, or suggests possible projects. Which sentences in the letter could be recast as more reader centered and persuasive? If the writer begins several sentences with *I*, suggest alternate sentence structures and emphases for those sentences.
5. After making a brief outline of the letter, check the sequence of information against the sequence suggested earlier in this chapter. Draw wavy lines under information that seems out of order, and draw arrows showing where it should be moved. Which paragraphs seem to cover too many topics? Where could the writer better connect one sentence to the next?

6. Which information in the letter should the writer emphasize or make more accessible for the reader, and how might the writer do this?
7. Where does the writer's tone seem too casual or overly formal? Suggest word changes in these places.
8. Go through the letter sentence by sentence and word by word, identifying elements that seem verbose or unnecessary. Make sure each element contributes something new or does rhetorical work. Apply the six techniques for creating more concision. This is especially important if the letter is longer than one page.

Résumé—Rhetorical Situation

As your résumé will accompany and be read in conjunction with your letter of inquiry, its rhetorical situation is basically the same. The audience will be the letter's addressee and anyone else this person asks to review the two texts or passes them along to. As with the letter of inquiry, your main purposes for writing the résumé are to persuade the audience to respond to your inquiry with information and, ultimately, to agree to sponsor a service-learning writing project with you.

Although both the letter of inquiry and résumé have informative purposes, the résumé is less concerned with informing readers about the course, assignment, and possible projects and more concerned with informing them about you—your educational background, your relevant job experience, your achievements and qualifications. In addition to these purposes, the résumé has the more persuasive purpose of convincing readers that you are qualified to help them produce a professional and effective set of texts.

In the more typical context of a job search, résumés usually have two types of audiences—a human resources staff person or manager who does the initial screening, and a manager or decision maker in the more specific area of the job. Both audiences will be busy and will likely be evaluating numerous résumés; this is a highly competitive situation. The human resources staff member will likely serve a screening function, eliminating irrelevant or unqualified résumés based on a specific set of criteria. These criteria can include whether the résumé has a specific objective statement, whether the résumé contains particular qualifications, and whether the résumé is polished and error free. Sometimes résumés are scanned and searched for keywords as a screening technique (we explain design strategies for electronic, scannable résumés below). By sheer necessity, the person screening résumés will likely only spend a minute or less on each one before putting it into either a "reject" or an "accept for further consideration" pile. Because this reader will be looking for specific pieces of information and will be reading only to determine in a general sense whether or not you're qualified, he or she will scan rather than peruse the résumé. If your résumé is passed to the manager or committee supervising the specific job search, it may receive more attention but will still be read and assessed quickly. At this stage the evaluation might be more comparative; your résumé might be measured against other résumés and more specific criteria.

Because your situation is a little different from applying for a job, and because you may be writing to a relatively small organization, you may be able to send your letter and résumé directly to the appropriate decision maker. In addition, your situation in this class is not as competitive as in a job search, although you are competing for the reader's time and energy.

Because of the way your audience will read and assess your résumé, it must be accessible and memorable. Readers must be able to find the information (e.g., qualifications) they're looking for quickly and easily, and this information should be specific and impressive enough to help readers remember you. You don't want to run the risk of blending in with other applicants in the job search. The following sections recommend specific strategies for achieving both of these qualities.

Résumé Parts and Formats

Most résumés by college students contain most of the following basic parts, often in this order from top to bottom.

- name and contact information (be sure to include an email address)
- objective
- education
- work experience
- relevant skills
- honors and awards
- activities.

The major headings of Kristianna's résumé, Figure 4.4, show a variation of this sequence. To keep her résumé to one page, she doesn't include honors and awards (as they seem less relevant than the other information).

Figures 4.4 and 4.5 are variations of two common types of résumés. The former, Kristianna's résumé, follows a more traditional, *chronological* format. After the objective and education section, she overviews her work experience, complete with dates, in reverse chronological order. This type of résumé, used by many professionals, highlights the writer's work history.

A second type of résumé, called a *functional* résumé, is designed to highlight relevant work experience and skills (Lay et al. 609). In this way it is more specifically tailored to the job at hand. In Figure 4.5, Heather, a Penn State student, focuses on two of her work experiences that most vividly show her qualifications and on her related skills from coursework. Some functional résumés revolve around headings about specific skills such as client service, sales, and writing and editing. This approach is often used by applicants who do not have a consistent work history or who have gained much of their relevant experience in unpaid positions.

For this assignment and to apply for most jobs, we recommend that you use a combination of these two types of approaches, as both Kristianna and Heather do. Along with overviewing her recent work history, Kristianna includes special sections on computer skills and community service experience. Although Heather's

Kristianna Hope Fallows

888 Chipley Court — 407.677.8888
Winter Park, Florida 32792 — khf2@pegasus.ucf.edu

Objective

To apply my writing, computer, and office skills to a collaborative service-learning project that meets a local community service organizationís communication needs.

Education

B.A., English (specialization in Technical Writing), expected Spring 2002
University of Central Florida Orlando, FL

- Coordinated three major small-group writing projects, including a set of technical instructions for a mechanical engineering lab.
- Wrote and designed several common genres of professional communication, such as memos, letters, instructions, progress reports, a feasibility report, and a proposal.

Work Experience

Administrative Assistant for Audit Services, Sun Banks — Orlando, FL — 1999-00

- Prepared several audit reports per week
- Organized several department functions such as the annual Christmas celebration
- Ordered supplies, electronically filed audits, and performed other general office duties

Insurance Clerk, ABC Life Insurance Company — Heathrow, FL — 1997-98

- Maintained and improved configuration of company database
- Helped train more than 15 new employees
- Produced numerous informational reports using Microsoft Word and Excel

Data Entry Assistant, Northwood Staffing Services — Maitland, FL — 1997

- Entered client information and other data into databases for several companies, including ABC Life Insurance

Computer Skills

- Word Perfect, Microsoft Word
- PageMaker, Microsoft Publisher
- Data Entry: 11,000 kph
- Microsoft Exel
- Adobe Photoshop
- HTML

Community Volunteer Experience

- Race for the Cure and Adopt-a-Mile Benefits for the Susan G. Komen Breast Cancer Foundation, Zeta Tau Alpha Sorority
- MS Walk, Walk America, March of Dimes Walk-a-Thon

Figure 4.4 Traditional résumé

Heather Pierce
333 W. Nittany Ave. #3
State College, PA 16801
(814) 867-3333
hp3@psu.edu

Objective

To apply my strong communication and design skills in aiding a local community service organization produce a needed document or set of documents.

Related Work Experiences

Earth Day Coordinator, Penn State Earth Day Celebration 1998

- Designed and wrote more than 10 newspaper ads, flyers, and press releases publicizing the event
- Designed a comprehensive website promoting the event and explaining its goals
- Wrote proposals resulting in more than $4,000 of funding from various local businesses and university departments
- Coordinated the information tables, children's activities, dance and musical performances, and the sound system during the celebration
- Supervised more than 50 volunteers

Assistant to Education Director, Stroud Water Research Center 1996-97
(field station of the Academy of Natural Sciences of Philadelphia)

- Created more than 20 flash cards and posters to be used in educational programs and the White Clay Watershed Association's Stream Watch
- Helped administer summer programs focusing on stream ecology and the importance of trees to a healthy stream ecosystem
- Collected and analyzed seedling data for a tree shelter restoration project

Education and Related Skills

B.S. in Environmental Resource Management 2000
Minors in International Agriculture and Science, Technology, and Society
The Pennsylvania State University, University Park, PA
GPA: 3.81
Courses: Technical Writing, Argumentative Writing, Speech Communication
Computer Skills: Microsoft Office Suite (including Word, Excel, PowerPoint, and Publisher), Dreamweaver (web publishing)

Other Awards and Activities

Dean's List (5 semesters)
University Scholars Program
Pennsylvania Association for Sustainable Agriculture Conference
Sierra Club

Figure 4.5 Functional, skills-based résumé

résumé focuses on relevant skills, these are conveyed under subheadings of specific jobs (with dates). This résumé also includes a more traditional *awards and activities* section. Most college students do not have extended work experience and don't apply to jobs for which this is a qualification; therefore, showing a continuous record of employment is not crucial. We recommend that you focus, instead, on your most relevant jobs and arrange them in hierarchical order from the most to the least relevant and impressive. To conform to readers' expectations, you should probably overview your education before work experience unless the latter is particularly relevant. If your work experience is scant, you may want to focus on any work positions you held (e.g., lab assistant) or major projects you completed in college; this would require an extended education section. Now we turn to a more detailed discussion of each of the sections of the résumé.

Objective

Some résumé screeners look for this first. The objective statement should be concise yet informative. It should tell your readers the type of position (including level of permanency) and type of company you are targeting. Instead of saying that you want this job to give you experience or expand your skills, tell the reader what you will bring to the job. Kristianna's résumé illustrates this reader-centered move. For this assignment, the description of the position can be more general, as you aren't applying for a designated job. Keep in mind that this can be a tricky part of your résumé. We have seen many student résumés that include problematic objectives. Some are very vague, along the lines of "to acquire an entry-level position in a growing company." Others are far too specific, as in "to become assistant director of advertising at Acme Products, Inc." As you identify your job search goals, strike a careful balance between giving your reader a sense of the kind of job you'd like and showing flexibility about possible positions and duties. Many professionals consider objectives to be passé, as so many of them say so little. Discuss this concern with your instructor as you work on your résumé.

Education

This section should include the name of your college or university, your degree and major, any minors or areas of specialization (especially if relevant to the job), and your expected date of graduation. Lead with information that will be most important to your readers—in most cases your degree and major but in some cases the university. Provide your cumulative and/or major GPA if it's impressive or if it's specifically mentioned in the ad to which you're responding. You may also want to include a short list of relevant coursework. Because Kristianna hadn't yet taken many upper-division courses in her major, her résumé instead briefly describes some coursework-related skills under "education." The length of the education section should depend, in part, on how much work experience you have; if you have a substantial amount of relevant work experience, this section can be shorter. If, on the other hand, you don't have much work experience, you can expand your education section to include descriptions of major projects and coursework-related skills.

Work Experience

Sometimes labeled *related* or *relevant* work experience, this section should list your skills and achievements under subsections for the jobs you include. Bulleted lists will make your achievement statements more accessible. Lead with your position and company rather than the place or date (see Figures 4.4 and 4.5). Kristianna and Heather emphasize their previous job titles as subheadings using italics and boldface, respectively. As we stated before, you will probably be more persuasive if you list these subjects in hierarchical rather than reverse chronological order; remember that your readers will be scanning your résumé from top to bottom.

Skills Section

This section, which could alternatively be placed after education depending on its importance, is your chance to reinforce especially relevant and impressive skills, some of which you also show in your achievement statements. Remember that some résumé evaluators look for particular keywords or qualifications. In addition to computer skills (see Kristianna's résumé), you could emphasize foreign language, lab, or other relevant technical skills. Use a descriptive heading for this section.

Honors and Awards/Activities

These sections, which could be combined, are usually included to display in list format the job candidate's quest for excellence and well roundedness. Few employers want to hire one-dimensional or seemingly uninvolved employees; sometimes employers look to this section for interview questions. With this said, it has been our experience that many students include too much information in this section, listing every award and activity they can remember. Include only a sample of the most relevant and impressive ones, as Heather does in her résumé. In most cases you don't need to include dates with this information, though you may want to mention the number of semesters you made the honor roll.

Keywords

Résumés designed to be scanned electronically rather than read by a person generally include a keywords section that lists terms indicating key qualifications for the job. You don't want your résumé overlooked because the computer couldn't find enough relevant qualifications captured in keywords. Keywords (usually nouns) can include *HTML* and other computer skills, technical skills, communication skills, and previous positions held. Take care to use the terms of your audience, particularly those mentioned in the job ad. Only include terms that capture your qualifications, of course.

Invention—Relevancy

Make sure the information you include is as impressive and relevant as possible. This is not to say that only technical skills are relevant. Think, too, of what we call transferable skills and qualities. Review a list of all of the paid and volunteer positions you've held, and think carefully about what responsibilities they involved. Think about what you learned from each one and how each experience shaped you as an employee. Then review the list of words below to see how they match up with your experiences. Applicable to almost every job, these include the following:

- supervisory skills
- client service skills
- organizational skills
- enthusiasm
- strong work ethic
- teamwork skills
- project management skills
- initiative
- trustworthiness
- communication skills

Rather than listing these skills and qualities as you would computer or language skills, try to show them in action through your descriptions of educational and work-related achievements. The statements in Kristianna's second bulleted list, for example, convey her organizational skills. In the next list she highlights her supervisory experience. The fact that she was promoted from a part-time to a full-time job at ABC Life Insurance Company demonstrates her initiative and trustworthiness. In addition to her communication skills, Heather's bulleted statements show her ability to raise money, to supervise a complex event with numerous volunteers, and to communicate with different publics through different media.

For this assignment, of course, you will want to focus as much as possible on writing and communication skills such as speaking, designing, editing, and supervising others' writing. You should also include more specific elements of your writing experience such as the types and genres of texts you have written, the media and programs with which you have written, and the audiences to whom you have written. Because Kristianna is a technical writing major, she includes a list of genres in the education section. Heather's work experience section is persuasively focused on communication skills and achievements, such as designing a website, writing proposals, and creating posters. In addition, Heather wisely conveys the range of the texts she has produced.

Perhaps this is your first upper-division writing course and you haven't done much writing in your job. In this case, first determine the types of writing, speaking, and electronic communication you have done for other types of courses. Along the same lines, think through the daily activities you have performed at work, however small or informal; you might be surprised how many of them involve communication with coworkers or clients. Short press releases and memos are examples of professional writing. Often seemingly unrelated jobs such as those in the service industry require transferable skills. Although working as an insurance clerk doesn't seem all that related to writing for a nonprofit community service organization, Kristianna tapped into that experience to show her project management, writing, and supervisory skills.

Memorableness

Given that your résumé will, in most cases, be compared against many others that list similar qualifications, the memorableness of your résumé is crucial to its success. One way to make your résumé memorable is to include short sections describing major projects from coursework, jobs, or community service. A brief description of an individual or collaborative honors thesis or a senior design or research project may be more impressive than a list of relevant courses. Such a description—including a title, purpose, main tasks, and main findings—would be an effective way to show rather than tell your qualifications as well as give your audience something concrete to associate with you. Kristianna might have replaced her two bulleted items under education with a more detailed description of the set of technical instructions she designed and wrote. The first major section of Heather's résumé constitutes an excellent example of what we're suggesting; she describes in detail her roles and accomplishments around a single major project or event—the Penn State Earth Day Celebration. Through her thick description of this event, we get a strong sense of several of her qualifications.

Another way to create a more memorable résumé is to turn your short, bulleted statements of duties into what we call achievement statements. Sometimes, you can do this by quantifying your work. Instead of the statement "Wrote several fundraising proposals," you could tell readers that you wrote four comprehensive proposals that generated more than $4,000. Instead of saying that you assisted in new employee training, you could say that you helped train more than 15 new employees. Kristianna could have been more specific in her first statement in the work experience section of her résumé. Another way to pitch your duties as achievements is to present them in terms of specific projects or major tasks you completed or helped complete. Emphasize any special responsibilities you were given.

Effective achievement statements include other kinds of detail as well. As the sample résumés show, such statements typically begin with parallel verbs, often in past tense. Instead of using general, somewhat passive verbs such as *performed* or *worked,* use more specific and dynamic action verbs. The list below gives some examples of more powerful verbs.

- analyzed
- coordinated
- designed
- developed
- edited
- evaluated
- installed
- managed
- organized
- planned
- presented
- programmed
- sold
- trained
- wrote

For the most part Kristianna and Heather use such words in their statements. They might replace *created* and *produced* with more precise verbs, however. The rest of your achievement statements should also be concrete and specific.

Rather than stating that you worked with clients daily, for example, tell readers the specific things you did for clients. Heather's achievement statements do a particularly good job of providing details. Notice that she lists the activities she coordinated at the Earth Day Celebration and gives the name of the water protection program for which she created materials. A reader is more likely to remember a candidate who "produced flash cards and posters for Stream Watch" than a candidate who "produced educational materials for an environmental program."

Document Design and Delivery

As we discussed in Chapter Three, accessibility and retrievability are crucial elements of a technical or professional document's arrangement. This is especially true of résumés, given the speed at which they are assessed. Three elements that can increase your résumé's accessibility are bulleted lists, informative and prominent headings and subheadings, and easily discernable blocks and columns of information.

We have already discussed the major sections of a résumé—objective, education, work experience, awards and activities, and so on—and most major headings correspond to these. If you include a major section that describes more specific qualifications, make its heading more specific, as Kristianna does with *Computer Skills* and *Community Volunteer Experience.* You can ensure the prominence of your major headings by putting them in a larger and different font, boldfacing them, and/or separating them with blank space. Take a look at the two sample résumés: Kristianna makes her headings stand out by boldfacing them and designing blank space above them; Heather separates them from the rest of the résumé by putting them in a separate column. Both writers use a larger font size and a sans serif font that differs from the serif font of the main text.

Sans serif fonts, or fonts that do not have horizontal lines on the letters, are commonly used in professional documents for headings and text accompanying visual displays. Examples of sans serif fonts include Arial, Helvetica, and Tahoma. Because the serifs may help guide the reader across the page, serif fonts—for example, Times New Roman, Palatino, and Garamond—are typically used for the main text in printed documents. Kristianna uses Helvetica for major headings and Times New Roman for the rest of the résumé. Heather uses a Helvetica–Times New Roman combination.

You should visually emphasize your name at least as much as major headings. In addition to a larger font size, Kristianna and Heather use horizontal lines across the page to make their names stand out.

Your subheadings will probably be even more specific than your major headings, as they convey specific education and work experiences, such as the different job titles you've had. Notice how both Kristianna and Heather use their specific job titles as subheadings in their work experience sections. Many résumé writers make the mistake of beginning these subheadings with the names or locations of their employers or the dates of their employment, information that is probably not as important as what they did there. Your subheadings should also be easy to find and therefore might require boldfacing, italicizing, or some other variation in typography. We suggest that you only emphasize the most important part of the subheading, as Kristianna and Heather do in their résumés.

Whatever design decisions you make about the typography of the headings, subheadings, and the rest of the text, make sure you implement them *purposefully* and *consistently*. Using too many variations in typography can make your résumé look like a direct-mail advertisement and create an inelegant ethos. Major headings don't need to be in all caps, bolded, italicized, and in a larger font size, for example. Additionally, overusing typographic cues can detract from the emphasis such cues will have for readers.

If the person assessing your résumé cannot tell where parts of the text begin and end, she or he is less likely to find and remember specific pieces of information. Therefore it is important to manipulate spacing (along with headings) to designate sections visually. You can begin by creating space between sections. We also recommend that you begin the text of a section immediately after the heading for that section and that you use indentions sparingly (see Kristianna's résumé). In addition to creating blocks of information, align like items whenever possible. Heather's résumé, for example, aligns headings, the main text, and then the dates. Kristianna's résumé similarly aligns places of employment and dates of employment in columns.

The physical presentation of your résumé will play a crucial role in creating your ethos for the audience. In most cases your audience is meeting you for the first time through this document. Just as evaluators sometimes screen résumés by looking for mechanical mistakes, they may also screen using elements of delivery such as paper quality, print quality, font, and spacing. After all, if a candidate doesn't show attention to detail in her or his own job application materials, how can the employer expect her or him to do so on the job? Make sure you use high-quality bond paper and a laser printer. Most readers prefer white paper, but very light grey and beige may also be acceptable. In addition, use generous margins, as some readers will be turned off by a résumé that looks cramped; not only is a cramped résumé aesthetically unappealing, it also shows poor organizational skills.

Electronic Résumés and Web Résumés

More and more job searchers and employers are shifting from traditional to electronic résumés. As Mike Markel explains in *Technical Communication,* many organizations can't read the sheer number of résumés they receive and therefore scan electronic résumés into databases that they can later search for particular skills (thus the importance of a keywords section). The job searcher has the advantage of more quickly transmitting the résumé to an employer; electronic résumés are often sent in email messages or as email attachments. Print versions must be completely scannable.

Although electronic résumés require a more careful attention to keywords, their design must be much simpler to facilitate electronic transmission and scannability. Figure 4.6 shows an electronic version of Kristianna's résumé. You'll notice how much plainer it looks. Instead of using multiple design elements such as columns, rules, boldface, italics, and multiple font types and sizes, electronic résumés should be written in ASCII text, that is, using only basic letters, numbers, and punctuation marks.

This résumé illustrates some of the other design requirements Markel points out as well (467).

- Use a sans serif font such as Arial or Helvetica for better text clarity.
- Use the space bar instead of tabs for moving text (as tabs may convert to the settings on the reader's program).
- Put text in a single column, and make the margins extra wide.

Kristianna's electronic résumé is written in 12-point Arial font. Because she can't set off headings with bold text and a larger font size, she uses all caps for the major headings. The bullets have been deleted, the text has been taken out of columns and flush left, and the margins have been slightly expanded. As a result, the text is slightly longer, running more than a page, but this doesn't matter in electronic format. Kristianna has also added a keywords section, reinforcing communication, computer, and genre-related terms for which her employer will likely be looking.

More job searchers are also creating web versions of their résumés and posting them on Monster.com and other web job boards. Like print résumés and unlike electronic résumés, web résumés can include multiple design elements. You'll still need to transform the design elements of your print résumé, however, to conform to web conventions (which we overview in Chapter Seven). These include using mostly sans serif font, using different font sizes more than bold or italics and using tables and rules. A web résumé can also include some additional design features such as color (though your color scheme should be simple and high contrast), figures, links, and navigational tools. Unlike print résumés and like electronic résumés, precise length is not really an issue. You don't want the résumé to require too much vertical scrolling or any horizontal scrolling, however. Design with a fairly small screen size in mind. Here are some more specific design suggestions.

- Include links very selectively, as most links will take the user out of your site (thereby increasing the chance that they won't return); common links include your email, a university, a current or past employer, and a website showing some of your work.
- Limit the figures you include to icons or other small, simple elements; these should have clear rhetorical functions and not break up the résumé's design.
- Use tables to create columns.
- If your résumé requires some vertical scrolling, include anchors that enable the user to jump up and down from section to section.
- Avoid including a picture of yourself or irrelevant personal information, such as your age or marital status, as this may cause you to be discriminated against and may detract from the professional ethos of your text.

KRISTIANNA HOPE FALLOWS
888 Chipley Court
Winter Park, Florida 32792
407.867.8888
khf2@pegasus.ucf.edu

Objective: To apply my writing, computer, and office skills to a collaborative service-learning project that meets a local community service organization's communication needs.

EDUCATION
B.A., English (specialization in Technical Writing), expected Spring 2002
University of Central Florida, Orlando, FL
Coordinated three major small-group writing projects, including a set of technical instructions for a mechanical engineering lab.
Wrote and designed several common genres of professional communication, such as memos, letters, instructions, progress reports, a feasibility report, and a proposal.

WORK EXPERIENCE
Administrative Assistant for Audit Services, Sun Banks
Orlando, FL, 1999-00
Prepared several audit reports per week
Organized several department functions such as the annual Christmas celebration
Ordered supplies, electronically filed audits, and performed other general office duties

Insurance Clerk, ABC Life Insurance Company
Heathrow, FL, 1997-98
Maintained and improved configuration of company database
Helped train more than 15 new employees
Produced numerous informational reports using Microsoft Word and Excel

Data Entry Assistant, Northwood Staffing Services
Maitland, FL, 1997
Entered client information and other data into databases for several companies, including ABC Life Insurance

COMPUTER SKILLS
Word Perfect, Microsoft Word
Microsoft Exel
PageMaker, Microsoft Publisher
Adobe Photoshop
HTML
Data Entry: 11,000 kph

COMMUNITY VOLUNTEER EXPERIENCE
Race for the Cure and Adopt-a-Mile Benefits for the Susan G. Komen Breast Cancer Foundation, Zeta Tau Alpha Sorority
MS Walk, Walk America, March of Dimes Walk-a-Thon

KEYWORDS
Writing, professional communication, document design, collaboration, computer database, report, instructions, proposal, Excel, HTML

Figure 4.6 Electronic, scannable résumé

Writing Workshop Guide for Résumé

In this writing workshop, as in the first, base your responses on the interests and needs of the writer's audience—a representative of an organization deciding whether or not to approve a service-learning project. Remember that the following questions may capture the criteria your instructor will use to evaluate your résumé.

1. Putting yourself in the place of a busy agency staffperson, do a one-minute scan of the résumé. Then turn it over and write down both your general impressions and what you can remember about the writer. Ask the writer if you remembered everything he or she wanted you to recall. What could the writer do to make her or his main qualifications more immediately apparent and memorable?
2. Critique the résumé's arrangement, including the order and accessibility of information. What information should be moved around so that the items move from the most to the least relevant and impressive? How could the writer create more easily identifiable chunks of information with blank space? How could the headings and subheadings be more prominent and informative?
3. What font types would be best for clarity and for distinguishing headings from the main text? Where could the writer use boldface or other font variations to emphasize crucial pieces of information? If the writer uses too many visual cues (in number and type), suggest a way to simplify this. Where could the writer be more consistent with spacing and fonts?
4. Scan the résumé again, this time focusing on the relevance of the information to the job and the rhetorical situation. Write question marks beside places that seem to convey information unconnected to the qualifications for the job.
5. Focus now on any descriptive statements in the résumé. Draw wavy lines under statements that could sound more like achievements and could convey qualifications (writing or work related) more specifically and actively. Give suggestions for improving these statements. Where could the writer quantify or more specifically describe?

Making the Final Selection

When you have completed your letter of inquiry and your résumé and have sent them out to your target agency, you will move into the next phase of choosing a project. We recommend that you follow up with a phone call to your contact person a couple of days after you expect the materials to arrive. Let your contact person know that you are serious about working on a project with the agency and that you are willing to follow through on your commitments. Learn as much detail as possible about the exact types of writing the agency needs. Develop a sense of deadlines and due dates. Do as much as you can to assess the agency's level of enthusiasm for working with you.

If your instructor expects you to work in collaborative groups for this project, you'll need to convince some of your classmates to join in on the project with you. To facilitate this process, we recommend something we call *pitch day*. This is a predetermined class period during which each student has a brief amount of time to pitch her or his project and ideas to the class. While this is happening, someone should record the names of the agencies under consideration so that they are visible to the entire class. After all the pitches have been made, you can begin to shift around, talking to others about common interests, exciting possibilities, and projects that sound especially viable.

As you begin to form groups, compare your schedule with those of the other possible teammates. You might have one person write down each person's weekly days and times available to meet outside of class and/or visit the agency. Although at this point you're just checking for basic compatibility of schedules, you may need to find another group if you won't be able to meet with the others on a regular basis.

We have used the pitch day approach for several years and have found it to be an effective way for students to group themselves. In case your project is not chosen by other students, be on the lookout for other projects that mesh with your background, values, interests, and skills. Come prepared to negotiate and collaborate. This is a class period you don't want to miss.

Activities

1. Check your mail to find examples of correspondence—invitations to apply for credit cards, requests for contributions to nonprofit agencies, and the like. Apply the criteria listed in the letter of inquiry workshop guide to one or two. Think about the aspects of these letters that you find persuasive, offensive, engaging, or dull. Develop your own checklist for effective correspondence based on this analysis.
2. Visit job search websites online. Find several sample résumés produced by professionals in your field. Evaluate them in terms of the criteria in your workshop guide. Based on your analysis, make lists of effective and ineffective résumé-writing strategies specific to your own area of expertise. Think about such details as the order of sections, the amount of detail included in discussions of job experience, education, and so on. Also, make some notes about effective and ineffective attributes for electronic scannability. Write a memo to your instructor and classmates listing strategies you have learned about résumé design from this activity.
3. In a memo or a post to the class listserve, reflect on how you adapted your letter of inquiry and résumé to conform to the organization's interests and values. Describe additional revisions you would make to the document if you were targeting a similar job in the corporate world.
4. Find a letter of application and/or résumé you have written in the past, perhaps when you applied to your current employer or university or for a scholarship or award. Analyze ways in which you would improve that text based

on what you've learned in this chapter. Make notes about these improvements, and share them with the other members of your group. (You might also keep these notes to refer to the next time you are called upon to write such a document.)

Works Cited

Lay, Mary M., Billie Wahlstrom, Carolyn D. Rude, Cynthia L. Selfe, and Jack Selzer. *Technical Communication.* 2nd ed. Boston: McGraw-Hill, 2000.

Markel, Mike. *Technical Communication.* 6th ed. Boston: Bedford/St. Martins, 2001.

Williams, Joseph M. *Style: Ten Lessons in Clarity and Grace.* 6th ed. New York: Longman, 2000.

Chapter 5

Refining Your Project

In the last chapter we discussed strategies for choosing an organization with which to work and acquiring a general sense of the project you'll pursue with that agency. In this chapter we'll give you some suggestions for clarifying the details of your collaboration with that agency, including selecting the kinds of documents you'll produce and creating a plan for accomplishing your work in a time frame that's suitable to you, your instructor, and your agency representatives. We will introduce you to an important document genre in technical and professional communication—the **proposal**—and give you suggestions for using this writing assignment to help you carefully plan the work you'll do throughout the course term or unit.

Assessing Strengths and Comparing Schedules

Whether you're working in a collaborative group or on your own, the first step to refining your semester project is assessing your strengths and skills. In Chapter Six we recommend a group-management strategy that involves regular board meetings, or somewhat formal group discussions during which one group member takes notes and records your shared decisions for future reference. Depending on your instructor's approach, these notes may fit into a **field journal,** a document genre we'll discuss in the next chapter. The group members' responses to the strengths survey below would be a good topic for your first board meeting. Identifying your individual and collective strengths will not only aid you in getting to know each other, but will also help you to begin thinking about how your collaboration process might work, what kinds of projects you can take on together, and even how you might distribute tasks and leadership responsibilities. All of this is useful invention for the **management section** of the proposal you will write.

Begin this assessment by making a list of the abilities and skills that you know you could bring to the project. What skills and areas of expertise did you emphasize in the résumé you sent to the organization? What writing, writing-related, technical, and transferable skills have you developed in coursework, an internship, jobs, or your work with student or community organizations? What have teachers and peers told you were your strengths? Now rank those strengths from 1 (low) to 5 (high) in each of the following areas. Don't rank the areas in relation to one another.

1. Researching, including web, library, and/or field research
2. Planning and organizing a document
3. Drafting under pressure
4. Revising
5. Editing
6. Designing tables and figures
7. Experience with community or campus service work
8. Experience writing relevant genres
9. Designing print and/or web documents, experience with design software
10. Leading a group
11. Coordinating group activities
12. Accepting direction from others
13. Following group plans for meeting times and deadlines
14. Offering constructive criticism
15. Accepting constructive criticism
16. Managing conflict
17. Listening
18. Previous coursework related to the subject
19. Specialized knowledge or training related to the subject
20. Work experience with this organization
21. Access to necessary resources like computer equipment
22. Giving oral presentations.

Now, to assess your group's combined strengths, compare your scores with those of your teammates, possibly placing names of group members next to their highest-scoring areas. Be honest about your own strengths and weaknesses, but also be sure to indicate areas in which you'd like to develop your skills. For example, if you usually hate giving oral presentations in front of groups but want to improve this skill, let the group know so that you can plan for such an opportunity. We'll discuss managing individual responsibilities more in the next chapter. For now your goal is simply to get a sense of what kinds of projects you will and won't be prepared to handle.

After you've considered your combined skills, discuss your schedules in more detail. We understand that one of your greatest concerns about participating in a service-learning project is finding time to meet and work outside of class. In Chapter Six, we'll discuss ways to complete a range of tasks given your varied schedules. At this point you might construct a table showing each group member's committed and open time slots during the week. This can help you determine where your board meetings, agency time, and other group activities outside of class might fit. One or two time slots during the week might be sufficient. Also list any group members' upcoming trips or other expected absences to plan for alternate group times.

Identifying Possible Projects

Once you have a rough sense of what you can do and when you'll likely be able to do it, it's time to start thinking about the specifics of your project. As you move in this direction, consider (or reconsider) what you know about the organization's texts and their needs. You may want to review your agency profile. Consider what your agency contact person has told you in response to your letter and/or initial visit. Does this person already have a specific project in mind for you to pursue? Did she or he give you a list of possible projects that would be useful to the organization? Are the organization's expectations written in stone, or is there flexibility regarding the kinds of projects you might produce?

Next, consult your research about the organization's existing documents. What can you determine about the organization's document needs and expectations from reviewing these materials? Have you identified any texts, such as a website, that need major revision and updating? If your organization has requested that you produce texts for a specific project such as an upcoming fundraiser, what can you learn from the texts used in similar previous efforts? Doing this research will enable you to go into the next meeting with a more informed and persuasive ethos.

Don't let your invention about possible texts stop here, though. Your contact person may not recall all of the documents involved in a service or campaign, or you might identify a new writing need yourself. In the next section we provide an overview of the range of document types to consider.

Considering Various Genres

The genres we describe below are certainly not the only ones you might produce, but they constitute a fair representation of the types of projects our students have done. We provide more detailed descriptions of some of the major genres in Chapter Seven and other chapters. Later in this chapter, for example, we'll explain more about the rhetorical situation, parts, and conventions of proposals. Chapter Eight covers progress reports, and Chapter Nine discusses evaluation reports and other more formal reports. Because the genres vary in length and complexity, not all would make adequate projects by themselves, and some might require considerable assistance from agency personnel. Most of our students' projects have involved one complex text along with one or two shorter supplementary ones. If you're working as a one-person team, one text of medium length and complexity might be enough. Keep in mind, though, that the number of words included in a text does not directly correlate to its complexity; sometimes it's more difficult to present an important idea in a shorter document, and some types of document have more complex design elements than others. We've

OTHER VOICES 5.A Arthur Padilla

Arthur Padilla is executive director of the Sexual Minority Youth Assistance League in Washington, D.C.

Agency Mission Statement: As a youth service agency serving the metropolitan area of Washington, D.C., including Maryland and Northern Virginia, our mission is to support and enhance the self-esteem of sexual minority youth—any youth (13–21) who is lesbian, gay, bisexual, transgender (lgbt), or who may be questioning their sexuality—and to increase public awareness and understanding of their issues.

The Sexual Minority Youth Assistance League (SMYAL) staff and volunteers find ourselves writing on a daily basis. Much of the writing we do deals with financial concerns. Most frequently we must produce program reports and their accompanying descriptive narratives. These narratives are crucial to contract monitoring and ongoing communication with long-term funders. This writing also plays a crucial role in maintaining a historical record of SMYAL's program successes and failures.

Everyone on staff is also expected to write a brief piece for the agency newsletter each month. The newsletter is another important element in the communication with funders. However, in these texts we must also address the interests of our clients. This dual audience challenges the staff to be more creative and much less technical in their writing.

The most challenging of all the writing, however, tends to be the information needed for brochures, fact sheets, and other educational/information materials that are needed for large scale, general community education. The process is usually a focused effort to take a large amount of theoretical and research based writing, distill it into laymen's terms, then present it in a format that is both compelling and informative. It would seem that this is the most used and distributed information from any nonprofit group, yet it appears to be the least addressed issue for staff training. This type of information is necessary for small print media and social marketing campaigns, brochures and information sheets, website information, and for fundraising.

Student writers involved in service-learning activities with our agency could help with a range of writing tasks. They could work with staff members to design brochures or fliers for upcoming events and programs. They could be responsible for designing and editing one edition of our newsletter. Working closely with agency personnel, they could draft narrative descriptions of program developments and/or update information on our website. Any of these activities would benefit our agency and would give students an opportunity to learn a lot about how nonprofits work, what youths are concerned about, and how important effective technical and professional writing is.

divided the documents below into four categories according to their primary purposes, although some have multiple, overlapping purposes.

Group One: Promotional Materials

- **Press releases** are short text blurbs designed for release to newspapers, radio and television stations, and other media. They most often describe major news or upcoming events. For example, a group of University of Arizona students wrote press releases to advise the community of a new mayor's initiative to recruit mentors for local youth. Though these documents had to be quite short to ensure that they would be published or broadcast, they also had to be carefully crafted. The students had to find ways to quickly capture the interest of their target audiences and to give them the information they needed to consider participating in the program. Other students have written press releases about major fundraising events and about changes in organization leadership. (A few of our students have also written longer newspaper stories about upcoming agency events.) Press releases are also often written to offer an organization's position on a recent legal action or controversial event, but you should probably leave this kind of release to your organization's legal staff.

- **Billboards, posters,** and **flyers** often entail even fewer words than press releases. When Bill Wood was charged with helping his local Big Brothers/Big Sisters chapter to recruit male college students to serve as mentors for boys, he knew that one of the best times to reach this audience was while they are driving. He therefore developed a slogan and design that would capture the attention of drivers in the campus area. Although it didn't involve much writing, it did take a lot of research, planning, and creativity. These types of texts can be good supplements to a larger text in a project.
- **Fundraising letters** or **packets** are often vital to a small organization's fundraising plan. You may decide to write letters to request donations of money, goods, or services. When two University of Central Florida students worked with a local tutoring program for at-risk children, they drafted a series of letters requesting that community members provide funds to sponsor individual children, that local businesses provide materials for arts and crafts projects, and that an area entertainment attraction supply complimentary tickets for the children to take a field trip.

Group Two: Informational Materials

- **Brochures** are another good option for service-learning projects. These promotional and informational documents are critical to the success of many programs and can often be based on existing materials. Some organizations develop small **booklets** instead of brochures. Students can offer fresh and creative perspectives about designing such texts. You may remember our previous discussion of the Penn State students who worked with their campus health center to create brochures about sexually transmitted diseases. They took the medical information provided by their contacts at the center and translated it into student-appropriate language. They laid out the brochures so they would be appealing to students and the information would be easy to access.
- Many organizations seek help in creating or updating and revising **websites** about their work in general or to promote particular programs or products. One group of University of Florida students employed their professional writing and design skills to design a site for the English department's writing

internship program. Their work enabled interested students to obtain detailed information about opportunities for advancing their training.

- **Fact sheets** are generally one-page documents that provide lists of facts or pieces of information. You may have seen them in your doctor's office or in your university advising center. Like billboards, they're designed to attract a reader's attention. Fact sheets are typically used to alert readers about a change in policy, an urgent situation, or other information that could be important in their lives. A group of technical writing students at Penn State designed a set of fact sheets for Centre County Youth Services to teach parents about the warning signs of drug use.
- **Newsletters** can make excellent projects for several reasons: They involve both writing and design, they usually have clearly defined audiences, and their parts are usually easy to split up among group members. In our experience, most organizations are eager to find help with these projects. Newsletters may include feature stories, staff columns, announcements, calendars, and advertisements. One group of students actually launched a newsletter for the membership of their campus Asian-American Student Association. They researched the membership's interests and concerns. Two of the students continued to produce the newsletter long after the semester ended.

Group Three: Technical Documents

- Many agencies need help in creating **instructions, procedures,** or **manuals** for tasks performed by office personnel or volunteers. Some also need instructions written for clients or to accompany products they produce. Instructional documents usually include many illustrations, such as action-view drawings or photographs. A University of Florida student in Advanced Professional Writing created a list of instructions for entering new prescriptions into the database at the pharmacy where she worked. This document helped decrease mistakes by pharmacy techs and waiting time for customers. Her instructions, which were adopted by her company's pharmacies district wide, provided a meaningful community and business service.
- **Online documentation** and **tutorials** are usually web or CD-rom texts that help clients use a new product or system. Documentation is usually used as a reference resource; tutorials guide a user through a process step by step. Often written as hypertext, these documents contain even more complex design elements than print instructions. Authors need both experience with the subject of the instructions and technical ability in web production. A group of advanced technical writing students from Penn State worked with the campus computing center to create online documentation for a new web-based course discussion program called CourseTalk. Some students wrote documentation for the instructors, and others wrote it for the students. In both cases they field tested the documentation with potential users.
- Most campus and nonprofit organizations must on occasion submit **proposals** for funding, other needed resources, or even permission to pursue a project.

Because of their complexity and the detail they require, we don't suggest proposal projects for beginning students. Advanced business or technical writing students, though, like the ones who wrote the sample proposal you'll see later in this chapter, sometimes choose to write such documents. In this case the students wrote two short unsolicited proposals requesting funds for a no-kill animal shelter in Orlando. Proposal writing may be a more viable option if you are working with an organization that already has targeted a funding agency and has written an initial draft of its proposal.

Group Four: Business and Managerial Documents

- **Form letters** are staple documents in almost every organization. You may be asked to help your agency create or revise letters for such purposes as thanking contributors and advising job applicants of their status. A University of Arizona student connected with the American Cancer Society office in his hometown, and he had an opportunity to review and revise several of the agency's letters. His work freed up staff members to concentrate on the business of working with clients and providing services.
- When organizations are in the process of reviewing and reassessing their services or products, they often use **questionnaires, surveys,** or similar tools to gather information from their clientele. One of this book's authors, Melody, worked with a medical anthropologist to write an assessment report on the Tuscon AIDS Project's safer-sex education programs. When they finished the report, she and her coauthors agreed that they needed to create a questionnaire to gather more information about the program's target audience. Because this audience ranged in age from 16 to 23, Melody believed that her technical writing students that semester might have better language and content instincts than did she and her agency coworkers. After some research on safer sex and on designing questionnaires, the students rewrote the questionnaire to make it more audience-centered and precise.
- Every organization—nonprofit, business, or campus group—must regularly produce **reports.** This broad category of managerial and technical documents includes annual reports, financial reports, research reports, sales reports, feasibility reports, and many more. Your contact person may ask you to draft or revise this kind of document. This typically involves research and analysis as well as writing, so it requires a lot of support from your contact person. You may even choose to take on a project of your own as did one Penn State University student. She wanted to investigate the reasons for low student participation in the campus Catholic Student Association. After interviewing students and staff members and circulating a survey, she produced a recommendation report that the organization used to renew their recruitment efforts.
- **Program assessments** are critical to groups who receive funding or authority from outside their immediate control, and this includes just about every group. A highly specialized type of report, program assessments are often full of grids and graphs that represent the measurable results of agency efforts.

They often contain lists of objectives, descriptions of activities, and summaries of effects. A University of Central Florida student created such a document for a local outreach program for teens at risk, helping her agency secure ongoing state funding for the program.

Discussing Projects with the Contact Person: A Second Visit and Trip Report

A crucial component of planning a successful project is deliberating about possible texts with each other, your instructor, and your contact person. We strongly suggest that your group make a **site visit** at this point so that all of you can meet your contact person and the other agency staff, become more familiar with the agency, and negotiate the details of a project. This negotiation will involve balancing the organization's needs with your group's interests and strengths and with your instructor's requirements for the assignment.

Rather than just showing up, make a special appointment for the meeting, and tell the contact person what you want to accomplish. You might even specify the kinds of projects your group is thinking about ahead of time. Clarify, too, that you are interested in learning more about the agency and its specific writing needs. Allow enough time for the entire group to tour or observe the organization and meet members of the staff.

Also before the meeting, choose one or more group members to take detailed notes to be used for a field journal entry or a trip report (see our discussion of this genre in Chapter Four). As the sample trip report in Appendix A illustrates, this second trip report can help give your instructor a clearer picture of the projects you're negotiating (in case he or she needs to intervene) and can help you document important information about the agency and the project that can be incorporated into your proposal. Finally, you should prepare by generating a list of questions you need answered about the agency's services, texts, and resources.

Conducting the Meeting

Make sure you arrive at the meeting on time. We suggest you walk or ride to the organization together. After introducing the group members and reemphasizing your interest in the agency and its work, give the contact person a chance to ask clarification questions about the parameters of the project and what her or his specific role will be. Make sure you have consensus answers to these questions. You might even bring a handout that provides both your contact information and a list of the project's requirements (including dates). Let them know you'll be writing a proposal and a progress report for them and the instructor.

Next, ask your questions about the agency and its texts. If the contact person suggests possible projects, ask to see examples of similar documents (e.g., the texts from last year's fundraising campaign). Here are some factors about which you should probably inquire:

OTHER VOICES 5.B Nathan Trenteseaux

Nathan Trenteseaux is communications director for the United Way of Alachua County, Florida.

Soon after a group of University of Florida students selected United Way as their sponsor for a community writing project, I asked them to meet with me so we could discuss possible writing tasks. I explained to them that one of our most important ongoing projects is recording success stories of local individuals who have benefited from services provided by a United Way agency.

Each year, I ask the United Way agencies to send me names of clients who have been positively affected by the agencies' work. This year, I received 6–8 names. It was then my mission to have the students contact the people, interview them, and write the success stories.

Unfortunately, it is not always easy to track down people and get them to tell us their stories. Sometimes we have problems contacting the person, and sometimes they want to keep their experiences confidential. In many cases I have to play "phone tag" with the contacts before finding a convenient time to do the interview.

I assigned the students two stories each. I first sat down with them and gave them an overview of the success stories and the agencies involved. Then I gave the students copies of last year's campaign brochure so they would have an idea of how we write the stories and how they appear in the brochure. I also worked with the students on what types of questions to ask during the interview.

I explained to the students that we like to keep the stories fairly short and full of good quotations. The stories begin with background on how the individual or family came into contact with the agency and then go into the things the agency did for them and how this affected their lives.

I must say that everything went well. Once they had examples of stories and were able to contact the people, they had no problem writing the stories. I was always available to answer the students' questions and to offer suggestions for revisions. We met on a weekly basis to give updates, answer questions, and go over drafts.

I was extremely pleased with each student's effort and quality of work, and I'm glad that they learned something about United Way agencies and their positive effect on people in our community. With the success stories, others will learn about these effects as well.

Sitting down with the students and offering them encouragement and guidelines on how to interview and write the stories was key to their success. I also tried to keep a business casual atmosphere during meetings. I wanted them to feel at ease and free to talk to me in a conversational manner about the project or other concerns. Coming into a professional setting can be intimidating at first.

I look forward to the opportunity to work with the students again in the future.

- sample projects and texts, including identification of the most needed or pressing
- resources needed and available for producing texts
- people involved in producing texts
- organization's expectations for quality of content and appearance of texts
- organization's time frame.

As you receive answers to these questions, you'll want to assess how well the projects fit your interests. Don't be afraid to reinforce your group's suggested project at this point. The negotiation will likely involve some compromise on your and the agency's parts.

After you work out the most promising project and gather as many needed resources as possible, you'll want to discuss logistical details such as how often and when your group will visit the agency, the best way to communicate with the contact person (e.g., email or phone), and a tentative schedule for accruing feedback from the contact person.

In most cases the organization will have more than enough possible projects from which to choose. Occasionally, though, despite students' efforts to select a viable organization, the organization's personnel won't know or be able to effectively articulate what they want. Maybe they haven't worked with student writers before. Maybe they work on a limited number of projects that the staff can currently handle. If you find yourself in this situation, you can take more initiative in suggesting a project, being careful not to overstep or overcommit. This will probably require some extra research and an extended dialogue with organizational personnel and your instructor. It may also mean that you'll receive less directive feedback from your contact person over the course of the project. If the organization seems unable to come up with specific project ideas because it is disorganized, however, consider switching to another organization that responded favorably to a group member's letter of inquiry. Ask for your instructor's guidance at this point. A group of Blake's students decided to work with a student environmental organization on campus, even though they didn't know much about the organization. Their excitement about the organization's stated mission caused them to overlook the fact that the organization itself was in disarray and only sporadically active. The group was concerned but decided to propose a project with the organization anyway. They later regretted it. As group member Ari Luxenberg explained at the end of the project, "They [the organization] had no clear goals or organizational infrastructure and were incapable of making decisions or giving our group any guidance as far as the project was concerned." As this group experience warns, you probably shouldn't base your decision to work with an organization solely on shared interests.

Evaluating Your Project Idea

Once you've had a group site meeting and negotiated a project, run your idea through the following checklist to ensure that it meets everyone's requirements and expectations. This group exercise may seem redundant, but it can help you avoid much frustration during the next few weeks or months.

- *Is your planned project writing oriented?* Sometimes when students get excited about working with an organization, they lose sight of the goal of improving their writing. Similarly, agencies may be so excited to find that students are willing to work with them that they will request other kinds of help. One group of students offered to produce documents for a Tucson wildlife protection agency, but they were asked to do much more, including taking care of animals and maintaining facilities. Their contact person didn't have a clear sense of the purpose of the project, so they had to reinforce the assignment requirements. You may volunteer to do additional service for the agency on your own; just be sure to focus on a writing project for your course.

- *Does the project involve enough work to keep your entire group busy for the allotted time?* Sometimes agency personnel who dislike writing or have not actually written documents like those they're asking you to produce overestimate the complexity and difficulty of writing tasks. Avoid the last-minute scramble for more writing assignments that one group of students faced when they discovered that their local Easter Seals office only needed five very similar versions of one donation request letter.
- *Can you realistically complete the project in the allotted time?* While it's critical to be able to stay busy and produce a significant writing project at the end of the term or unit, it's also important to be sure that you can bring all of the pieces together by the deadline. A University of Central Florida student agreed to produce promotional materials for an Orlando charter school but was dismayed when she realized that the school needed far more assistance than she could offer in one semester. Her early efforts were so scattered that she had a hard time creating a larger coherent project to submit for her final course portfolio.
- *Do you have the skills, training, and abilities needed to complete the project?* If you completed the previous assessment exercises, then you've already given this question some thought. Be sure not to commit yourself to using software or equipment that you haven't already mastered or don't have time to learn. Select a project that matches your strengths, as did the University of Arizona environmental engineering students who applied their skills from a recent engineering course to a water resources project.
- *Are the documents you plan to produce needed?* Sometimes an organization contact person is so eager for assistance with writing projects that she or he will suggest project ideas before thinking through them thoroughly. In your visit you might ask the contact person who will read the documents, how the documents will be distributed, and what readers will do with the information presented in them. One student discovered quite late in the semester that the veteran's services agency with which he was working did not have a clear plan for using the proposal he was writing for them. He had to improvise a new plan at the last minute to make his project work.
- *Do you and the agency have the necessary resources to use, produce, and distribute your texts?* Again, sometimes a contact person's eagerness to obtain assistance or to help you with your class project may cause her or him to suggest projects that have not been carefully planned. Some may not understand everything that is involved with producing or using a particular text. People who don't know much about the web may not understand that they must have space for posting their sites as well as staff resources for maintaining them. Some may not have the funds available for creating desired documents. They may not know how expensive colorful brochures can be or recognize

that advertising on one billboard can cost several thousand dollars per month. This information came as a surprise to Bill Wood's contact person at the Big Brothers/Big Sisters office. To see his efforts pay off, Bill did some extra work soliciting donations and discounts for ad space.

- *Do you understand exactly what your contact person is expecting?* Making contact with a local organization through service-learning can be a mutually beneficial process, but it can also go from pleasant to frustrating if communication becomes shaky. You'll want as clear an idea as possible of what the agency representative is expecting. You may want to ask the representative to show you a sample of a document of the same basic quality and approach they're expecting to ensure that you're on the same page. You'll want to address such concerns as the length of the document, the expected level of technical information, and the expected final appearance. Get these matters straightened out up front to avoid the frustration faced by a group of Penn State students who wrote proposals for a local AIDS service organization. Toward the end of their project they were surprised to learn that their contact person had expected them to do extensive health care research on their subjects and to cite this in their proposals. Needless to say, they spent a few long nights at the library.
- *Does your contact person have the authority to assign you this task?* It's important to become as familiar as you can with the social context of your work. To avoid stepping on someone's toes, find out about the organization's chain of command. One group of students got into a complicated situation in the process of revising brochures for several programs at a local children's charity. Though their contact person (the director of the agency) authorized them to overhaul the documents, she failed to let the individual program directors know about this decision. When the students contacted these directors with their suggestions for substantial changes in the promotional materials, the directors were offended and frustrated. The students had to garner the goodwill of these collaborators before their writing and design suggestions were accepted.
- *Do you have the authority and access to information to complete all pieces of the project?* Keep in mind that many nonprofit agencies providing services to clients deal with sensitive and confidential issues. You may not have the legal or ethical right to access certain types of information that seem relevant to your work. If you face such a situation, ask your instructor and contact person about ways to compensate. One group of University of Arizona students working with a program for Latina at-risk teens ran into this problem when they were told to review client case files to get ideas for promotional materials. Ultimately, the problem was settled when the contact person agreed to summarize the information in the files, omitting such personal details as names and addresses.
- *Does the contact person understand precisely what is and is not your responsibility?* This question may seem redundant, but it is very important. Be sure that your contact person understands how much of the research you'll do and how much of the legwork for production and maintenance you'll provide. A group

of University of Central Florida engineering students who worked with a state wildlife and game agency found themselves in a tough spot when their contact person expected them to complete data collection and come up with a plan for distributing leaflets to fishing enthusiasts at state lakes. Be sure to make your abilities, commitments, and limitations clear to your contact person.

Many of the examples above may sound like service-learning horror stories. Don't get the impression from reading them that all projects are doomed to failure. That's certainly not the case. Despite all of the glitches the abovementioned students faced, they found ways to work out their difficulties and make impressive contributions to their agencies and communities. Often our students' best learning came out of such experiences, and we'll discuss how you can turn messy situations into interesting opportunities for improving your writing and management skills when we introduce the progress report assignment in Chapter Eight. You can avoid some frustration and save time, however, if you address problems with your contact person and instructor early or, if need be, negotiate another project. As we've noted before, a truly successful service-learning project must provide benefit to everyone involved. We've discussed many ways for making sure this happens—choosing your agency carefully so that it matches your interests and values, refining your project to take advantage of your skills. As you work on the project, be on the lookout for possibilities for turning this work into additional opportunities. One University of Arizona student was awarded a prestigious internship at his university's public radio affiliate partly as a result of his service-learning work there.

Moving Toward a Project Plan

Once everyone involved is satisfied with your project idea (which can involve several related texts), you can begin to develop a plan for tackling it, starting with dates and other logistical concerns. We recommend scheduling an initial group planning meeting in which you visually map out your project on a large calendar. You might draw one on an electronic whiteboard, a poster board, or a large piece of butcher paper. If you use paper, ask one member to convert the plan into electronic form to share with the entire group. Fill in the calendar with the following items:

- *Assignment due dates from your course syllabus.* Include the dates for the proposal, progress report, workshops, the evaluation report, project presentation, and any reflection assignments.
- *Important agency dates.* Using a different color, mark all meeting and workshop times with agency personnel, the agency due date, and the agency production date.
- *Group and individual scheduling.* Using yet another color, add information about weekly and other group meetings. When and where will the group meet? How many of these meetings will be at the agency? Your instructor may have a few days on the syllabus dedicated to group work. Mark also any

dates when members plan to be out of town, studying for major tests in other classes, or working extra hours at their jobs.

- *Fill in the steps of the process.* Keep in mind that you'll have to build in time for every phase of the document writing process. Whether you're doing a semester-long project or just one unit in the course, you'll need to plan carefully as you will likely have other class responsibilities and because due dates will not always match up with your process. Possible project phases include the following.
    1. writing the proposal
    2. doing necessary research
    3. planning the documents
    4. drafting the documents
    5. exchanging and commenting on the documents
    6. reflecting on the project
    7. planning your revisions
    8. revising the documents
    9. editing the documents
    10. putting the parts of the documents together
    11. transmitting the documents.
- *Using Chapter Six as a resource, plan your approaches to collaboration.* When and how might you split up tasks, work together, or hand off tasks from one person to the next? How do you plan to exchange and comment on documents?
- *Cover all the bases.* Email your plan to your teacher; then revise it in light of her or his suggestions. Double check your agency's deadlines and your teacher's expectations in terms of number of required pages, range of documents, and so on.

You can use much of this invention in drafting the management section of your proposal, the genre we discuss next.

Proposal—Rhetorical Situation

Now that you've met with your organization's contact person and tentatively worked out a project plan, it's time to formalize your plans in a proposal. A proposal is a request for an opportunity to pursue a project and is a critical component of many tasks in the workplace. This is a genre of writing that most professionals produce on a regular basis. One of your audiences for the service-learning project proposal will be your contact person and others at the organization involved with your project. You can think of this document as a contract between your group and your agency contact person(s). For these readers the proposal will confirm your assignment, clarify their roles, and outline important dates. Although this audience will likely keep your proposal on file, initially they may only skim the document.

The proposal is also a contract with the second audience, your instructor, who may not yet know the details of the work you have agreed to do. In addition to evaluating the quality of the writing in your proposal, your instructor will likely study the document closely to ensure that your proposed project meets the requirements of the assignment (e.g., scope, deadlines) and has been carefully thought through. If your proposal isn't detailed enough, especially about the implementation of the project, your instructor will likely have you revise it before you proceed. Thus, your proposal also has a persuasive purpose for this audience.

Like most proposals, yours will not only further specify and seek approval for the project, but also, once approved, serve as a reference document for you and your supervisors. For you, the proposal will function as a blueprint. The better the blueprint, the smoother the process and the better the product. Therefore, it is in your best interest to be as thorough and as detail oriented as possible. As a contract, the proposal will be a reference point against which your progress report and final product will be measured. For example, your progress report may refer directly to the timeline in your proposal.

Proposal Types

The proposal is one of the most diverse genres of technical and professional writing. Proposals can be long or short, formal or informal, external or internal, solicited or unsolicited, promising research or delivering goods and services. You might think of proposals as documents that offer to solve a problem. The complexity of the problem and the solution usually determines the complexity of the proposal. Relatively simple, easily implemented solutions may only require memos, letters, or emails; more extensive solutions may require longer, formal proposals.

As part of analyzing a proposal's rhetorical situation, it might be helpful to place it along a continuum from formally solicited to unsolicited. On one end we have proposals that are formally solicited by requests for proposals (RFPs). If a company, a government agency, or other organization needs a problem addressed, a product produced, or a service performed, it might issue an RFP to solicit solutions. Some RFPs set out very specific guidelines for the solution and the components of the actual proposal. On the other end of the continuum we have proposals that are not invited or solicited and, in some cases, address problems their audiences aren't even fully aware of. For example, an employee may discover a more efficient way to file student data and propose the new system to a supervisor who is content with the current system. As Markel points out, most formal proposals are written to audiences who at least have agreed informally to accept them (though not necessarily to approve them) because they require so much time and effort to compose (485). Many proposals fall somewhere between the two ends of the continuum. For example, some proposals are informally invited, and others are solicited through general requests or are written in response to general guidelines. Some proposals are written to audiences who recognize a problem but haven't yet sought solutions for it.

The degree to which a proposal responds to an audience's solicitation will certainly affect its shape. A proposal responding to a detailed RFP must adhere meticulously to the RFP's requirements and reflect its concerns. A proposal written to an audience who has identified a problem only generally or hasn't identified one at all will need to do more convincing in the problem section before providing the details of the solution. An internal proposal invited by your employer may not require as much detail about your capabilities and experience as one written in response to a publicly issued RFP.

The proposal for this assignment might be considered both solicited and unsolicited. Although you are being assigned to write the proposal, the specific community and writing problems it addresses have not yet been defined. Thus, you will need to convince your instructor that you are addressing an important, urgent set of problems before you convince her or him that your specific solution is desirable and feasible.

Proposals can also be categorized according to what Markel calls their deliverables, or "what the supplier will deliver at the end of the project" (486). Deliverables can be research or goods and services. Research proposals, common in the sciences, propose to study a problem and produce some type of research report of the study. Goods and services proposals promise to deliver some kind of product, service, or combination of the two. For example, an engineering firm might write a proposal to design and build a bridge, or a professional communication consultant might propose to conduct a set of workshops for improving communication within a company.

Your service-learning proposal and project may lead to both types of deliverables. You will discuss your service-learning "research" for the agency in a progress report and a final report for your instructor. More importantly, your project will produce a usable professional document or set of documents for your sponsoring organization.

Proposal Parts

Formal proposals (and, to some extent, short, informal proposals) are conventionally arranged around some version of the following set of sections.

1. introductory summary or overview
2. problem section, sometimes combined with the introductory summary
3. solution section, including a discussion of objectives that is sometimes a separate, preceding section
4. management section, including a discussion of the proposers' capabilities that is sometimes a separate section
5. cost or budget section
6. optional concluding section.

Because you likely will not be requesting funds for your project, your project proposal may not need a cost or budget section. If, however, you are working on a project that will require the purchase of supplies, you should include this in a brief budget section. You may indicate whether you expect to pay for these items yourselves or you expect the organization to supply them. Also, some nonprofit agencies can obtain tax credit and increased community funding when volunteers contribute a certain number of hours to their efforts, so it's good to estimate an approximate number of hours you expect to work on the project. Noting this in the proposal will help you to remember to keep track of your hours and to turn in a record of them at the end of the project. This will further benefit your organization.

In what follows we further describe the sections of a proposal in the context of your assignment, referring to the sample student proposal in Figure 5.1. Figure 5.1 is based on a proposal written by advanced professional writing students at the University of Central Florida. In this sample, three students propose to write two funding proposals and a shorter, related newsletter article for A Dog's Best Friend (ADBF) no-kill animal shelter in Orlando. Although we recommend that beginning technical and professional writing students not choose grant proposals as their major writing assignments due to the context-specific complexity of the genre, the student writers in this example had been exposed to technical writing and proposals from previous coursework and jobs. Although the proposals we include here and in the appendixes are for group projects, we realize that you might be proposing an individual one. This might simplify some parts of your proposal, such as the management section, but it might make other parts, such as the solution section, more challenging. You'll likely have to work harder to convince your instructor that you have the skills and initiative to carry out the solution by yourself. Your project will still be somewhat collaborative, of course, as agency contact people and your instructor will be involved.

Title Sheet

Even before the introductory summary, formal proposals begin with a title sheet that contains an informative title, the authors' names, the date, and the primary readers' names. The title in Figure 5.1 both indicates the general solution and tells readers the genre of the text.

Introductory Summary

Although you may think of this section as just an introduction, it has several crucial functions. In some rhetorical situations, the introductory summary will be the most-read section of the proposal. Busy managers and other readers may rely on this section to get a handle on what's being proposed and to decide where to turn next. The introductory summary sets the context for the rest of the proposal, telling readers the purpose of the text, previewing the problem and solution, and laying out the text's arrangement of subsequent sections. Managerial introductions, detailed introductions meant to give managers and other decision makers a full overview of the text, also preview the overall costs.

Remember that a purpose statement tells readers the rhetorical purpose of the text. The writers of the ADBF proposal in Figure 5.1 don't include such a statement, as the primary audience, the instructor, already knows the text's purpose. In other cases, however, the audience needs the immediate orientation provided by a purpose statement, either because they don't know the text's purpose(s) or they need to

Procuring Animal Care Funding for
A Dog's Best Friend Animal Shelter:
Project Proposal

By Team Buddy:
Maggie Boreman
Mark Dunn
Lori Phillips

Submitted to:
Professor Melody A. Bowdon, UCF
Director Charlotte Klokis, ADBF

February 10, 1998

Introductory Summary

A Dog's Best Friend (ADBF), the only truly no-kill animal shelter in Orlando, desperately needs more funding for animal care operations and shelter improvements. This nonprofit organization is supported wholly by donations and is run completely by volunteers.

The organization must supplement the individual donations its raises through its newsletter and web site with grants from foundations and other funding sources. Although the shelter's volunteers are aware of its needs, they do not have the time or expertise to research and write grant proposals.

Therefore, Team Buddy[1] proposes to produce two short grant proposals for ADBF, one for operational supplies and one for capital improvements. The project will involve meeting with the volunteers to prioritize and cost needs, identifying and researching promising funding sources, and writing and revising the proposals.

The remainder of this proposal will elaborate on the problems the project addresses, delineate the elements and steps of the project, and explain how we will carry out the project successfully.

Problem: Raising Funds for Operational and Capital Needs

ADBF is no-kill animal shelter (i.e., shelter that does not practice animal euthanasia) in Orlando with the following mission: to provide housing, food, and veterinary care to

(continued)

Figure 5.1 Team Buddy's project proposal

Figure 5.1 *(continued)*

discarded dogs and cats until they can be adopted by new, responsible pet owners who will spay or neuter them.

The growing number of stray or abandoned animals in the Orlando area has caused area shelters to become overcrowded, even as more and more animals are being euthanized by county animal control facilities. Thus, ADBF is operating beyond its capacity of 100 animals, which only compounds its funding crisis.

Funding and Writing Problem

ADBF is a nonprofit organization that operates solely through adoption fees and private donations by individuals (raised mostly through special events, letters, and the organization's newsletter and website). These funding sources are not meeting the shelter's operational and capital needs, outlined in the following lists:

Operational Needs

- dog and cat food
- cleaning supplies (e.g., litter, racks, pails)
- medical and veterinary expenses
- office supplies (e.g., file cabinets, copier)
- other supplies (e.g., cages, collars)

Capital Needs

- whole-house fan
- tile for floor
- new drainage system
- more kennels and runs
- van for transporting

Because some of the above expenses (especially capital ones) will be somewhat costly, ADBF needs a way to raise larger chunks of money. Such funding could be raised through grant proposals, but none of the current volunteers are able to write them. ADBF has no paid staff members; instead, the organization is directed and run by a handful of volunteers, four of whom run the physical shelter.

Although they are acutely aware of the shelter's needs and might also know of a couple potential funding sources, these volunteers have little to no experience researching and writing grant proposals. They are animal caretakers, not technical writers. The current volunteers wouldn't have time to write the proposals anyway, as they are already working overtime just to keep the shelter running.

Solution: Writing Two Grant Proposals to Animal Welfare Foundations

To solve the funding and writing problems, Team Buddy will write two grant proposals to national foundations specializing in animal welfare. The first proposal will request funding for operational expenses (including a copier), and the second will request funding for capital improvements and a van to transport animals. In addition, we will write an article for ADBF's email newsletter explaining the tasks involved in writing a grant proposal and asking for more volunteers to aid in this effort. Thus, our main objectives for this project are the following:

(continued)

- To identify about 10 or so potential foundations or other similar funding sources and to further research the two or three most promising ones.
- To produce two detailed and professional grant proposals out of our collaboration with ADBF volunteers.
- To secure at least one grant from this effort.
- To identify several existing or new volunteers who have some technical writing background and who could assist the core volunteers in writing future grant proposals.

The rest of this section will give a rationale for our solution and detail the four major tasks that it entails: 1) prioritizing needs and determining their costs; 2) identifying and researching promising funding sources; 3) writing and revising the two proposals; 4) writing supplementary materials, including materials to accompany the proposals and a newsletter article recruiting more volunteer proposal writers.

Rationale

We could have chosen to assist the shelter with several other communication projects, such as working on the design and content of its website and print newsletter or producing a set of informational and promotional texts for pet adoption days. According to the director of ADBF, however, grant proposals and other fundraising texts are needed more. Although we are not experienced technical writers, and proposals are complex texts, we will learn about proposals in class and will supplement this information with other reading. Further, we are getting proposal writing experience with this very assignment.

Most grant proposals go unfounded, and those that are successful usually only get one-time funding. Yet we will only be able to write two proposals, and many of the shelter's expenses are ongoing. To compensate for this, we will also try to recruit a group of volunteers to serve on a proposal writing committee. This group would compile a database of funding organizations and would write two or three proposals per year. We will write an article in the shelter's email and print newsletters proposing the committee and explaining what would be involved.

Prioritizing Needs and Determining Costs

The first task of our solution, and one that we have already begun (see list on p. 1), is to collaborate with those who run the shelter to specify further their operational and capital needs, prioritize those needs, and estimate the costs of those needs.

To arrive at some of the estimates, we may need to consult professionals as well, such as an auto dealership for pricing new and used vans or a contractor for pricing a new floor drainage system. We will clearly need to rely on the shelter's director and workers to prioritize their needs and decide which we should specify in our two proposals. Even

(continued)

Figure 5.1 *(continued)*

those we do not pursue can be pursued in future grant writing efforts. We will then try to target sources that will likely fund the most urgent needs.

Identifying and Researching Promising Funding Sources

The second major task, which we have also already begun and which will overlap some with the first one, is to identify and research animal welfare foundations and other promising funding sources. This will not only help us determine who our audiences will be for the two proposals, but will help us form a larger database of funding information for future proposal writers to use.

Using the Foundation Directory, we have already identified three potential grantees:

1) The Summerlee Foundation, which makes grants to alleviate fear, pain, and suffering of animals and to promote animal protection and the prevention of cruelty to animals.
2) The Company of Animals Fund, which funds spaying/neutering, adoptions, cruelty investigations, and rescue. The money must be used for emergency or ongoing care of companion animals.
3) The Edith Goode Foundation, which funds efforts to prevent cruelty to animals and to promote animal welfare.

Using foundation directories, we will find out more about each funding source, including its mission and values, the types of projects it funds and doesn't fund, short descriptions of recently funded projects, and requirements for proposals and funded projects. We may need to adjust the prioritized list slightly based on the requirements of the funding sources we target. For example, most foundations have a limit on the amount and only fund certain types of items or projects.

Writing and Revising the Proposals

After we get a better grip on our audiences and their requirements, we will be in a better position to begin drafting the actual proposals, which will likely include roughly the same parts as this proposal along with a cost section. Some foundations ask for shorter proposals, however. Before drafting we will make a comprehensive list of specific questions to which our audiences will likely expect answers. Also before drafting we will make an outline of the proposals' major sections and what they will cover; this will ensure conformity to requirements as well as coherence.

In drafting we will make sure that we answer the audiences' questions in as much detail as necessary. We will try not to include extraneous information. Drafting the cost sections of the proposals will require us to work with the shelter's accountant. For the management and cost sections, we will first draft any illustrations and then the accompanying text.

Once we have a complete, carefully written draft, we will workshop each proposal twice, once with our classmates and once with our contact person at ADBF. The former

(continued)

workshop will focus on the proposal's persuasiveness to its audience and its concision, coherence, and cohesion. The latter workshop will focus in part on how accurately it represents the needs of the agency. We will also give our instructor drafts for comment.

After extensive revising of our drafts based on the comments from the workshops and our instructor, we will closely edit the proposal, fine-tuning its style and physical presentation. This part of the process will ensure that the proposal is as error free and polished as possible.

Writing Supplementary Materials

Finally, after we have completed the proposal writing and editing process, we will need to write cover letters to transmit our proposals to the two funding sources. These should be fairly easy to write but must explain what the proposals address and must highlight their most important points. For this part of the project, we will consult a proposal-writing textbook about conventions of transmittal letters. We will also closely consult with ADBF's director in writing and revising the letters, as they will also bear her signature.

The second supplementary text we will write is a newsletter article recruiting volunteers for a proposal writing committee. In this letter we will explain the proposal writing process, overview the resources volunteers would have to work with (including a proposal writing textbook, sample proposals, and a database of needs and foundations), and explain how the committee would work. The article will be written as a short and informal "How to Volunteer" piece.

Management: Collaborating and Completing Tasks

Now that we have explained what the parts of our project will involve, we will explain how we will successfully enact them. This section will first explain who will work on the project and what their roles will be, how we will collaborate with each other and ADBF volunteers, and what resources we will draw upon. The second part will present our timeline for completing the project and its parts.

Team Roles and Collaboration

The three members of Team Buddy are Maggie Boreman, Mark Dunn, and Lori Phillips. Maggie adopted a puppy from ADBF and is already fairly familiar with how the shelter operates. Additionally, as a technical writing major she has written proposals for several course projects, one of which was business plan. Journalism major Mark Dunn has substantial writing and design experience, including working as an editor for the campus newspaper. He has also served as a volunteer in two local community service organizations. Lori Phillips, a mechanical engineering major, is part of a team who wrote a successful proposal for designing a solar car. She designed all of the figures and tables for this proposal.

Team Buddy's contact person at ADBF is Charlotte Klokis, its director. She has agreed to meet with us regularly as well as give us extensive feedback about our work. We will also be consulting Rachel Walker, the shelter's accountant, about the cost sections of the two

(continued)

Figure 5.1 *continued)*

proposals. Our professor, Professor Melody Bowdon, will supervise our work throughout the project as well. Here is the contact information for all involved in the project:

Table 1: Contact Information for Project Participants

Name	Phone #	Email
Maggie Boreman	407-677-8554	mbb2@pegasus.ucf.edu
Mark Dunn	407-677-2187	newsman1@hotmail.com
Lori Phillips	407-607-3338	lphillips@pegasus.ucf.edu
Charlotte Klokis	407-605-9595	klokis@hotmail.com
Rachel Walker	407-605-9595	rachelwalk@aol.com
Melody Bowdon	407-607-6500 ext.230	mab@pegasus.ucf.edu

As Figure 1 shows below, Maggie will be the team's organizational leader and main liaison to Ms. Klokis, whom she already knows. Maggie will call and remind the other members of group meetings, ensure that meetings are productive, and relay questions and messages from the group to Ms. Klokis and vice versa. Although all three group members will contribute to the research, writing, design, and revising of the proposals, Mark will supervise the writing process, meaning he will make sure members get their parts written and create a system for revising and proofreading each part. Mark will also be the group's main liaison to Professor Bowdon. Because of her research and design background, Lori will oversee the group's research of funding sources and design of all figures and tables to be included in the proposals. This latter task will also require Lori to be the one who meets with Ms. Walker, the accountant.

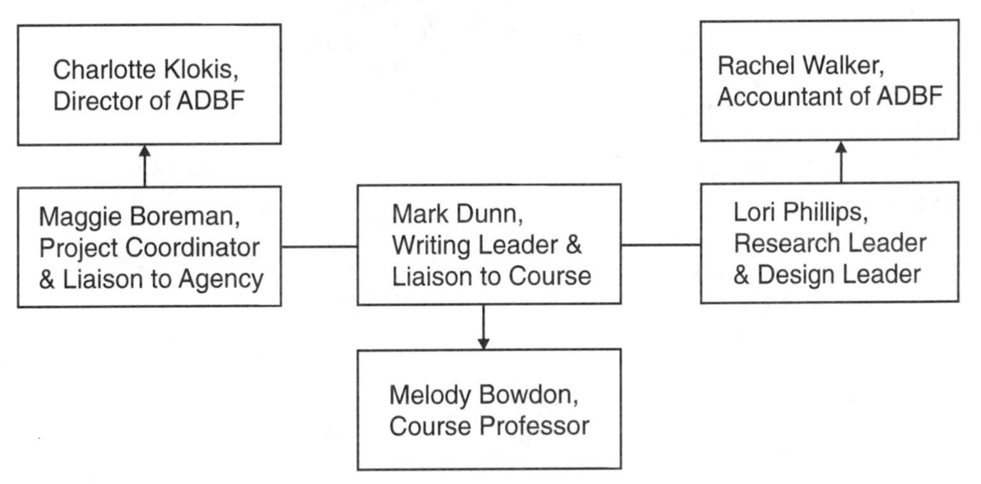

Figure 1: Organizational Chart of Project Participants

Meetings and Timeline

In addition to class time designated for working on our projects, our class time will have two standing meetings per week at the following days/times/locations: 1) Mondays,

(continued)

Figure 5.1 *(continued)*

10:30 A.M.–12 P.M. in our classroom computer lab after class; 2) Thursdays, 4–6 P.M. in the Library West computer lab. Meeting in computer labs will enable us to design, draft, and revise online, thereby saving time and effort. When we have less to work on, we may decide to conduct meetings in our course's synchronous chat room on the web. We also understand that we will need to hold additional meetings at certain points in the project.

Most of our contact with our contact person at ADBF, Director Charlotte Klokis, will be through email, all of which Maggie will send. Ms. Klokis has also agreed to meet with our group to review our work once at a week, on Mondays at 1:30 P.M. at the agency. Some of these meetings will require the entire group to attend, and some will only require Maggie, the agency liaison, to attend. On weeks that Ms. Klokis cannot meet, we will send her updated materials as email attachments. We will need to visit the shelter a little more toward the beginning of the project when we further determine the agency's needs and toward the end when we ask for feedback from our contact person, the accountant, and possibly others.

Figure 2 below is a milestone chart showing the project's major tasks, their timeframes, and the group members supervising them. You will also notice important due dates, some for course requirements, some for agency requirements, and some for both.

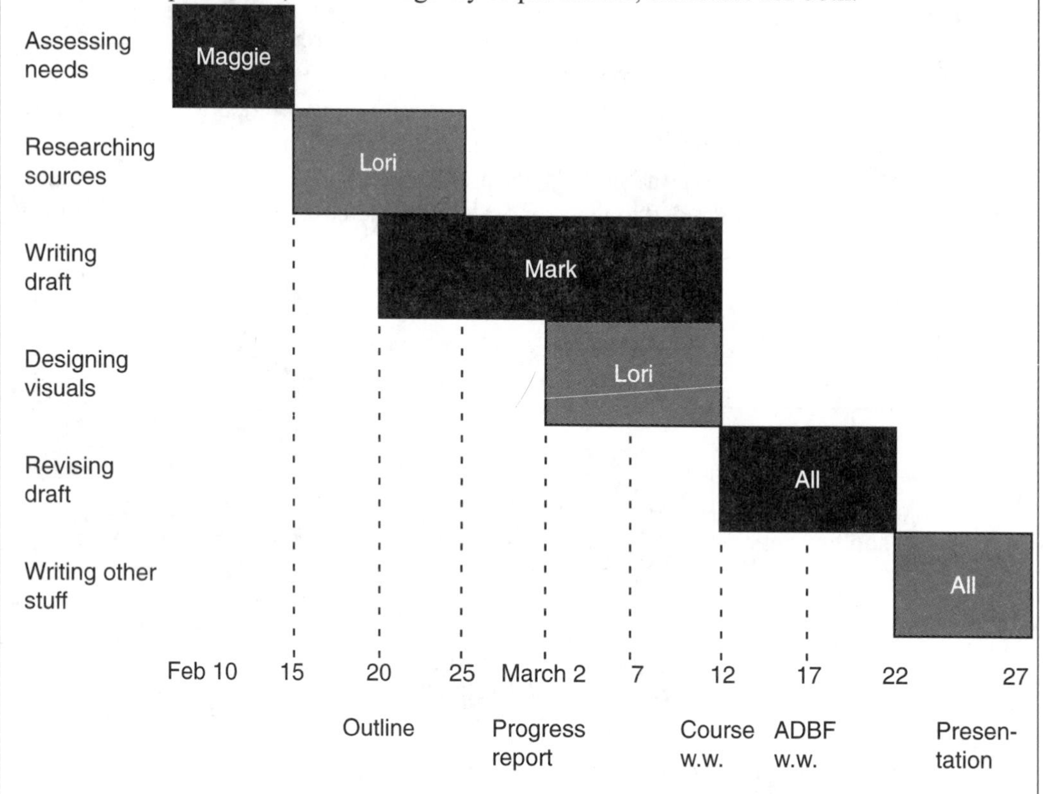

Figure 2: Gantt Chart Showing Major Tasks and Deadlines

(continued)

Figure 5.1 *(continued)*

As you can see, we have arranged our work to coordinate with course dates for the progress report, writing workshop, and presentation. This schedule will also enable us to complete the proposal before the agency's due date of April 1.

We call your attention to two other elements of our management plan shown by the Gantt chart above. First, part of our collaboration will be divide-and-conquer and part will be simultaneous. Although we will split up leadership responsibilities and the parts of the proposal to write, we will also meet together to revise each part. We will use our standing and class meetings to go over previously assigned work and to set new objectives and individual tasks. Second, we allotted a full ten days to revising the proposal based on the feedback from the two writing workshops. This should be ample time to solve any problems that might arise in preparing the proposal for delivery.

We allotted less time for writing the newsletter article because it is a shorter, simpler assignment and because we will draw on materials already gathered in our invention for the proposal.

Conclusion

In conclusion, we are confident that our shared commitment to helping stray pets, the eager cooperation of our agency, our writing experience and resources, and our project management skills will enable us to produce a successful grant proposal that will address an urgent community need. At the very least we will leave ADBF with a professional proposal template and an initial campaign to recruit future proposal writers.

[1]Consisting of three Professional Communication students at the University of Central Florida, Team Buddy gets its name from Buddy, an Akita/Scottish Deerhound puppy adopted by one of us from ADBF.

be reminded of it (them). Also, the proposal's purpose is signified, to some extent, by the title and forecasting statement. The sample proposal instead begins by previewing the two problems—the underlying community problem and the organization's writing problem—as well as the solution and its major parts. By presenting these parts, the writers give the audience a more complete picture of the solution and, at the same time, preview the major subheadings in the solution section.

Problem Section

In this section you will address two related problems: (1) the larger community problem(s) that the organization is trying to address in their writing efforts; (2) the organization's writing problem and its more direct exigence. Although the first of these is important, partly because it helps generate the second, the writing problem should be the main focus of your proposal. In other words, your proposal should primarily address how you will help the organization meet its writing/designing/editing needs. Remember that your solution will focus on writing and how it will help the agency. In addition to being generated by a larger community problem and the organization's efforts to address it, the writing problem might be caused by conditions at the organization such as lack of staff, resources, or expertise to produce the needed texts. Let's look at the example in Figure 5.1 to clarify the difference between the two types of problems.

The section first explains the underlying community problem that has generated the organization's needs: The area's growing number of stray animals has pushed ADBF beyond its capacity and ability to fulfill its mission. Then the writers move to a more detailed description of the organization's funding problem, the more specific problem that their proposed writing will help solve. They end by describing the conditions at the organization (e.g., the already strained staff and its lack of writing resources and experience) that contribute to the problem and constrain the possible solutions for it.

As we have already explained, the length of your problem description (and especially its explanation of the significance) depends on how much your audience recognizes and understands the problem—the less the audience is aware, the more explaining and convincing you will need to do. Whether the proposal is unsolicited or solicited, however, its problem section needs to be specific, audience centered, and pointing clearly toward a solution.

We've found that many students have difficulties describing a problem or a set of problems specifically or locally enough. Problems are more specific than topics or even issues. In the context of a problem description, a problem might be defined as a clearly discernable and at least partially solvable set of circumstances producing negative effects. In the case of the ADBF proposal, these circumstances involved the agency's needs for funding and writing, the first of which the writers list in detail.

The purpose of such specificity is to demonstrate to the audience that you thoroughly understand the nuances of the problem from their perspective, with their concerns in mind. This is one of the proposal's persuasive purposes. A superficially defined problem will lead to a superficially defined solution. When drafting your service-learning proposal, you may want to run your problem description by your contact person at the agency to make sure you understand the problem, its causes, and its effects the same way she or he does.

One reason the problem description is so important is that it sets up the objectives and solution. As we discuss below, the coherence among these sections is crucial to the proposal's success. If the problem is the circumstances that produce negative effects, then the objectives can be thought of as more beneficial effects and the solution as the circumstances that can produce them. In describing the agency's funding needs and writing problem, the writers of the ADBF proposal create the rhetorical space for their specific objectives and solution. Their description of the writing problem even refers to the solution of writing grants, however generally.

Some proposals end the problem section with a list of objectives, and others place the objectives in a separate section between the problem and the solution sections. Most, including our example of the ADBF proposal, include the objectives in the solution section, however, as a way to announce what the solution will accomplish.

Solution Section

The solution section is the heart of the proposal; it describes in detail the scope, the parts, and the desirability of your proposed project. Its parts should include an overview, a list of objectives, a discussion of alternate solutions, and a detailed description of the solution's major parts or tasks. Although this section is primarily informative, it also has persuasive purposes—you must persuade your audience that you have thought out your solution carefully and that it is more desirable than alternatives.

As a longer section with multiple parts, this section might begin with an overview of the solution's scope and deliverables, as this section of the ADBF proposal does. Given its primary audience of the instructor, your proposal's solution overview should probably address, at least implicitly, how your solution fits within the parameters of the course assignment. Sometimes the explanation of the scope explains what the project does not entail as well as what it does. Remember that the proposal is also a contract between you and the agency. Here you make explicit just what the agency can and cannot expect from you.

A list of the solution's objectives—or specific, measurable outcomes—is also sometimes included toward the beginning of the solution section. Whereas goals can be general, objectives must be specific and concrete. The bulleted objectives in the sample proposal, for example, specify the number of foundations the writers will research and the number of proposals they will write and submit. As we have already suggested, the objectives should flow directly out of the problems.

Just as the problem section should focus on the agency's writing problem rather than the larger community problem it addresses, so should the solution focus on your writing tasks rather than the work your deliverables will help the agency accomplish. The parts of the solution in Figure 5.1 clearly reflect the tasks of the group's writing project. The writers begin to establish a credible ethos through the detail of their solution subsections, taking the reader through every anticipated step of their writing process from researching the subject matter and audiences to workshopping and revising the documents. The writers also explain how and why they will perform each task, anticipating any questions their audiences (especially their instructor) might have.

Another way you can strengthen your ethos and instill confidence in your readers is by demonstrating that you have done your homework related to the solution. In describing the texts you will produce, you should draw on the information you have gathered in your preliminary research. Under the subheading "Identifying and Researching Promising Funding Sources," for example, the writers describe this task as a continuation of the research on foundations that they have already completed.

Management Section

This section, perhaps more than any other, is where the proposers inspire (or fail to inspire) their audience's confidence in their ability to make their solution a reality. Proposal writers can show that they understand the audience's problem and propose a desirable, well-thought-out solution, but if they cannot convince their readers that they are capable of implementing the solution, their proposal will not succeed.

The management plan typically includes a description of the project personnel and their roles as well as an outline of the project's schedule. Basically, then, this section answers the readers' questions about who and how. Like the cost section, the management section typically contains several illustrations that help the audience visualize the plan and make its details more accessible.

The *who* subsection of the management plan explains exactly who will carry out the proposal, what their specific roles will be, and what their major qualifications are. Some proposals elaborate on the proposers' capabilities in a separate section toward the end of the document; longer, formal proposals might even include minirésumés of the main personnel involved. In our sample proposal, the writers lead with short descriptions of their qualifications and resources. The writers also explain who will be working with them from the agency. Because its audience will likely use it as a reference document, this proposal includes a list of contact information—a helpful, reader-centered addition.

The writers include an organizational chart (drawn in Word) to supplement their verbal description of the group members' roles, particularly leadership roles. This part of the management plan should explain who will do what, who will be in charge of what, and how each team member will participate in the actual writing and revising of the proposed documents. The group members' leadership roles should correspond with their individual strengths. For instance, if someone has more experience and expertise in design, she or he might be in charge of those elements. If you don't have teammates, then the organizational chart could focus on you, your instructor, your contact people, and perhaps their supervisor(s).

The *how* subsection is even more important, as it lays out the solution's major deadlines and will likely be referred to in subsequent progress reports. Because your proposal involves complex collaboration, you will need to describe your plan for this collaboration, perhaps referring to the concepts discussed earlier in this chapter or in Chapter Six. The ADBF proposal writers explain in detail when and where they plan to meet with one another and with their contact person on a weekly basis—information their instructor will definitely want. Once again, if you're not working as part of a team, you only need to describe your strategies for collaborating with your instructor and agency personnel, although you should still mention obtaining feedback from in-class workshops.

The heart of the management plan is the project timeline or Gantt chart. A Gantt chart is basically a timeline in the form of a bar chart with the bars running horizontally. This type of chart can show the duration of each project phase as well as when phases overlap. Like the example in Figure 5.1 (created using the table function in Word), the major tasks on the vertical axis of the chart should correspond with the major tasks described in the solution section. The horizontal axis shows the dates on which each part of the project is expected to begin and end. The ADBF Gantt chart also includes two additional features: the deadlines for the major texts involved in the assignment and the name of the group member in charge of each major task. This is an exemplary chart not only for its detail, but also for its clarity. Although it conveys a great deal of information, the chart is easy to read and includes only necessary verbal and visual elements; it is uncluttered. The dotted lines connecting the segments to the dates and the alternating shading of the segments make horizontal and vertical grid lines unnecessary. Also eliminated were the names of the months before each date.

Beyond providing their audience with a detailed timeline and description, the writers of the sample proposal explain why this timeline is feasible and why they allotted the time they did to certain parts of the project, especially to the revision process. In our experience, student groups rarely give themselves enough time to revise and edit; this is time-consuming, meticulous work that usually requires simultaneous collaboration. Be sure to build it into your plan.

Cost or Budget Section

Although it is not always relevant to a project proposal for a course, most proposals include a cost or budget section that presents in table form the proposed costs of the project or the funds for which the writers are asking. Some proposals break down the costs according to the project's tasks and according to the personnel involved. If you include a cost or budget section in this or another proposal, rather than letting the tables speak for themselves you should also include a verbal explanation of the costs and why they are reasonable or necessary.

Accessibility

Although most proposals include context-setting statements, headings, and lists, these accessibility strategies are even more important in longer, complex documents. As you will recall from Chapter Three, setting the context for a document involves letting the reader know what it will contain and sometimes previewing its major parts. The introductory summary of the ADBF proposal ends with a straightforward forecasting sentence that does just this. Because of their length and detail, the solution and the management sections contain more specific forecasting elements of their own. The solution section presents a numbered list that forecasts the subsequent subheadings.

Although a list of subsections is probably the most common forecasting device, it is not the only one. In "Techniques for Developing Forecasting Statements," Markel discusses other forecasting elements based on the journalistic prompts of *what, where,* and *why*. The first of these types focuses on the scope and overall content of the document; the second points readers more specifically to the most important parts of the remaining text; the third attempts to strengthen the audience's motivation by explaining the rationale behind the text. Each of these types, then, forecasts a different aspect of the text that follows it.

The headings in the ADBF proposal do more than indicate conventional proposal sections such as "solution" and "management"; they also tell the audience what, specifically, each section is about. The solution heading, for instance, reads "Solution: Writing Two Grant Proposals to Animal Welfare Foundations." The subheadings in the major sections are also informative and well crafted, indicating specific tasks in parallel grammatical structure (e.g., *prioritizing, identifying, writing,* etc.). Notice, too, how the heading levels are easily discernable, how each level is consistently rendered, and how each heading or subheading is part of the same visual block as its following text.

Functional Redundancy

Keith Grant-Davie, in "The Strategic Use of Redundancy in Document Design," explains how structural and informational repetition can serve important reader-based functions in a technical document such as a formal report or proposal. Redundancy can accommodate different workplace readers who may focus on different parts of a technical document, read it out of sequence, or enter it at various access points. "[I]n order to be robust enough to survive the rough conditions of workplace reading," Grant-Davie explains, "reports also need the kind of structural redundancy that engineers deliberately build into vulnerable systems" (7).

Repeating key information in different parts of a report or proposal can serve several functions, including providing multiple access points to the same information, emphasizing key points, setting the context, and creating coherence among different sections (12–13). Repeating information in the cover letter, executive summary, and introduction of a report, for example, can accommodate different readers who will focus on different sections; managerial readers may focus only on the executive summary (a summary that should be complete enough to stand alone), while technical readers may be more interested in the more detailed main report and its introduction. The same principle can apply to the introductory summary and subsequent sections of a proposal.

The introductory summary of Figure 5.1 gives readers an overview of the sections to follow, even providing the tasks around which the solution section will be organized. These same tasks will also be repeated in the solution section's own forecasting, the solution section's subheadings, and the management section's Gantt chart. Although it may seem unnecessary, this repetition will help readers remember the solution's main parts and connect the parts of the proposal. In reading the proposal the first time or referring to it later, the instructor or another reader may skip the introductory summary or another part of the report; forecasting the solution's main parts in both the introductory summary and the solution section will ensure that the reader understands the solution's structure.

Verbal–Visual Integration

As we mentioned in Chapter Three, illustrations can serve a range of purposes relative to the verbal text: They can repeat it, complement it, supplement it, summarize it, illustrate it, and so on. This relationship, along with the illustration's importance to the text's audience and argument, might shape the extent to which an illustration is integrated into the text. If a table or figure is only supplementary and not crucial to the text's main argument, it might be placed at the end without much comment. A more important visual, such as a Gantt chart in a proposal or a drawing showing a step in instructions, should be integrated more thoroughly and placed as close to its verbal explanation as possible.

As a rule, tables and figures, however important, should rarely be left to simply speak for themselves. The following three-step technique is one way to integrate them.

1. Introduce them.
2. Point out their key features.
3. Help readers draw conclusions from them.

Introducing an illustration, which usually involves referring to its specific title, enables the writer to connect it to the preceding discussion, thereby maintaining cohesion. Because visuals contain information of differing degrees of importance, it is sometimes helpful to explain to readers how to approach the illustration and what to notice about it. Finally, this surrounding explanation might suggest what the reader should take away from the visual.

Figure 5.1 effectively demonstrates this technique with the Gantt chart. The paragraph preceding the figure introduces it and its main components. Then, after the chart, the writers spend three paragraphs pointing out its most important features and emphasizing its overall message—that the project's implementation has been planned out carefully to meet the course and agency deadlines. The second paragraph after the Gantt chart also elaborates on how and why the group scheduled the project the way they did. This method of verbal–visual integration also provides functional redundancy by offering readers some choices for retrieving information.

Style Focus: Coherence

Coherence—the flow and logical connection among parts of a text—may be more important to proposals than any other technical genre. Proposals are conventionally straightforward, top-down documents. Proposal readers do not expect any surprises but expect their questions to be answered as soon as they arise, and they expect the main points and structure to be previewed up front.

In addition, proposals typically make a **problem-solution argument** that necessarily builds on itself as it goes along. The underlying causes lead to the problem, which prompts the objectives, which generate the solution, as shown below.

Each part of this argument should lead logically and smoothly to the next. Indeed, may project proposals are rejected by evaluators because the proposed solution and procedures are not well suited to the objectives or because of some other coherence glitch. You may recall from Chapter Three that this is rhetorician Kenneth Burke's notion of form—the creation and fulfillment of desire in the audience. In *Rhetorical Grammar,* Martha Kolln also discusses coherence in terms of reader expectation. "Active readers . . . ," explains Kolln, "fit the ideas of the current sentence into what they already know: knowledge garnered both from previous sentences and from their own experience. At the same time they are developing further expectations" (42).

We recommend that you make a detailed outline of your proposal before you begin to draft. This outline should include all headings and subheadings as well as transitions connecting the sections. Including transitional elements in an outline is a good way to ensure that your divisions are arranged in the most reader-friendly order; it can also make your own writing process easier.

We now turn to a fuller lesson in creating sentence-level coherence in a proposal. Not only should each section of the proposal be tightly connected to the preceding and subsequent ones, but each sentence should also move the reader in a smooth and logical progression. To illustrate the following three strategies, we will refer to the first page of the ADBF proposal, reproduced in a slightly modified form as Figure 5.2.

Maintain a Focus and Create a Logical Progression of Ideas among a Group of Sentences

Although we do not believe that every paragraph should have only one topic indicated by an opening topic sentence (such a rigid approach doesn't account for the contingencies of the rhetorical situation), we do recommend limiting the topics you present in a paragraph to keep from overwhelming or confusing readers. This is especially important in technical and professional communication, given the hectic environments and quick-reading strategies of many users.

One technique for identifying places where you might begin a new paragraph or rearrange ideas requires that you determine the rhetorical purpose of each sentence in a passage, asking, "What is this sentence doing for readers, and how is it contributing to the passage? (You can also do this on a larger scale with paragraphs.) Then write its purpose beside it in the margins. Not only can this help you identify sentences that could be deleted or combined with others, but it can also help you identify topic shifts and the rhetorical relationships among sentences. The sentences in a paragraph should topically "hang together"; their relationships should be clear and they should move in a logical, easily recognizable progression.

For an example, let's look at the first three paragraphs of the "Problem" section in Figure 5.2. The writers originally wrote these as one paragraph. The three paragraphs move from a description of the larger community problem addressed by the agency to an explanation of how this problem has created an overcrowding and funding crisis for the shelter to a discussion of the agency's writing problem, on which the rest of the section elaborates. The writers decided to dedicate a separate paragraph to each of these rhetorical moves, calling particular attention to the focus on the writing problem with a subheading.

Create Given–New Information Chains

Creating a given–new information chain involves connecting a sentence or a paragraph to a previous one by beginning with a familiar idea. This can help give readers a framework within which to place new information and show them the logical relationship between two pieces of information. As we mentioned in the style section of Chapter Three, Kolln uses the term *known–new contract* to signify the writer's obligation to connect each sentence to what has gone before and fulfill the reader's expectations (43–44).

Figure 5.2, a modified version of the first two sections of Figure 5.1, provides some examples of this technique. In the first sentence of the second paragraph, we can see that the first part of this sentence (including the main subject and verb)

Introductory Summary

A Dog's Best Friend (ADBF), the only truly no-kill animal shelter in Orlando, desperately needs more funding for animal care operations and shelter improvements. This nonprofit organization is supported wholly by donations and is run completely by volunteers.

The organization must supplement the individual donations its raises through its newsletter and web site with grants from foundations and other funding sources. Although the shelter's volunteers are aware of its needs, they do not have the time or expertise to research and write grant proposals.

Therefore, Team Buddy proposes to produce two short grant proposals for ADBF, one for operational supplies and one for capital improvements. The project will involve meeting with the volunteers to prioritize and cost needs, identifying and researching promising funding sources, and writing and revising the proposals.

The remainder of this proposal will elaborate on the problems the project addresses, delineate the elements and steps of the project and will explain how we will carry out the project successfully.

Problem: Raising Funds for Operational and Capital Needs

ADBF is no-kill animal shelter (i.e., shelter that does not practice animal euthanasia) in Orlando with the following mission: to provide housing, food, and veterinary care to discarded dogs and cats until they can be adopted by new, responsible pet owners who will spay or neuter them.

The growing number of stray or abandoned animals in the Orlando area has caused area shelters to become overcrowded, even as more and more animals are being euthanized by county animal control facilities. Thus, ADBF is operating beyond its capacity of 100 animals, which only compounds its funding crisis.

Funding Needs and Writing Problem

ADBF is a nonprofit organization that operates solely through adoption fees and private donations by individuals (raised mostly through special events, letters, and the organization's newsletter and web site). These funding sources are not meeting the shelter's operational and capital needs, outlined in the following lists:

Operational Needs

- dog and cat food
- cleaning supplies (e.g., litter, racks, pails)
- medical and veterinary expenses
- office supplies (e.g., file cabinets, copier)
- other supplies (e.g., cages, collars)

Capital Needs

- whole-house fan
- tile for floor
- new drainage system
- more kennels and runs
- van for transporting

Figure 5.2. Sample from revised proposal

mentions the donations the organization must supplement. This ties the sentence to the previous one, which ends, in part, on the topic of donations. You can also create this type of direct chain by repeating key terms or by using pronouns with clear antecedents (i.e., nouns they refer to) in the previous sentence. Given–new information chains do not always involve the repetition or echoing of key words, though. We can find another example in the first paragraph under the "Funding Needs" subhead. Here the second sentence begins by referring to the funding sources listed at the end of the first one.

Joseph Williams suggests checking the need for more given–new chains by first underlining the first few words of each sentence in a passage and then going back and determining if the ideas expressed in these beginnings are new to readers. If they are new, they might need to be better connected to previously discussed ideas (107–108). We are not suggesting that every sentence in a passage should follow the given–new structure. This structure is less important when new ideas seem to hang together as part of a familiar topic to readers. When you are presenting complex technical information or trying to represent a causal relationship or a process that isn't obvious, however, this approach is important.

Add Transitional Words or Phrases to Reinforce Organic Cohesion

Transitional elements can clarify connections among sentences or larger elements for readers, making it easier for them to discern topics and to follow a progression of ideas. Of course, different transitional words signify different types of relationships, such as similarity, contrast, addition, clarification, illustration, summary, spatial order, and cause–effect.

Because you are probably quite familiar with transitional words, we won't list them here. What is important to remember is that they should be used not to create surface coherence on their own, but to reinforce already existing logical coherence. Williams calls the logical connection among and flow of ideas "organic cohesion" (115). In other words, just tacking a transitional word between two disconnected sentences or ideas does not create coherence. It can even increase reader confusion.

To illustrate the supplementary function of transitional words, let's look at the third paragraph in Figure 5.2. The transitional word *therefore* cues the reader to a cause–effect relationship between the two sentences. The connection between the two sentences is already apparent through their meaning, however; the first sentence describes the agency's inability to write needed grant proposals, and the second one overviews the writers' plan to do just that.

Writing Workshop Guide for Proposal

For this writing workshop, try to think like your instructor. First make a list of the questions she or he will want answered; then evaluate the proposal from a skeptical stance. Which questions does the proposal leave unanswered? Which questions could it answer more convincingly? After this initial reaction, consider the following steps.

1. Begin with the introductory summary. How thoroughly does this section overview the agency and proposal for a managerial reader? Which parts should be more or less specific? How could the forecasting be more helpful?
2. How well does the problem section focus on the organization's *writing* problem? Draw wavy lines under places that seem to depart too much from this focus. How could the writers better emphasize the problem's significance (from the readers' perspectives), more specifically explain its causes and effects, and more strategically point toward the writers' objectives and solution?
3. Now turn to the objectives in the solution section. Are they listed in parallel fashion as specific, measurable outcomes? How could they address more closely the problem and its effects?
4. How could the solution's opening better explain its scope and forecast its main parts? To what extent do these parts revolve around a writing project and its process? Where might the writers discuss how the solution is preferable to alternative solutions?
5. As you read through the solution section, write question marks beside places where you have questions about the solution's parts, rationale, or implementation (i.e., questions about what, why, and how). Draw wavy lines under sections that get away from writing and related activities.
6. Read through the solution section again, this time deleting unnecessary words, restructuring sentences with expletives, and combining sentences where appropriate.
7. How could the writers tighten the connections among the problem, the objectives, and the solution? Where could they make these relationships more explicit?
8. Which parts of the proposal could be more accessible? Where could the writers use lists or break up paragraphs? Which headings and subheadings could be more helpful and parallel? How could the headings and subheadings better correspond to forecasting elements?
9. Turning to the management section, write question marks beside places that could answer the instructors' questions more specifically. How could the writers more persuasively present their qualifications and resources? What else should the writers include about their collaboration? Where could they strengthen reader confidence by explaining the rationale and advantages of their schedule?
10. Take a closer look at the organizational chart, timeline or Gantt chart, and any other visuals in this section. How could the writers make them more informative and professional? How could the writers better integrate them into the text by placing, introducing, and explaining them? Make at least two suggestions for improving the design (including clarity) of the Gantt chart, in particular.
11. Write a *t* by places in the text that need transitional elements to connect them to previous or subsequent information. Identify the type of relationship each transition should indicate.
12. Write a *g* beside new information that could be more smoothly introduced through given–new information chains.

Activities

1. Reflect on your initial impressions of the power dynamics among the members of your group, your instructor, and the person/people with whom you will work at the agency. Write a journal entry in which you explore how these dynamics might affect your decisions as well of those of the group.
2. Consider ways in which the proposal assignment and the larger project position you in multiple, perhaps conflicting roles. Create a visual representation of these roles, presenting complexities and overlaps. List possible strategies for negotiating these roles.
3. The relationship among a problem, an objective, and a solution of a proposal can be encapsulated in this scheme: You will address problem X by achieving objective Y through implementing solution Z. Try condensing your problem, your objective, and your solution into a few words, and then fill in the preceding scheme. Write a memo to your group members suggesting ways to clarify these relationships.
4. Here's an exercise to help ensure the coherence of your proposal. Cut your initial draft into pieces, paragraph by paragraph, but only cut straight lines. Then mix the pieces up and give them to a classmate to see how accurately and easily she or he can put the text back together. This process will likely highlight parts that need rearranging or transitions.
5. Create a teamwork calendar for each member of your group, using an online scheduling tool or a calendar function in a word-processing document. Agree to share responsibility for updating the calendar, and commit to discussing changes in it at the beginning of each joint work session.

Works Cited

Grant-Davie, Keith. "The Strategic Use of Redundancy in Document Design." *Studies in Technical Communication.* Ed. Brenda R. Sims. Association of Teachers of Technical Writing, 1996. 3–15.

Kolln, Martha. *Rhetorical Grammar: Grammatical Choices, Rhetorical Effects.* 3rd ed. New York: Longman, 1999.

Markel, Mike. *Technical Communication.* 6th ed. Boston: Bedford/St. Martin's, 2001.

———. "Techniques of Developing Forecasting Statements." *Journal of Business and Technical Communication* 7.3 (July 1993): 360–366.

Williams, Joseph M. *Style: Ten Lessons in Clarity and Grace.* 6th ed. New York: Longman, 2000.

Chapter

Managing Your Collaboration

In our teaching of service-learning writing courses, we've found that the collaborative emphasis of service-learning projects can lead to students' greatest challenges and best learning. Many experienced students groan when they learn that a class requires extensive group work. Who doesn't have bad memories of the slacker who never holds up her end of the work, or the bossy person who insists on doing everything his way, or the couple of people in the group who talk only to each other and never have anything to contribute to group deliberation? We certainly have our own stories to tell about collaboration. As we wrote this section we reflected on a bad experience we had in graduate school ten years ago when we sat down and tried to write a proposal for a conference together. At the time, we agreed on the major points we wanted to make and were intellectually in sync, but when we sat down at the computer to quickly put together a *one-page* document that was due the next day, we learned that even the best of friends can run into serious problems when it comes to writing together. We have different styles and different processes. We imagine texts differently from each other and picture their evolution along different trajectories. As best we can recall, by the time we finally finished that document, we were angry with each other and neither of us was proud of our final product. We can't remember whether the proposal was accepted or rejected, but we remember our feelings of surprise and frustration. Fortunately, we have since learned that through careful planning, concerted effort, and mutual commitment to a project, we can make a collaboration work. So can you.

We want to clarify up front that collaboration is hard work—it's as hard or harder than creating a document on your own—but if it's done correctly and in the right spirit, it almost always leads to better work than one person can do alone. Our students who have graduated and gone on to work as engineers, technical writers, scientists, business executives, and other kinds of professionals tell us that the writing they do in their jobs is almost always collaborative to some extent. When they write reports, they must often meld the work of several writers into a single seamless text. When they write letters to clients, they must keep in mind the ethos and the standards of their companies. When they write documentation, they might work with technical developers, marketing representatives, and prospective users to ensure accuracy, marketability, and user friendliness. When they write proposals, they might collaborate with cost, legal, and other experts. All of this is collaborative writing.

Though not as serious as in the workplace, the stakes are also high for collaborative course projects. In our experience, most student groups face challenges that, if not handled well, can compromise the quality of the final projects and the pleasantness of the collaborative experience. Members who don't pull their weight can be asked to leave the group, which can present a problem for all involved. Among the many student projects we've supervised, we've only seen a few groups fall apart in this way, but it can happen if members don't follow some basic guidelines.

STUDENT VOICES 6.A Comments on Collaboration

Sandy: Working on team projects can often be frustrating. Everyone has their own ideas and it can become time-consuming. On the other hand, brainstorming with a group of people often produces great ideas. My most important advice to collaborators would be to allow plenty of time for collaborating and to plan ways to deal with problems like attachments you can't open, etc. Plan for Murphy's Law!

Sam: Despite the fact that I am technically working alone on my project, I will certainly need the help of a number of other people to get it done. The rules for collaborating with classmates also apply to working with agency contact people. It may be even more important to do things like leaving plenty of time and so forth with this group. I'm the only one with a grade on the line.

Tabitha: I groaned when I heard the term "group project," but once I chose a partner with interests similar to mine, I felt better. Because this is a web-based course, we are primarily using the "divide and conquer" approach. Our schedules are pretty different, so the more stuff we can do and report back to the other on, the better off we are. The division of labor is a little tricky since there's only two of us—so we each have to wear many hats, but that's turning out well because we each get a chance to do a variety of kinds of work. Also—as much as I love my email, it's amazing how a phone call or two can clear up confusion.

Gia: I really believe in the "results-driven meetings" idea. I've found that if I take some notes when there's something I need to know (or need clarified) then collect these notes over a day or two, I can (1) either figure out what I need on my own (so as to avoid asking dumb questions I just needed a clear head to find answers to) and (2) avoid bugging my groupmates with a zillion emails.

Jose: Collaborative work always seems more challenging to me than it should. I tend to work better alone because I know exactly what my goal is and I do not have to worry about what goals another person or group of people has in mind. In the course of being a tech writing major, I have learned a great deal about collaborative work and how to form a common goal with a writing team. I'm getting better at using tools to facilitate teamwork. One of the most important for me is what one of our classmates called "icebreakers"—these are activities that help us to get to know each other.

Collaboration Challenges and Strategies for Handling Them

When we were writing the proposal described above, we faced three kinds of challenges. First, we had a *writing* challenge. We had to produce a tight, persuasive, and intellectually engaging document. It had to be 300 or fewer words long and include enough references to texts and theories to demonstrate that we knew what we were talking about. It also had to persuade our readers that we had a new and interesting point to make. This was a hard task, especially for a couple of relatively inexperienced students.

Next, we had a *personality* challenge. We had to negotiate between Blake's more linear, pre-planned approach and Melody's more exploratory one. (You can only imagine how long it took us to settle on those labels, and we're still not altogether satisfied with them.) We had to accommodate the strong personalities of two highly opinionated people with different ideas about what to write. Our fear about our lack of experience and our deep desire to get our paper accepted only magnified this challenge.

Finally, we faced *logistical* concerns. The proposal was due in less than 24 hours, and we each had other responsibilities to fulfill during that period. We had to make six photocopies of some sections, three of another, and take all of it to the post office to be postmarked.

Each challenge was enough to cause us frustration, but when all of the challenges were grouped together, we experienced some serious stress. As you face your own collaboration tasks, you'll find that these types of challenges compound each other throughout the process. Sometimes a personality difference creates a writing problem, or a logistical problem such as a tight deadline exacerbates a personality problem. In the pages that follow, we'll suggest a number of strategies for making your group's collaboration on all of these levels as smooth and successful as possible. We'll insert some examples from our own work and some anecdotes from student projects. But we'd like to begin with a few simple rules.

- *Allow adequate time.* While two or three or four heads are almost always better than one, they are not necessarily faster than one. Often the process of collaborating takes much longer than working individually. Don't save joint work for the last minute, thinking that you can divide it up quickly and complete it. Finish drafts as early as possible to leave plenty of time to accommodate each person's writing pace, habits, thinking processes, and so on.
- *Follow the Golden Rule.* It may seem trite or obvious, but treating your teammates as you'd like to be treated is probably the most important strategy you can follow. Assume that even the most confident person is sensitive about her or his writing. Remember that everyone has her or his own values and habits. Try to be aware of and sensitive to your groupmates' feelings and opinions.
- *Don't take things personally.* Remember that you are working together toward a final goal. This project, other schoolwork, and other life responsibilities may overshadow your colleagues' concern for your feelings from time to time. Try

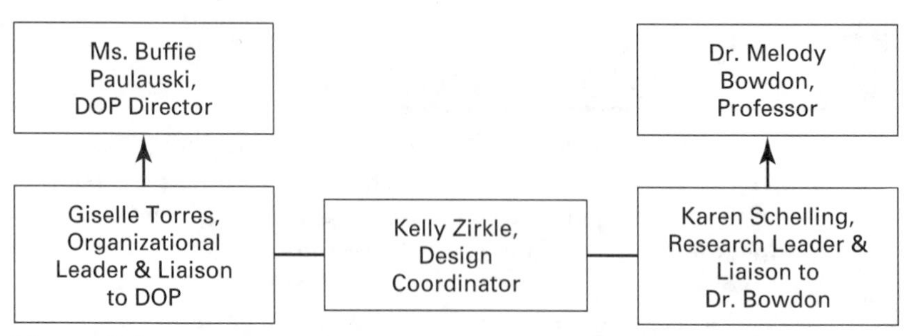

Figure 6.1 Collaboration planning chart

not to take bad moods, brusque comments, low enthusiasm, or tepid first reactions too seriously.

- *Don't assume that group members can read your mind.* We've seen several groups become stressed mainly because the members didn't communicate with each other honestly. Though it may seem completely obvious to you and one or two others that one group member is not carrying her part of the load or is making everyone miserable by railroading things, that person may not be at all conscious of the problem. The person who shows up at the library ten minutes late for every meeting may not recognize how frustrating those of you who have made the effort to be punctual find this pattern. Though it may sound obvious, we suggest that you remind yourself regularly to just *tell the person.* Most people don't want to be disliked by their group members and will make an effort to change their behavior if they are alerted to problems.
- *Create a visual representation of your collaboration.* It can be quite helpful to diagram your relationships as you begin to work on a joint project. Don't forget that you are not collaborating only with your team members; you are also working with one or more agency representatives. Further, you will be collaborating with other members of your class through peer review and class presentations and discussions. Your instructor will also be providing input and guidance. As you design your collaborative plan and model, figure these characters in as well. See the simple chart one group used to represent their collaborative responsibilities in Figure 6.1.
- *Approach these concerns in a sophisticated way.* Together apply the ideas you'll find throughout this chapter. Learn as much as you can about your joint goals and ideas, and look at them from an intellectual standpoint as well as a practical one. Develop a framework for thinking about collaboration that will apply to your work beyond this course. Develop your own set of strategies and policies for collaboration.

Types of Collaboration

As we've noted, most students have one or two disastrous collaboration stories to share before they start their service-learning classes. Our negative joint writing experience described above is a typical example: we came to the computer, sat down together, and expected to quickly produce a seamless text. Fortunately, through the years we've learned that this is not the only kind of collaboration available to us. There are at least three distinct models of collaboration as well as combinations of these. Your group will likely rely on all of them before you finish your project.

Simultaneous Collaboration

As you probably guessed, in this approach group members take on one task together at one time. Although such document-sharing technologies as MOOs and NetMeeting may allow you to do this kind of collaboration from different physical spaces, most people engaging in simultaneous collaboration are working in a room together. This can happen in all kinds of spaces—a computer lab with several terminals, someone's office or apartment with everyone huddled around one computer, a regular classroom, or a library workroom. Simultaneous collaboration is most appropriate for work at the beginning or the end of a project phase. If your instructor gives you time in class to work together, consider saving some of the duties below for those periods. Specific activities that can be accomplished through simultaneous collaboration include:

- *Planning.* Your final product will be better and you will save time if everyone's voice is included in the planning of the project, whether it's about research, writing, design, or revision.
- *Outlining.* If you're deciding what to include in a document, you'll want to involve more than one person to avoid leaving anything out. You might have everyone send preliminary lists of inclusions to one person for compiling, but we advise that you hold at least one group session to finalize the outline.
- *Determining group guidelines.* Although many writing activities are best done solo, it's important that you work together to define a general group style and ethos. We'll describe specific strategies for accomplishing this later in the chapter.
- *Processing feedback.* If you're receiving input from members of a peer group, your instructor, and/or your agency contact person, we suggest that you make an effort to process this information together. Comments on your documents can often be interpreted in several ways, and it's helpful to have more than one pair of eyes looking at them together to determine what they mean and how important they are. You may want to make copies of responses or divide several responses up among your group for reading, but decide how to act on those ideas together.
- *Putting together the final product.* After a long process of producing a major project, it may seem that the last few steps are the easy part. But we've found that some of the most difficult and frustrating experiences of the entire process come at the very end. It never fails that a printer will malfunction or the photocopier at the print shop will jam or there will be some kind of last minute decision to be made about the appearance of the portfolio or the cost of binding or reproduction. For this reason, it is critical to involve all group members at this stage. Everyone should have a chance to

assess the product's contents and appearance. Together you're less likely to forget to include something, overlook a misspelled name, or put things in the wrong order. Meeting simultaneously also ensures that everyone signs off on the project, which prevents placing the onus of final responsibility on one or two members.

- *Presenting your project.* If you've ever waited until the last minute to rehearse a group presentation, then you know how important it is to plan and practice together to ensure coherence and professionalism. Although your actual presentation will probably be sequential—one person presenting at a time—your preparation should be largely simultaneous.

"Divide-and-Conquer" Collaboration

Some parts of a project don't need to involve more than one person. These activities can be assigned to individual group members and then processed by the group as a whole. When you use this approach, be sure to distribute the work fairly and to avoid duplicating efforts. The following kinds of activities fit well in this approach.

- *Research.* Typically only one member of your group needs to go to the library or onto the web to find a piece of information or to locate a source. Some people are considerably more effective as researchers when working alone.
- *Composing/Drafting.* Very few people can write effectively while several people are talking to them. Most of us need quiet time and a clear mind to generate good texts. We recommend that you work individually to write initial versions of most texts, keeping the ethos and decisions of your group in mind.
- *Correspondence.* You may choose to give one member primary responsibility for corresponding with your agency contact person or your instructor. This will help you to avoid duplication or contradiction. It will also simplify things for your correspondent, as she or he will not have to puzzle over which member to contact with a question or problem. The group member who takes on this responsibility must commit to keeping all other members up to date about messages. It's a good idea to copy the group on any emails or letters that you generate.
- *Miscellaneous details.* A complex writing project involves many small details such as photocopying and collating. It can also involve formatting the page numbers or headings in the text or delivering the final product to the agency. Assign these tasks to individuals, but be sure that no one is given more than her or his share.

Sequential Collaboration

In this approach collaborators take turns working on the same document or project. One person may be responsible for starting the project or laying the foundation, and other members must add to or review that initial work. As we'll discuss later in this chapter, communication and text-sharing technology can simplify this process. Sending a document as an email attachment, for example, is much simpler and faster than meeting your partner at the library at 6 P.M. on Saturday to hand-deliver it. Activities that are best accomplished through sequential collaboration include:

- *Early revising.* Though we've advocated that individuals take responsibility for generating early versions of texts, we suggest that early revision, which involves evaluating global concerns such as content and arrangement, be done sequentially. You might pass an early draft among the members of your group and have each person make comments about the shape of the document and add missing elements. We specify that we are referring to *early* revising because, as you'll note in several places in this chapter, we believe that the overall process of revising is a group responsibility.
- *Preliminary editing.* Again, it's important for everyone to be involved in editing a document, but even with a very tough skin, it's hard for some of the most experienced and confident writers to be present for and participate in the editing of their own texts. Likewise, most of us hate to mark up someone's text while they're watching us. We suggest that you split up the first round of editing, perhaps have different group members focus on different aspects of the text in their editing passes. The sequential approach to this process will also mean that the group members see the document with fresh eyes; if a document is new or you haven't seen it in a few days, you will be better able to assess its strengths and weaknesses.

Combination Approaches

Although we've categorized certain parts of the collaboration process as particularly suited to sequential, divide-and-conquer, or simultaneous models, the reality is that *most processes require a combination approach.* Even if one person is primarily responsible for composing a first draft, for example, he might consult with the rest of the group several times during that process. Although a research task may be assigned to one member, the group might help her brainstorm sources to pursue and keywords with which to search. Although a group might design a document grid collectively, one member might draw the final version to distribute.

In the process of writing this book, we have followed each of these models at one point or another. Table 6.1 shows our collaboration system for the first three chapters. After planning the chapters together, we each took the primary responsibility for drafting certain chapters, agreeing to do most of the research and writing separately. (Later chapters required us to draft sequentially, with one person writing a segment of the chapter and then sending it to the other for completion.) Then we exchanged the chapters and made revisions to each other's work directly on electronic copies of files, saving them under new names to preserve the old copies. We then exchanged chapters again so that the original author could see and respond to the changes made by the reviser. After we were confident that the drafts were complete and fairly polished, we sent them to the publisher and our series editor for review, and the drafts went to outside reviewers with expertise in technical and professional writing. After receiving feedback from our editors

Table 6.1 **Chart of author's collaboration scheme for first three chapters**

	Planning →	Drafting →	Work-shopping →	Planning Changes →	Revising
Chapter One	Both, simultaneous	Melody, then Blake	Series Editor, Reviewers	Both, simultaneous	Blake, then Melody
Chapter Two	"	Melody, then Blake	"	"	Blake, then Melody
Chapter Three	"	Blake, then Melody	"	"	Melody, then Blake

and reviewers, we returned to simultaneous collaboration, working together to synthesize suggestions and formulate a revision plan. Revision and editing responsibilities for each chapter were then split up.

More than any one combination of models, we advocate a careful, thorough planning of your collaboration. We also recommend considering the following factors that shaped our own collaboration.

- *Interests.* Occasionally one of us was more enthusiastic about the subject of a chapter than was the other. In those cases, it was easy to choose our combination approach. We planned simultaneously, gave one writer the responsibility for composing, and then revised and edited sequentially.
- *Abilities.* Similarly, in some cases, one of us had more experience or knowledge about a topic than the other.
- *Schedules.* We wrote the book on a fairly quick and tight timeline. Sometimes it was simply more practical for one or the other to write a first draft or tie up loose ends. This factor also determined which of us would travel to the other's town for work that required simultaneous collaboration.
- *Equity.* Though we didn't bother to keep track of it on paper, we kept an informal running tally of whose turn it was to add page numbers or drop off a package at the post office. This allowed us to each occasionally have responsibility for simple tasks that didn't require much thought and sometimes to be able to finish the writing or editing without worrying about the next step.

The rest of this chapter presents strategies that apply to your group regardless of the combination of models on which you decide. Following these strategies, based on our own observations and research on industry teams and student groups, will help ensure that your process is productive, smooth, and successful.

Characteristics of Effective Teams

In *Team Players and Teamwork,* Glenn Parker identifies several characteristics of effective teams in business. According to Parker, members of effective teams do the following (33):

OTHER VOICES 6.A Summer Smith

Summer Smith, PhD., is an assistant professor at Clemson University. She directs the technical writing program at the university and coordinates a program-wide service-learning effort.

Whole-Class Collaboration on a Single Document

If your class is working together to produce one substantial document, the task of coordinating everyone's efforts can seem overwhelming at first. One of my students remembers the experience this way:

> When I stepped out of the first class this semester, I was truly scared! How could 24 people collaborate to write a report?!! But somehow we did it, and along the way, we learned a lot about working together, being professional, and accomplishing a task.

This student's class collaborated to write an annual report for a governmental agency in charge of recycling and solid waste collection in the county. It was the agency's first annual report, so the students also worked together to research the audience's and client's needs and write a proposal for the content, organization, and design of the report. Thousands of copies of the report were printed and distributed by the agency.

In this class, just as in a small group, effective collaboration depended on collective decision making. Here are some strategies that can help your class make decisions together.

- ***Short arguments.*** When the class needs to make a decision, everyone's voice should be heard. Short arguments are brief email or bulletin board messages arguing for a position on an issue such as how groups should be formed, what should be covered during class time next week, what factors should influence the style of the class's document, or how the class could improve its collaboration. Everyone in the class should write a message, and everyone should read the messages before a class discussion or vote.
- ***Presentations and progress reports.*** Your class may divide into small groups that research different questions or write different sections of the document. Weekly oral progress reports from each group to the class and presentations of completed work (such as a research report or a section of the class's document) provide an opportunity for groups to share ideas and recognize conflicts or overlap in their work.
- ***Storyboarding.*** To produce your class's document, small groups or individuals will probably write separate sections that need to fit together without contradictions or excessive repetition. Storyboarding allows everyone in the class to share ideas about all sections of the document. Begin storyboarding when you are deciding what information to include in your section. Write your ideas on a sheet of paper and post it on the wall of your classroom with a blank sheet for comments. Everyone in class "walks the walls," reading each storyboard and writing comments. Continue posting storyboards throughout the writing process as you develop drafts.

1. Share a clear purpose.
2. Participate in open discussions.
3. Listen carefully to one another.

4. Engage in civilized disagreement.
5. Arrive at consensus decisions.
6. Maintain open communication.
7. Establish clear task assignments.
8. Share leadership roles.
9. Develop effective external relations.
10. Engage in periodic self-assessment.

Recall the last highly successful group project in which you participated. How many of the above characteristics were true of your group? In what follows we elaborate on some of these characteristics and how to achieve them.

Establish a Group Identity

The first and most fundamental group activity is to establish a group identity or team ethos. As we have already discussed in the last chapter, this involves getting to know the other team members—their backgrounds, values, skills, and communication styles—and giving the team a name. By calling themselves the Zoo Crew, the students who worked with the Central Florida Zoological Park, for example, reconceptualized themselves as a unit and made such details as email subject lines and group correspondence simpler.

On the task level, this part of the process involves establishing a collective mission and set of objectives. Research on dysfunctional teams has shown that one of the main reasons for their dysfunction is the lack of a shared mission. Part of your mission should be to produce a better product as a group than any of you could as individuals. When determining group objectives, remember that objectives are specific, measurable outcomes. Your objectives might be course related (e.g., to earn an "A" on the assignment), agency related (e.g., to recruit 20 new volunteers for a new program with brochures), and text related (e.g., to improve the design and thereby the usability of two brochures).

Distribute Leadership Roles Strategically

To distribute leadership roles strategically, you will draw on each group member's project-related strengths, based on your previous assessment activities. Along with comparing each group member's strengths, you should identify the leadership roles the project will involve. These might include the following:

- Group coordinator—in charge of group communication, coordinating group meetings, and setting agendas.
- Lead liaison to agency—in charge of correspondence with the organization's contact person and other involved members.
- Lead liaison to course—in charge of correspondence with the instructor and other class groups.
- Design leader—in charge of designing the documents, producing visuals, and integrating visuals into the text.
- Invention leader—in charge of research for the texts and of developing the style sheet.

- Writing head—might be split up according to the main parts of the texts; if you were producing a set of instructions, for example, one person could be in charge of the introductory sections, another the step-by-step instructions, and another the troubleshooting and warning elements.
- Revision leader—in charge of preparing the texts for writing workshops, synthesizing feedback, and implementing changes.

Now it's time to divvy up the roles, keeping in mind that effective teams take advantage of each group member's expertise and experience. The group member with the most layout or computer experience should be the design leader, for instance, and the person with the most technical knowledge of the texts' subject should be in charge of research. Because your group will probably not have seven or more members, each group member will obviously have to assume more than one of the leadership roles mentioned above. Consider that the drafting and the revision will probably be the two most time-consuming elements, and leadership in these areas should likely be circulated among members.

We are not suggesting that each leader should be solely responsible for a task, only that she or he be the director or primary facilitator of that task. One of the most important characteristics of a leader, in our view, is the ability to facilitate the participation of others. Rather than micromanaging or simply taking over the work of other team members who are struggling, a leader finds ways to enable their production and raise their standards. If your instructor follows our suggestion, she or he will require that each group member do some of the writing and that all of these tasks—especially revision—involve some simultaneous collaboration.

Determine Clear and Fair Task Assignments

Along with distributing leadership roles, this task is the nuts and bolts of group planning. Before you can determine who will do what, you'll need to decide what collaboration model(s) will be most productive and efficient for your group. It should be obvious by now that we recommend a combination, using simultaneous, divide-and-conquer, and sequential approaches for various phases of the project. Consider one another's schedules and writing preferences when devising your plan.

You'll then need to discuss specific tasks and assignments for the divide-and-conquer parts of the project. These assignments should be as clear and precise as possible. In addition, everyone should participate in this decision making. In our experience, tasks assigned to absent members are seldom completed on time, if at all.

An equitable distribution of tasks is not only crucial to the group's morale, but also to its success. In our observation of student groups, we have found that groups that benefit from the full involvement of all members produce higher-quality products than those that have one or two members doing all the work. There is no room for slackers or control freaks in service-learning groups. We have often seen high-achieving students take over the project at the revising/editing stage. What usually happens, however, is that the student grossly underestimates the time and effort

OTHER VOICES 6.B Sheila Cole

Sheila B. Cole is the quality assurance manager for Fiserv Orlando. She has worked in the financial service industry for more than twenty years. This was her first experience with a service-learning project.

Successful Collaboration with a Non-Profit Organization

I took Dr. Bowdon's professional writing course during a condensed summer term. There was a lot of material to cover, a lot of new concepts to learn, and our service-learning project was due in only six weeks.

During the second class, I was chosen as our project team's agency liaison, responsible for keeping in touch with our non-profit organization. To get the ball rolling, I sent an email to our agency the day after my selection. I am so glad that I didn't wait. The early communiqué impressed our agency contact, Jessa, which got us started in the right direction, and her quick response showed me that she, too, would be responsive to our needs.

Jessa became an active member of our team, and communication became an integral part of our collaboration effort. When Jessa clarified or changed a requirement, I was responsible for sharing that information with the rest of the team. When a question arose from a team member, I made sure that the message made it to Jessa and that she was aware of our deadline for a response. The arrangement worked well for both the agency and the team. Jessa didn't have to be concerned about whom to contact with suggestions or issues, and the team's tasks stayed on target which, with our short-term constraints, was exceedingly important. The process allowed us to move continually forward with our project.

Even though we exchanged phone numbers, e-mail became the team's primary means of communication. This worked well for our agency contact, too. When Jessa was out of town on business, she was still able to respond to our messages in a timely manner.

Despite the short semester and conflicting schedules, our project was completed on time. We learned a lot from our service-learning experience, and we made new friends and important new contacts. Without a doubt, successful collaboration was the key to our success.

required at this stage and, as a result, ends up with a mistake-ridden, only somewhat coherent text that doesn't represent the group's efforts fairly. We repeat: Everyone must be involved in the writing and revising of the project texts.

You may not be able to anticipate all tasks, and you may want to adjust task assignments as you go along to ensure equity. For these reasons, we strongly recommend clarifying at the end of every meeting what each person is expected to do before the next meeting and putting that information in writing, perhaps in your field journal, an assignment we explain in detail at the end of this chapter.

Facilitate Open Communication

Open communication among all group members is crucial to the success of a writing group. It is not something that just happens, however, but must be actively nurtured by your group. You can start by establishing communication channels and protocols outside of class. As we'll discuss later in the chapter, this could include a group email list or an electronic bulletin board. For any system to work, group members have to use the group's communication channels consistently, of course. You should commit to checking your email or the bulletin board daily for questions, concerns, reminders, and so on. You should also commit to responding promptly to one another's correspondence (just as you would your instructor's) and to notifying other group members promptly if, on a rare occasion, you are unable to meet a group deadline or attend a group meeting. Unless you have an emergency, however, you should never miss or even show up late to a group meeting; not only is this extremely disrespectful, but it burdens other group members with extra work and leaves you out of the loop.

It is especially important to communicate about problems that arise. Relationship problems, such as personality or ego conflicts, can be especially difficult to discuss. Addressing the problem, however, if done diplomatically and supportively, can be the first step in resolving it. At the most basic level, creating a supportive, open environment involves listening to each other. You can show your teammates that you are actively listening to them by not interrupting, giving them verbal and nonverbal signals (e.g., shaking your head, saying "uh huh"), paraphrasing or summarizing their ideas, asking questions, and taking notes on what they say.

Encourage Active Participation and Consensus

In addition to establishing procedures for open communication, your group should establish meeting procedures that encourage active participation and collective decision making. You might, for example, take turns giving input at the beginnings and endings of meetings. The group member in charge of the work being discussed should take the lead in ensuring that the other members actively contribute. Some group members might feel more comfortable contributing by email or some other electronic forum; whenever possible, create multiple channels for contributing.

Remembering that some disagreement is a sign of a healthy group, don't be afraid to offer differing opinions in your group meetings and other communication. Group members who never disagree are not invested enough in the project and/or do not feel comfortable voicing their opinions, sometimes because one overbearing member is making all of the decisions. You may remember from our discussion of rhetoric in Chapter Three that ancient rhetoricians saw deliberation with others and identifying alternative courses of action as necessary steps in determining the best course.

Run Results-Driven Meetings

As you have probably already discovered, integrating group meetings and exchanges into your already busy schedule can be a difficult task. Despite this difficulty, many student groups don't take advantage of meetings adequately and, as a result, have to schedule more meetings. Making results the bottom line of your meetings can help you avoid this problem.

- One way to run a results-driven meeting is to *set an agenda.* Group members should discuss ahead of time what to address in the meeting and what work to bring. We recommend that at the end of each meeting the group set a tentative agenda for the next one. Once the project is underway, meetings should revolve at least partly around reporting results of each person's work in progress. Assigning group members to bring their results can prevent them from procrastinating and can enable you to take full advantage of the time you have together. Group meetings should be used for simultaneous collaboration, such as group planning and revision. With this in mind, you should treat assignments for meetings just as seriously as course or agency deadlines.
- A second way to make your meetings count is to *establish procedures for resolving conflict and arriving at consensus decisions.* This is not to say that part of your meeting shouldn't consist of open discussion of differing opinions and courses of action, only that you don't want to end the meeting without coming to a decision that will guide your future work. You can't let disagreements be debilitating. We recommend democratic voting after each group member has one last chance to state a position. If you are at an impasse, you can always ask for your instructor's opinion or agree to table the issue for a short period of time if that's logistically feasible. A solution may be clearer to all the members in a day or two. In arriving at a decision, each group member should be prepared to compromise—this is one reason you should avoid becoming emotionally attached to your ideas.
- As a group, you should *keep a record of decisions.* Once decisions are reached, at least one person in the group should note those decisions for future reference. As we've noted above, this record can be part of your field journal. This record might also be kept in an electronic forum such as a bulletin board.

Your collaboration outside of meetings should also be at least somewhat results driven. In addition to discussing your ongoing work via email, for example, you might exchange and comment on drafts, about which we offer more advice below.

Assess Your Work

Though we dedicate Chapter Nine of this book to project assessment and evaluation, we believe that this process must be a constant part of your project. You can see that we've provided guidelines for assessment throughout the text in the form of writing workshop guides and end-of-chapter activities. In addition to the assessment your instructor assigns you, however, you'll want to assess your group's interpersonal dynamics and progress periodically throughout the process. This will enable you to discover and solve problems before they become serious roadblocks.

Strategies for Managing Drafts

Because your group will probably divide writing tasks for at least part of the project, you will need to establish procedures for exchanging drafts, giving each other feedback, and making your texts consistent and coherent. We discuss below some strategies for managing your work on drafts from the planning stages to the final stages of editing. We also suggest some technologies for facilitating this process.

Create (and Revise) a Style Sheet

Style sheets are reference texts that document the choices or rules for a text style, including grammar, mechanics, word choice, and sentence structure. They might indicate, for example, whether to use a comma before the final element in a series, how to represent numbers, or whether passive or active voice should be primarily used. Style sheets often record design rules as well, such as how to distinguish among different levels of headings and how many spaces to indent.

Professional communicators use style sheets when writing and editing mostly to ensure consistency and visual coherence. Style sheets are especially useful for collaborative projects in which different writers draft different parts of the text. Creating a style sheet early in a project can help a writing team clarify its style and document design choices. Individual group members can then make sure their drafts conform to the style sheet, saving the group editing time later. Once they reach the editing stage, the group members have a reference point against which to check their work.

We should emphasize here that style sheets are malleable instruments that writers modify and add to throughout a writing project. Your group won't always be able to anticipate all of the stylistic and design decisions that you will face. As you develop additional guidelines for the style sheet, make sure you share and discuss these with your teammates so that everyone's writing style is consistent. We discuss style sheets in more detail and show you an example of one in the next chapter, where we take you through the recursive steps of executing your project. You can find another sample style sheet in Appendix B.

Provide and Track Specific, Helpful Feedback

Another recommendation for managing your group's writing process is providing each other specific, helpful feedback, not just at the end but throughout the process. Group meetings are ideal venues for doing this, of course, but you might also share and comment on drafts electronically. Although you can certainly send drafts via email or a bulletin board, you may also have access to text-sharing and commenting software. Regardless of the method, group members engaged in divide-and-conquer collaboration should be well aware of one another's work and how it is progressing in relation to their own.

If part of your collaboration is sequential, that is if you are transmitting drafts on which to work one at a time, it might help to determine a procedure for tracking the changes each person makes and comparing one version of the text with another. You may decide that some of the changes a group member made should be reversed, for instance. Commenting, tracking changes, and comparing, which are also useful at the revision and editing stages, are features of most word processing programs. In Word, the commenting function is under the "Insert" column on the toolbar, and tracking and comparing are located under "Tools."

Commenting on your peers' drafts can be a tricky task. Unless they are also training to be teachers, writing students are rarely trained on how to approach this important process. The first step is to create a group environment that encourages critical feedback that is given in a constructive manner. Agree to offer clear, strong, and specific criticisms. Remember that the term *critical* doesn't necessarily mean negative. A critical comment is simply one that is clearly engaged with the text and offers analysis and suggestions for improvement. Besides being critical, your comments should be encouraging. When we comment on student papers, we try to write some of our comments in a coaching tone, giving the writer specific questions to think about or offering a couple of suggestions for improvement. Here are some additional recommendations for commenting.

- Start with positive comments that point out the text's strengths.
- Make both line-specific comments and summary comments.
- Write summary comments in a numbered list, from the most to the least important.
- Offer suggestions and identify problems.
- Ask questions.
- Comment on the text, not the writer.

We also suggest that you tell your peers what you want from them, preferably in writing or via email. Figure 6.2 shows an example of such an email that one student sent his proposal writing group when giving them a draft of his work. Notice the specificity of his concerns and the request for feedback.

Allot Extra Time for Collective Revising and Editing

In our experience, the most common task-oriented mistake that student writing groups make is underestimating the time it takes to revise and edit their work, especially to make it consistent and coherent. Often someone in the group will generously but perhaps naïvely volunteer to do this alone, asking the other members to send their parts of the project to her or him to compile.

But revising and editing a document is a multi-step, problem-ridden, time-consuming exercise that requires the resources of the entire group. You will undoubtedly experience some kind of file, computer, or printer problem as you put the parts of the text together and integrate illustrations. After you receive feedback from writing workshops and, perhaps, usability tests, you will need to develop a revision plan and then actually implement the changes, some of which will require more invention. Even at the final editing stages, you group will need to take a number of passes through the text, focusing on different elements and levels in each one. At the very least, you might use your style sheet to edit for the following:

- correctness
- accuracy
- completeness
- stylistic consistency
- design consistency

Subject: My requested response criteria

Previous Thread
Next Thread
Close

Reply
Reply Privately
Quote
Download

Message no. 236:
posted by Joseph Dale (jcd61518)
Mon Nov 13, 2000 20:44
My target audience will understand the terms and budget, but class peers probably have no basis to critique these areas. My request for feedback targets style and organization areas which do not require specialized topic knowledge.

I want to know about style and content in five areas and request that you ignore spelling and punctuation in my proposal. 1. The major issue I am struggling with is the problem statement. Did I define the problem clearly? If you think I did, then please paraphrase it for me. If not, a simple no is not sufficient; please comment on what seems fuzzy in the problem statement.

2. The second major issue I am struggling with is the who cares part of the problem. Did I explain why my project should be funded? Is it persuasive?

3. On the initial read without any reasons other than instinct, what sentences or areas just do not feel right or are fuzzy? I want to target areas where I assumed too much knowledge on the part of my target audience. If it is fuzzy to you, it will be fuzzy to them.

4. What headings or topics seem out of sequence or place? I want to target the organization and flow of the proposal. If you stumble in one area or have to reread a previous or subsequent area to understand a topic, then I want to revise it to explain to the reader what seemed implied or obvious to me.

5. Specific style issues that I have a real problem with are metaphors and nominalizations. Please look for areas where I slipped out of first person voice or used bad metaphors. These two style issues are the major tools I will use to establish rapport with the grant makers.
Previous Message
Next Message

Previous Thread
Next Thread
Close

Figure 6.2 Student's email requesting information from group

- verbal–visual integration
- coherence among the document's parts.

Creating a coherent document and editing for consistency are often the most time-consuming parts of the editing process. They are also often the most crucial. Even if its parts have been well written, a team document can still come across as disjointed and unprofessional if its parts do not smoothly "hang together" verbally and visually. This coherence is especially important when the text will represent a larger organization. The organization's ethos is on the line. As we've noted above, we hope that part of this process will be the synthesis of comments from your peers, instructor, agency contact person, and possibly even such additional audiences as projected users or a professor in your department with extensive expertise in the subject area. Interpreting and negotiating what to do with this potentially contradictory input can be time consuming and difficult.

Even if your group revises as you go along, you should probably allocate at least a couple of weeks for revising and editing. Once again, allow plenty of time for simultaneous collaboration during this time period. Revision decisions are group decisions, and revision and editing are group tasks.

Use Collaboration-Enhancing Technology

As we've noted earlier, communication technologies can facilitate and simplify a wide range of collaborative activities. Before you begin to rely too heavily on these tools, however, keep in mind that some group members may not have access to them. You might want to check with your campus computer services office to find out what programs are available in campus labs. Even if everyone has access, some tools will be more practical and familiar than others. And of course no matter how up-to-date our summary of the tools in this section may be, by the time you are reading and using this book, they may well be enhanced or even obsolete. What we want you to take away from the discussion below is a willingness to experiment with technology; this can sometimes make an inconvenient process doable or even turn an activity that might otherwise feel like drudgework into something enjoyable. Use the ideas below to guide your experimentation.

Communication technologies fall into two categories—**asynchronous** and **synchronous.** Asynchronous tools allow you to collaborate sequentially. They do not require that all students be online at the same time. Some examples of this kind of tool include:

- *Email:* This is our collaboration tool of choice. We prefer using email to talking on the phone for most of our work on documents because it allows us to send at any time and helps us not to hedge our opinions. We regularly attach Word documents to emails and send them back and forth. We sometimes use the "tracking changes," "compare," or "save versions" features to see exactly what each of us has done with the texts. As we've suggested before, if you choose to use this method, you'll want to be sure that everyone in your group uses an email program such as Eudora that allows them to send and receive attachments easily. Also make sure that everyone uses an up-to-date antivirus program.

- *Bulletin boards:* Your group may choose to use a web-based bulletin board program for sharing and recording your plans and ideas. This tool stores a threaded discussion on a web page. The major advantages of the web board are that it maintains all of your postings in one space and that it is often searchable by the subject and the writer's name. Many comprehensive online course tools such as WebCT include built-in discussion features. Your instructor can set up a private space for your group to store your conversations.

Synchronous or real-time collaboration requires that you and your collaborators be online at the same time. You may be dialing in from distant locations, but you must all be working simultaneously. Some examples of this type of tool include:

- *Chatrooms:* A chatroom is a virtual space for conversation among group members. Participants can read each other's comments and post immediately visible responses. Your group may choose to enter an established chatroom that is part of your course tools or to create one of your own using another service provider. One advantage of the chatroom function of most school-affiliated programs is that they often maintain a log of the conversation.
- *Group conferencing software:* Programs such as NetMeeting allow collaborators to work on documents together at the same time. The author of a text can give and take away control over the document, allowing the collaborator to edit or comment on a text from a distance. With a fast enough connection and the right equipment, users can also introduce video and audio components, which allow them to see each other's faces and hear each other's voices during the exchange. This kind of software works best when only two or three people are involved.
- *Whiteboards:* Most conferencing software and course programs include a whiteboard function. This is a virtual space where you can work together to draw diagrams, make notes, and record information. This function also works best with a small number of people. The results can be saved in word-processing documents for future reference.
- *MOOs:* MOOs are basically text-based virtual environments. The MOO is an enhanced version of the MUD, or multiple user domain, an early version of today's chatrooms that was developed by online gamers. MOOs add another element to the mix—they allow users to create virtual worlds, including buildings, landmarks, and objects. The two *O*s in MOO stand for *object oriented.* Authorized builders can use text to create offices and homes for their characters, which are represented by user names. Your campus may have a MOO or may subscribe to one owned and managed by another institution. This kind of space allows you to store your documents in a virtual workroom and interact together as you work on them. It can host many users at one time. Logs of MOO sessions can be a part of your group's work record.

Keeping It All Together: The Field Journal

At several points in this chapter, we've mentioned the field journal. This is an ongoing record of your work on the project. Whether you are working in a group or as an individual, we recommend that you maintain such a log. It can serve several functions, including these.

- *Record keeping:* Recording such things as mini-progress reports, updated timelines, group decisions, and plans of action in your field journal will help you stay on track as a group. It will also give you a record of all the work you put into the project.
- *Collaboration:* Working together to summarize what you've done during the course of a week or of the entire project can help create unity as well as shed light on conflicts. It is certainly better to deal with conflict early than to watch your project fall apart late in the game.
- *Reflection:* The field journal is an excellent place to store your personal reactions to what is happening in your group and with your project. Unlike most of your other assignments, the field journal invites you to write about frustrations or triumphs in an informal way. As you'll recall, reflection is one of the critical elements that distinguishes service-learning projects from other real-world projects.
- *Future reference:* Your teacher may ask to keep a copy of your field journal as a resource for future students. If a group in a future class wants to work with your agency, they might get a sense of how to approach it from reading your field journal. They could learn more about the kinds of work that actually go into a complex project or pick up strategies for managing the workload and dealing with collaboration problems.

Field journals can take many different forms. Your instructor may assign a particular format or allow you to design your own as a group. Our students have experimented with many approaches, but most fit into one of the following categories.

- *One journal compiled by the entire group.* Some groups like to create a single document that meets their standards for joint documents and has no clearly identifiable single author. For this kind of approach, they might compose each entry together at the beginning or the end of regular meetings and insert important documents as they are produced.
- *One journal with multiple authors.* Other students prefer to take turns keeping up with the field journal, recording notes from meetings and decisions from correspondence. Most entries identify the writer's name to document his or her contribution.
- *A compilation of separate journals by each group member.* You may also choose to keep individual records and reflections throughout the project and then put them together in one journal at the end of the term. This can be an especially interesting approach if you include in a cover memo some reflection about the differences among individuals' perceptions of the same events or processes.

Figure 6.3 shows a portion of a field journal kept by two University of Central Florida students, Cat and Kattie. You can see that it includes a wide range of items, including

- references to correspondence
- group decisions
- task assignments

Field Journal

9/29 Met for first time regarding organization selection for our project. Narrowed down to two organizations (Coalition on Donation and Outreach Love). Decided on date that we will pick one based on interaction with contacts. Created a milestone chart to have specific dates for group meetings and deadlines.

9/30 Kattie tried to contact CR of Coalition on Donation to find out about specific projects they may need assistance with. She was out of town. Left message.

10/06 Kattie tried to contact Carol Rumsey again. Left message.

10/10 Cat talked with HG, Outreach Love Volunteer, regarding possible ideas for our project. Received a copy of "Help Wanted" flier for tutors that is currently in circulation. Heather listed many needs of Outreach Love, but upon clarification of our requirements she referred Cat to contact CW for more information regarding tutor and textbook needs.

10/11 7:30pm–10:00pm We met to finalize organization selection. Opted for Outreach Love because of inability to speak to someone at Coalition on Donation and pending deadline. Worked on first rough draft of proposal. Updated milestone chart. Focused on list of potential needs and benefactors, as well as creating the introduction and objectives portions of Outreach Love proposal.

10/17 Cat talked with CW regarding the tutor and textbook needs. She explained our desire to design marketing tools for tutor recruitment and organizational information for potential benefactors. C referred us to BG for project approval. B is currently on vacation.

2:00pm–4:30pm We met to finalize the first draft of proposal one. Utilized feedback from Melody and peers. Gained more focus on scope of the project. We narrowed down donation needs and brainstormed marketing possibilities. Updated milestone chart. Added "Scope of Project" to the proposal.

Assigned responsibilities for next week. Cat will research newspaper/ newsletter advertisement costs. Kattie will research printing costs. Kattie will also research the potential of marketing tutor opportunities to University of Central Florida students.

10/18 Kattie called copy center to get price quotes for various types of copies for budget purposes.

10/20 7:15pm–8:15pm Met in class to discuss "plan of attack" regarding deadlines. Prepared questions for BG regarding organizational history.

(continued)

Figure 6.3 Sample from students' field journal

Figure 6.3 *(continued)*

10/21 Cat called BG to discuss scope of project. B provided possible contacts for research materials. She was very excited and happy about us helping her organization. Scheduled a meeting with her for Monday, October 25th at 7:30pm.

11:30am Cat called Kattie regarding project update, meeting with Beverly and reconfirming group meeting for this evening.

Cat called *Orlando Sentinel* regarding cost for placing a classified advertisement.

7:30pm–9:30pm Met to discuss organizational history. Discussed meeting with B. Discussed resources for supplies and production of brochures and fliers.

10/25 7:30pm–9:00pm Cat met with BG regarding organizational history and scope of project. Discussed several aspects of what we hoped to do. B helped narrow down the needs of Outreach Love as pertaining to our project. Our primary focus will be to market the program to recruit tutors. If time permits, we will assist with the solicitation of educational resources for the program.

9:15pm–10:30pm We met to create rough draft of the organizational history for peer review.

10/31 2:00pm–5:30pm Met to finalize organizational history and to complete our final draft of the proposal. Updated milestone chart. Contacted BG regarding specific questions on Outreach Love and our project.

11/1 7:30pm–9:30pm Met to complete cover letter for proposal. Updated field journal. Began work on the brochure for the Outreach Love marketing strategy.

11/3 Worked on progress report in class. We reevaluated our timeline and due dates.

11/8 7:30pm–9:30pm Met to complete rough draft of progress report. Added text to brochure. Updated milestone chart.

11/10 Met before class to exchange progress report with the team for Children's Home Society. Met with Melody regarding status of project.

11/15 7:30pm–8:30pm Met to finalize the progress report.

11/17 7:30pm–9:15pm Used class time to develop form letters for tutor requests and supplies.

11/28 12:00pm–6:00pm Updated all graded materials to be turned in with final project. Completed form letters, fliers, advertisement, and worked on brochure. Minor glitch with computer and brochure file.

12/01 Class time. Updated progress report. Developed new brochure due to computer glitch.

12/04 3:00pm–6:30pm Worked on brochure. Assigned final responsibilities. Kattie will work with brochure graphics. Cat will obtain video for presentation and final photos for brochure and PowerPoint.

12/06 6:45pm–9:30pm Completed brochure. Printed final documents. Developed presentation. Created PowerPoint for our presentation.

- discussion summaries
- occasional glitches in the process.

Their journal also demonstrates each type of collaboration and explains how they made those approaches work. Depending on your instructor's expectations, you may also choose to include other items that Kattie and Cat put in their extended journal, such as

- actual copies of correspondence (letters, memos, emails)
- "before" versions of documents
- research materials—printouts from web searches and transcripts of interviews
- informal notes from discussions with the instructor and agency contact persons
- meeting minutes.

If you are assigned to keep a field journal, do not put it off until the last minute. Use this task as an ongoing opportunity to reflect on your progress and work through problems and concerns. Keep it up as you move into the process of creating your documents.

Activities

1. Draw diagrams of at least three different collaboration approaches your group could take. Keep in mind that most groups combine different models in their approach.
2. As a group, decide on a set of criteria for evaluating your task-level progress and your relationship-level interaction. Then look at some of the writing workshop guides in earlier chapters of this book. Based on some of the criteria covered in them and on your own sense of effective writing and collaborating, design a guide to assess your progress. Compare your guide with those of two other groups in the class to find ways to modify it.
3. Practice using the "commenting" and "tracking changes" functions in Word by having each person in the group add to the agenda for the next group meeting. Figure out a way to track each person's contributions.
4. Create a set of collaboration ground rules for your group. They can include all kinds of concerns. You might agree to avoid certain negative behaviors, prohibit the use of certain words, limit debate time on certain subjects, or build rewards into the process. Agree to follow these rules and to gently remind each other of them in the event of a slip-up.

Works Cited

Bebee, Steven A., and John T. Masterson. *Communicating in Small Groups: Principles and Practices.* 3rd ed. New York: HarperCollins, 1990.

Lay, Mary, et al. *Technical Communication.* 2nd ed. Boston: Irwin McGraw-Hill, 2000.

Markel, Mike. *Technical Communication.* 6th ed. Boston: Bedford/St. Martin's, 2001.

Parker, Glenn M. *Team Players and Teamwork: The New Competitive Business Strategy.* San Francisco: Jossey-Bass, 1990.

Chapter

Executing Your Project

Now that your service-learning project is underway, it's time to begin to create the documents you committed to in your proposal. In this chapter, we'll offer you more focused invention advice and introduce you to some basic processes that technical and professional writers often undertake when developing documents. Through the discourse analysis, style sheet, and document grid assignments, you'll have an opportunity to plan your documents to meet your agency's and readers' needs. Producing a style sheet will be the primary focus here because this tool is so commonly used by writers and editors and will greatly simplify your collaboration. After this, we will present some guidelines for producing a few common agency texts that we haven't already covered. This chapter doesn't provide you with step-by-step directions for producing every possible kind of agency project. Instead, our philosophy is based on the idea that you can effectively adapt certain rhetorical and design principles to a range of situations and tasks. In the pages that follow, we will offer you guidance for doing just that.

Gathering Information

You've already gathered some materials from the agency to write your proposal. You will undoubtedly need to gather more documents from the agency and other sources to write and design your final project. Assessing what your agency can offer will help you determine how much web, library, and field research you need to do on your own. Some larger agencies, such as the American Cancer Society and United Way, may have small libraries of resources on which you can draw. The national websites of such organizations can also be good sources of information. A group of students developing a patient information packet about American Cancer Society-sponsored services was able to draw on already-published brochures, program fact sheets, and the local and national websites throughout their writing process. These sources even provided the students with possible graphics such as photographs and logos, making their job largely one of synthesizing and adapting.

Other organizations, especially newer and smaller ones, may not have much material on hand. A group of students designing fact sheets for a local chapter of Surfrider Foundation ran into this problem. They needed information about shoreline problems in North and Central Florida, but the organization could only provide a national website that focused on West Coast issues. This meant the students had to get an early start on their research, which depended largely on the websites of such organizations as the Florida Department of Environmental Protection, the Florida Internet Center for Understanding Sustainability, and the Florida Sierra Club. These sites pointed them to valuable evidence of local coastal system problems, from beach erosion to saltwater marsh dredging, and to current efforts to address them. The students could have also taken fuller advantage of their position as college students by consulting other campus organizations that focused on local environmental issues as well as other professors on campus who were studying them.

Beyond possibly providing you with valuable information sources, gathering materials from your agency will aid you in the next assignment we discuss—the **discourse analysis.** Studying these materials will help you understand the agency's conventions and expectations for its texts and their designs. As we suggested in Chapter Four, you will want to study materials as similar to your own service-learning project as possible. These materials may provide you with multiple design options, particularly if you are revising an existing text or creating a type of text the agency frequently produces. Or you may be producing a document without precedent at the agency. Because the local chapter of the American Cancer Society hadn't produced a patient information packet before, it could not provide the student writers with a sample design. Therefore, this group turned to similar client packets produced by United Way and other agencies for design ideas, finally settling on a folder containing a set of multisized color-coded information sheets. Even if you have past models from the agency, we recommend obtaining similar texts from other organizations to develop a fuller repertoire of design options. Perhaps you are designing a brochure, and all of the agency's former brochures take a three-panel form and use graphics in the same way. If your agency contact person has given you the freedom to experiment with design, you might want to look at four-panel brochures by other agencies, especially if you have much information to include. You may want to improve, also, upon the agency's standard design conventions concerning font type and placement of graphics.

Analyzing Your Discourse Community: The Discourse Analysis Memo

In your invention for the proposal, you have already written an agency profile—a sketch of your agency and its goals and values. At this point in your project, after you have solidified your project plans and are gathering additional materials, we recommend that you learn more about the discourse community for which you are now writing, as well as the specific rhetorical task you now face. In Chapter Three we discussed the importance of analyzing your roles as writers in relation to your coworkers and others in your organization, but you haven't yet had much of an opportunity to study the agency you will represent. This assignment gives you such an opportunity, asking you to think of your agency as a discourse community and to analyze its characteristics and writing conventions.

You may remember from Chapter Three that a discourse community is a network of people defined by a body of texts and practices who share a common interest, specialized knowledge, and ways of communicating (e.g., forums, styles). Because the members of your agency share a community concern, mission, ways of communicating, and specific texts, they constitute a discourse community. Depending on how much the clients interact with the agency, these clients might also be viewed as part of the discourse community. As a discourse community, your agency has designated forums or textual sites for communicating, including, perhaps, a website, a newsletter, meetings, and an email list. These forums follow specific conventions or customary practices, from the silent rules that define acceptable behavior in a meeting to the ways members address each other in email to the arrangement of formal reports.

In the remainder of this section we present several sets of questions, most adapted from James Porter's forum analysis, to guide your analysis of your service-learning assignment. You may find them especially useful in determining the conventions of genres we don't focus on in this book, such as instructions, newsletters, and websites. A discourse analysis will not only aid you in planning your work, but parts of it might also be adapted for future texts you write as part of the project. For example, part of the organizational history you compile might be used in an agency website or brochure you are creating. Part of your discourse analysis can be used in your style sheet, the next assignment we discuss in this chapter. These questions will also help you begin to examine the ethics of the agency's texts, particularly as they relate to their audiences. You may remember from Chapter Three that the ethics of texts can be measured by how well they fulfill their duties to the audience, how well they embody communally defined ideals, and how beneficial their effects are.

For this invention assignment, we suggest that you write a three- to four-page report (in memo format) to your instructor analyzing the discourse community of your agency. Focus on those questions that will supplement most usefully what you already know about the agency. To make your report more accessible, you'll want to use headings and subheadings.

Beyond your sponsoring agency for this project, the questions below can be fruitfully asked of any organization and its discourses, from academic disciplinary organizations to professional trade organizations to companies to government agencies. Performing a discourse analysis can be a valuable invention exercise for your future writing-related work, especially when you are still learning and adapting to the culture of a new workplace. Learning to be an effective technical or professional writer means more than mastering a set of writing skills; it also means becoming acculturated into a network of interlocutors, particularly learning their ways of communicating and their discourse conventions. The discourse analysis can help you do this.

Organization's History

You may have already collected materials such as informational brochures that answer some of the following history questions. You may need to interview your contact person or another agency representative for others.

- Why, when, how, and by whom was the agency started?
- Has the agency been known by other names during previous time periods?
- What was the agency's first formal mission?
- How is the agency funded?
- With what other local, regional, and national organizations is the agency affiliated and how?
- What are the agency's primary accomplishments? (Include here awards, honors, and statistics about the agency's successes in serving clients.)
- What niche does the agency fill in meeting community needs or concerns?
- What kinds of publicity has the agency received from local television, radio, and print media?

Discourse Community

You probably already know some of the discourse community's characteristics, such as its mission, values, and major players. Use these questions to expand that knowledge.

- Who are its leaders and members? How would you map their roles and describe their interpersonal dynamics?
- What are their backgrounds and areas of expertise?
- How are these leaders chosen?
- What are its members' preferred means of communication with each other?
- What other kinds of texts does the agency produce?
- Who supervises and participates in producing agency texts? How and to what extent do they supervise writing? What are their expectations?
- To what extent does the agency involve clients in the production and evaluation of its texts?

Discourse Conventions

To answer the following questions about the conventions of the discourse community, you will need to study the sample texts you have gathered, especially past examples of the major text you are producing. If you are producing a text that's part of a larger forum, such as a section of an annual report or an article for an agency newsletter, you'll want to focus on that forum. If you are writing a set of instructions for an organization, you might study other instructions to assess the formality of their tone, complexity of their sentences, and formatting of steps and visuals. Common technical and professional genres such as instructions, proposals, reports, brochures, and newsletters tend to follow basic generic patterns of arrangement, style, and design. Such patterns are not formulas or templates, however, but general guidelines that discourse communities adapt and refine, and from which they depart. Therefore, you'll want to study the organization's own versions of various genres.

The following questions basically ask *who, what,* and *how* of the discourse community's texts. The more specifically you can answer the *how* questions—which focus mostly on arrangement and style—the further along you will be in your invention for the style sheet.

To whom are the texts addressed?

- What kinds of audiences do the agency's texts address?
- What are these audiences' needs, interests, and values?
- How much does the audience know about the texts and subjects?
- How would you describe the audiences' relationships to the agency?

What do the texts discuss?

- What topics are most valued by the agency's members and audiences?
- What kinds of arguments do the agency's texts employ? (Logical, emotional, ethical?)
- What kinds of sources do the agency's writers consult for information?
- How do the texts reflect the agency's values?
- How do the texts portray their clients/audiences and their needs? Do you see any problems with these depictions?

How do they write the texts?

- What design patterns do the agency's texts follow? Does the agency use templates or standard designs? Consider forms, color, fonts, spacing, and use of graphics.
- What are the main modes of arrangement used? What are the main devices used to create accessibility? Consider headings, lists, paragraph size, and so on.
- What style or documentation guides are followed?
- What specialized terms are used?
- How formal is the style of the texts? What stance do the texts take toward their audiences? What ethos do the texts project?
- What patterns in sentence type and length can you identify?
- What patterns in tense, person (first, second, third), and voice (active, passive) can you identify?
- What are the most effective elements of the agency's texts? What are the least effective?

Figure 7.1 shows an abbreviated discourse community analysis produced by a group of students writing a set of "success story" profiles about United Way agencies to be included in a United Way newsletter and brochure. Although this

To: Professor Blake Scott
From: Team Helping Hands
Subject: Analysis of United Way Discourse Community and Conventions
Date: March 2, 2001

Background of United Way

The purpose of this memo is to inform you of the discourse community of the United Way in Alachua County. The discourse community of the agency, in our case, includes people in need of the various services and those willing to donate money to the United Way's Community Care Fund. The United Way allocates the donations to numerous organizations—28 to be exact. These organizations range from Big Brothers/Big Sisters, which aids in "helping children succeed," to ElderCare of Alachua County, which aids in "promoting health and wellness."

There have been several brochures produced annually in the past. They contain the same basic format. These include a brochure produced by a volunteer at a local public relations agency with three success stories written by volunteers and a list of all agencies. The list includes a short description, a number where they can be reached, and a website.

What United Way Writes About

The documents that we are being assigned are promotional in nature. They have to perform many jobs.

1) They must portray the organization in a positive way.
2) They must be appealing to a reader.
3) They must contain relevant information. Because the audience for these documents contains people of all backgrounds and socioeconomic statures, they must not make any group of people feel alienated. They also must perform different functions for the organization.

The brochure, which is the major concentration of our project, must inform those who need help of the many services offered by the United Way. It contains stories of success from people who have received help from the United Way. Interviews are conducted with people who are willing to share their positive experiences with the organization. The implied claim made by these stories is that the organization can help anyone, including the reader, in a similar way. An important thing to do when making such a claim is to create similarity between the people in the stories and any potential reader. The brochure does this by giving quotes made by the people featured in the story and

(continued)

Figure 7.1 Discourse community analysis

Figure 7.1 *(continued)*

including pictures of them. To make these appealing to the reader, they are made relatively short and the paragraphs are separated into small blocks. Because the target audience may have doubts that they can be helped, keeping the information short and believable is important.

Some other documents produced by the United Way are informational fact sheets. These are targeted to people who may be potential supporters or volunteers. They must convince people to help support the United Way. One example fact sheet is a simple bulleted list of statistics produced by a few United Way organizations. For instance, one line says that Alachua County Healthy Kids supplied insurance for more than 2,500 kids. These impressive bits of information show donors that their money will benefit the community. Another sheet gives the scope of a donation by listing what certain amounts can do. It lists the effectiveness of donations from one dollar to twenty dollars. This not only tells the reader that the money is being used well but also lets a potential donor know what an appropriate donation might be. The United Way lets its actions speak for themselves by putting their statistics and stories on their promotional material.

Who Writes It

The writing is split up between a few staff members and volunteers. Most of the volunteers are professionals from the community who are good enough to donate their time, writing or editing various brochures and information sheets. People staffed at the United Way also do a lot of the writing. Nathan, their communication director, handles everything from writing the success stories for brochures to handling special events that the United Way sponsors. He expects new text to emulate brochures from the past; therefore, we will be using old brochures as guidelines. Almost all of the writing we will be doing will be completed outside the United Way. This will be done either in a group or will be split up and done individually and then edited and revised as a group. We communicate within our group by either emailing or calling each other. We meet weekly at the United Way to communicate with the director and let him know that things are proceeding smoothly, or to get more information from him.

To Whom Do They Write

The majority of the texts are aimed at people potentially in need of help. This is demonstrated in the success stories published in their brochures. For a person with mixed feelings about being helped by the United Way, these brochures can inspire the motivation needed to seek help. This is done by demonstrating the success of other people in similar situations. The brochures show a wide variety of different organizations, each aiding a different need. This helps people relate to one another, accept their problems, and seek help. Their secondary audience is the volunteers and potential donors. By showing all the people that have been helped by the United Way, the general community becomes more enthusiastic about donating money because they know it is going to good use.

(continued)

Figure 7.1 *(continued)*

How They Write It

The standard design conventions for the success stories in the brochure are

- AP style
- Approximately 250 words each
- Times New Roman font
- Left justification
- Agency title in larger and different type font

The primary rules of AP style are one space after every period, and having the source of a quote stated after it. Sentence lengths vary and seem to depend on the most effective way of writing it. Furthermore, the majority of the text is in present tense and third person.

The arrangement of information is that, first, the background information of the person and/or organization is presented. Next, the client's problem is stated, and the reason for the visit to a particular agency is given. Then comments from the client follow, stating how that agency helped and why he or she chose it. These paragraphs are short, containing one to three sentences.

The stance writers take relative to their audience is like that of a news reporter writing a human interest story. The writers are telling a story to the audience. Therefore, complicated words are avoided so that the scope of the audience is enlarged. Also, a persuasive voice is used to convince the audience that their experience with a United Way agency was truly a success.

memo doesn't go into the history of the agency's local chapter, it does begin with more general background information about the community and its texts. The memo is then organized around the three general categories of *what, who,* and *how.* Subheadings might have made the information under each more accessible. The writers are fairly specific about the conventions of past success stories in United Way brochures, and this goes a long way toward the development of their style sheet later on. Beyond the list under the heading "How They Write It," the writers identify patterns in sentence structure, the parts and arrangement of the profiles, and tone. This section on style could be even more thorough, as it should comprise a good chunk of the memo.

Now that you have thoroughly analyzed the agency as a discourse community, its texts, and its ways of communicating, you are better prepared to focus on the specific rhetorical situation of the major text you will produce. At this point, we return you to the invention questions in Chapter Three. In fleshing out the answers to these questions, pay particular attention to the intertextuality of your document, that is, how your document relates to other texts by the agency. How does the website that you're producing supplement and expand brochures and other print texts about the agency, for example? On what previous texts should the volunteer training manual you are producing draw?

Creating a Style Sheet

In the last section of your discourse analysis memo, you documented stylistic patterns in the agency's texts. The next document your group will produce—a style sheet—will draw on this section to specify the stylistic and design features of your service-learning project. If some of the agency's conventions need to be improved, don't include them as guidelines on your style sheet. If you make major changes, you should consult your contact person to discuss them.

Some organizations have their own comprehensive style manuals or guides, which thoroughly document the style, usage, grammar, mechanics, documentation, and design conventions for the major types of texts they produce. Other organizations, such as your university's press office, rely on standard professional style guides, such as the *Associated Press* (AP) *Stylebook*. Style sheets are shorter and less comprehensive than style guides as they are developed for a specific text and a specific group of writers. Even if your organization follows a style guide, not all of its guidelines will apply to your text, and your text may involve additional considerations. You will speed up your style sheet creation process if you begin with the organization's style guide, however.

As we mentioned in the last chapter, style sheets are designed to ensure consistency in the style and design of a document. Style sheets document stylistic and design choices where more than one option is a possibility, such as whether to write out "percent" or use the "%" symbol. They also specify rules relevant to the particular text (e.g., spelling of subject-specific terms) as well as rules that the writers/editors would otherwise have a difficult time remembering (e.g., when to use apostrophes). Figure 7.2 shows a style sheet developed by two students in a professional editing course. The students, Leila and Lisa, designed a newsletter for the University of Florida Community Campaign, a campaign that raises support among UF employees for charitable organizations (including United Way, Environmental Fund for Florida, and National Voluntary Health agencies) in Alachua County. Figure 7.2 is the final version of their style sheet, which started as a simple table with just a few items.

As Leila and Lisa's example illustrates, style sheets can cover guidelines about numerous areas, including these:

Grammar and Mechanics

- capitalization
- punctuation (often divided into multiple sections such as commas, hyphens, and quotation marks)
- abbreviations/acronyms
- spelling of special or field-specific terms
- numbers

University of Florida Community Campaign Newsletter Style Sheet

Design

Style	Adobe PageMaker, Newsletter, 11"x17" double-sided, 4-pages, double-column front and back, single-space, left justify. White space sets off written text from graphics and photographs and separates articles.
Layout	Left and Right, 2 Columns with .2" space between columns. Paragraph style—.1 L, .1 R, .2 first line, 1 space after headings, no space between paragraphs.
Type & Font	Body Text: Arial, black, 10 pt., 13 pt. leading Newsletter Masthead: UFCC Logo, Parisian, 26.5 pt., 23.5 pt. leading, PMS 258 blue, centered. Headline: Parisian, 23 pt., auto leading, bold, PMS 258 blue, centered. Article Headings: Parisian, 18 pt. bold, auto leading, PMS 258 blue, centered, 2 spaces before, 1 space after. Writer credit: Arial 6 pt, 12 pt. leading, black, at end of article, right justified.

Grammar & Mechanics

Abbreviations	Spell out first use before using acronym; if same acronym is used in a different article, spell out first use again.
Capitalization	Use uppercase for academic department names and titles; use uppercase when referring to specific department or division; capitalize University of Florida and University when used alone as a reference to UF.
Commas	Use comma before "and" or "or" in a series. Use commas to set off direct quotes; place commas and periods inside right quote marks. Commas set off nonrestrictive clauses. Separate independent clauses joined by a coordinating conjunction with a comma.
Credit	Name and affiliation of writers and photographers providing written information or photographs are noted, 6 pt. black Arial font.
Dash	Leave one space before and after a dash (use em dash—not two hyphens).

(continued)

Figure 7.2 Style sheet for newsletter

Figure 7.2 *(continued)*

Hyphens	Hyphenate most compound adjectives that precede a noun but not those that follow a noun.
Lists	Use a colon when introducing a list with "as follows" or "the following."
Numbers	Use numbers for all quantifiable units of measure (including numbers under 10). Use numerals with percent (written out, not symbol—5 percent). Do not begin a sentence with a number—rearrange sentence. Decades do not take an apostrophe (1960s, 1970s).
Paragraphs	Indent .2" for new paragraphs. Leave one space after sentence period.
Quotation Marks	Use quotation marks to set off a direct quotation; place commas and periods inside end quote marks; use ellipses to reflect omitted words in quote.
Nonsexist Language	Use Chair (not Chairman); avoid using him or her, his or her, or sexist language; revise words or phrases that stereotype or unnecessarily call attention to gender.
Style	Informal, but professional, writing style; avoid passive voice sentences; use of first person acceptable. Written for diverse audience consisting of University of Florida faculty, staff, and administrators and UFCC agency representatives.
Visual Elements	
Photographs	Black-and-white photographs of individuals and events are included to increase reader interest. Standoff on text wrap set at .167". Captions below or to the side are used when needed to provide detail about the photographs—9 pt. Parisian, 10 pt. leading, PMS 258 blue. Photo credit in Arial 4 pt., 6 pt. leading, black, typed on vertical frame, standoff on text wrap reduced to .0. Stock photographs included as filler and for visual change are not captioned.
Charts and Graphs	Graphic displays of information are included to provide a pictorial display of information to improve readability and visual relief. Created in Microsoft Excel and imported into PageMaker. Framed with 1 pt. black line. Chart/Graph title in Arial, 12 pt. bold, black. Arial 5 pt. or SmallFonts 5.5 pt. or larger to label data points. Bars in bar graphs, slices in pie charts, etc. in PMS 286 blue and shades of blue.

(continued)

Figure 7.2 *(continued)*

Agency Listing	Two Column list. Agency affiliation centered and underlined, BellGothic, 10 pt., bold, black. Agencies numbered, left justified, BellGothic, 9 pt., bold. PMS 186 blue.
Page Numbers	Bottom center, American Typewriter, 9 pt., white font in PMS 286 blue oval.

- other quantitative figures (e.g., equations, formulas)
- citations.

Design Elements

- font type, size, and style (for headings, main text, visuals)
- levels of headings
- spacing (could be divided into leading, justification, indentions, space surrounding headings, spaces after periods and colons)
- emphasis (e.g., use of boldface or italics for key terms)
- page numbers
- tables (including title, caption, number alignment, design)
- figures (including title, caption, labels, design).

When style sheets are used as aids in the drafting process, they sometimes contain more complex usage guidelines, such as these:

Usage Guidelines

- use of nonsexist language
- use of first, second, and third person
- use of active and passive voice
- types of sentences (i.e., simple, compound, complex)
- tone and register.

In addition to guidelines about the above elements, useful style sheets often provide brief examples to *show* the writer/editor what the rule means. A rule about representing numbers might simply be expressed as "Ratios are expressed as 1:5." Similarly, if the rule addresses design, embody it in the example, putting the rule "Font size and type is 11 point Helvetica" in the type to which it refers. It often helps to include examples of grammar rules as well, especially when some writers may not be familiar with specific grammatical terms. If you purchased a style handbook or manual for your class, you may choose to connect your style sheet guidelines to it by indicating related page or section numbers.

Style Sheet Design

Because they are quick reference documents that writers use while drafting, designing, revising, and editing, style sheets must be *accessible, retrievable,* and *easy to read.* To aid accessibility, style sheets are often arranged as simple tables or at least in multiple

OTHER VOICES 7.A Leila Cantara

Leila Cantara is a student at the University of Florida and an employee in the university's Business Services Division.

Making the Style Sheet Assignment Work for You In Class and Beyond

As a nontraditional student, I have had the benefit of many years of work experience and several semesters of English composition courses. In Professor Scott's Professional Editing class I had the opportunity to put what I've learned to practical use. We learned to develop *style sheets* as tools to assist writers, editors, graphic designers, printers, and publishers in their collective effort to take a written product from draft to final publication.

My first style sheet was for a newsletter produced three times a year and distributed to staff, faculty, and administrators as part of the University of Florida Community Campaign. As work progressed on the newsletter, the style sheet grew from its original simplistic outline to a detailed listing of design and stylistic choices and guidelines. The end result convinced me that the original investment of time to develop the style sheet had paid off.

As my job at the University of Florida Business Services Division has diverse responsibilities, I must juggle multiple priorities under demanding time constraints. The use of style sheets has greatly improved my ability to manage time, delegate responsibilities, and create professional printed products.

My staff consists of student workers who are creative, talented, and often temporary. Training employees every other semester in the different writing styles, formats, layout, and printing requirements for the newsletters, annual reports, newspaper advertisements, and journal articles that we produce was very time-consuming. That created time problems, as frequent discussions with designers and printers were always necessary.

Using style sheets has made a marked improvement in our printed materials. The time required to prepare the initial style sheets for each project was minimal. I use a common layout for the style sheet, put in the required information, and adapt where necessary to incorporate unique requirements. My student staff quickly adapt to the different requirements of our diverse products, as the style sheets help us maintain consistency from concept to closure. I have also noticed fewer problems with designers and printers as the need to explore alternative layouts or ink colors has been eliminated.

My efficiency certainly has increased, but, more importantly, my stress level has decreased. I have been rewarded for my diligent effort to complete my education—not only because of what I learned but because I have been able to apply that knowledge in a practical and productive way to both a worthwhile community fundraising campaign and to my job.

columns (see Figure 7.2). If you use a table, grid lines should be minimal to aid clarity. In addition, the guidelines and examples are often arranged in a bulleted or numbered list. If Leila and Lisa had used lists, the specific information under each category would not blur together so much. We also recommend headings that stand out.

Whenever possible, keep the style sheet to one page so that it doesn't require flipping back and forth. Most style sheets are arranged alphabetically by category in the manner of other types of reference texts. Longer, more detailed style sheets might require multiple pages and a modified arrangement. Notice how Leila and Lisa's style sheet is divided into three major sections: design, grammar and mechanics, and visual elements. They divided it this way based on how they would use it to edit the newsletter. They planned to take one editing pass checking for spacing, font, and general layout; another checking for grammar, mechanics, and other style elements; and another for checking illustrations. You might develop another way to arrange the major sections; just be sure to base your decision on how you will use the text. You might want to put your style sheet on a web page or have a paper copy laminated as you will be consulting it on a regular basis.

Style Sheets and Templates

Some style sheets do not cover many design elements because these elements (e.g., font size and type, leading, indentions) are already saved in a template. Many of you will work with organizations that have templates or programs for creating particular types of texts such as newsletters and annual reports. Many such templates are available online. Others might be stored in agency word-processing or document design files. This is often the case at national organizations like the United Way and the American Cancer Society.

If you work with an organization that doesn't have saved templates for the texts you will produce, you can easily create your own with almost any word processing program. In Microsoft Word, for example, you can simply save a document as a template. Your sponsoring organization might currently use templates that need major revision. In either case, part of your contribution to the organization could be a template providing the agency with a design blueprint for future texts of that type. This would save them time and encourage consistency in future documents. Before you make changes to the template, you might present your contact person with a short memo proposing the changes and explaining your rationale. Designing or revising a template would be an especially valuable addition to a design-heavy document such as a newsletter, brochure, or set of instructions. Even if your group has a template with which to work, this template probably won't regulate every design element of your text automatically. You might still need to check certain elements of the text's spacing, for example. Such elements could be included in a style sheet.

Genre-Specific Style Guides

If your organization doesn't have a more comprehensive style guide, consider producing one as part of your service-learning project, especially if one of the texts you planned to produce falls through. To make such a task more doable, you might focus your guide on just one of the major genres (e.g., proposal, fact sheet, web page) that the organization regularly produces. You could also focus on either style or design. As a major part of their service-learning project, a group of

Melody's students produced a comprehensive web style guide for Downtown Orlando Partnership, a networking group for local business leaders that sponsors nonprofit agencies and performs charity work itself. Figure 7.3 shows this guide.

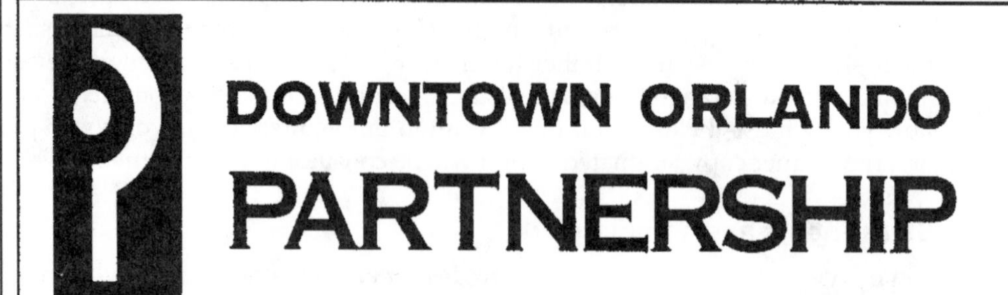

Web Style Guide

Prepared by Valerie Hall, Rick Hartig, and Bruce Hickman

The DOP Web Style Guide defines conventions used in page design, layout, and grammar and mechanics, including but not limited to navigation, document structure, and cosmetics. The style guide is intended to create a consistent environment for all persons utilizing the DOP web.

Table of Contents

(continued)

Figure 7.3 Comprehensive web style guide

Figure 7.3 *(continued)*

B. Production
   1. Document Searchability
   2. Navigation
   3. Hyperlinks
   4. Menu Pages
   5. Documents and Text
   6. Images and Icons
   7. Background Images and Colors
   8. Frames Navigation
   9. The Frameset Container
   10. The Frame Tag
   11. Navigation for a Site with Frames—Targeted Windows

II. Grammar and Mechanics
   A. Abbreviations and Acronyms
   B. Capitalization
      1. General
      2. Job and Position Titles
      3. Structures and Places
      4. Do Not Capitalize
   C. Punctuation
      1. Commas
      2. Dashes
      3. Hyphens
      4. Apostrophes
      5. Colons
      6. Semicolons
      7. Ellipses
      8. Periods
      9. Quotation Marks
   D. Numbers
      1. General
      2. Telephone Numbers
      3. Time
   E. Inclusive Language
      1. Sex and Gender
      2. Disability
      3. Race and Ethnicity
   F. Publications, Presentations, and Reports

(continued)

Figure 7.3 *(continued)*

I. WWW Style

A. Text

1. Web Reading Habits

If you are hoping to appeal to an external audience (including prospective businesses, investors and residents), keep in mind the ways in which web documents differ from printed ones.

- When people read copy on a computer screen, especially if they are browsing the Web, they tend to skim rather than read more carefully, line-for-line and word-for-word, as they might if reading a printed piece.
- Readers tend to skip from one web page to another, from one link to another, rather than reading a page from top to bottom.

2. Writing for the Web

Text written for the Web is most effective if it follows these general guidelines:

- Visitors to your site should know exactly whose site it is and what its purpose is when they view your home page. Otherwise, they may be lost or, worse, may leave the site entirely and never come back. This is accomplished with a link back to the site home page from every available page on the website.
- Outline, organize, subordinate: Use links to take readers into deeper levels of a topic. Think of upper-level pages as summaries or abstracts to whet the reader's appetite. Then use your links as a map of where they can go from there.
- Use subheads when your copy runs more than one screen in length, or break copy into more pages.
- Write short paragraphs and sentences (avoid complex sentence structures and jargon where possible).
- Speak directly to the reader, where appropriate (Web discourse is generally perceived as personal and informal).

3. Web Terms

E-world lexicon continues to change as rapidly as you can save your Word file. Here are some conventions to follow for now:

- Use "the Web" to refer to the entire network of the World Wide Web.
- Lowercase "web" when referring to a particular page or site.
- Use two words for the compound "web page" but one word for the compound "website."

4. Creating Emphasis

When we used typewriters, hitting the Shift Lock key to create full caps was one of two ways to create emphasis. The other was to underline. But reading more than a few words of full caps or underlined text is unpleasant. Computer users now have a number of options that can provide emphasis with increased readability. For example:

- Type of different sizes can indicate emphasis or levels of importance.

(continued)

Figure 7.3 *(continued)*

- A combination of two compatible typefaces, when used judiciously, can signal relative importance of copy blocks.
- Boldface and italic type can give emphasis to words and phrases and are much more readable than full caps and underlined type.
- Extra white space surrounding a copy block can help draw attention to it.

5. Spell Checkers

Sadly, spell checkers are fallible. Your word processing program's spell checker may see deer and register that it's spelled correctly, but if you intended to write dear, you could be in deep trouble. Use a spell checker as a first round of checking, but have a qualified person do final proofreading.

6. Typefaces

Some institutions (corporate and academic) prescribe two standard typefaces for routine use—one serif (with "feet") and one sans serif (without).

Here are four rules of thumb to guide daily word processing and document preparation.

- The smaller the type size, the shorter the line length.
- Recognize that the same point size in different computer typefaces does not necessarily mean that the letters will be the same height (see examples below).
- Serif typefaces (such as Times, Palatino, Garamond) tend to be more readable in long copy blocks and long lines than most sans serif typefaces (Helvetica and Arial are two commonly used sans serif typefaces), so text set in sans serif should be set in shorter lines or larger type.
- Note that, in most typefaces, the characters for the letter "l" and number "1" are different, so don't retain the old typewriter habit of typing lowercase letter l's when you mean the number.

B. Production

1. Document Searchability

As soon as a document is placed on the DOP web system, the search engine software begins to index the content, and it instantaneously becomes available through the search engine. Inserting search engine criteria in web documents increases the rank in search results, as well as displays that information in the search results page.

- All web pages are required to contain:
  - A properly documented Title tag
  - Informative Meta Keywords and Description tags.
- All Microsoft native documents and Adobe PDF documents are required to have the document properties fields completed.

2. Navigation

All web pages must be "treed" back to the DOP homepage. This is to ensure that when a user travels to a secondary page within a website, a path exists to return to primary pages

(continued)

Figure 7.3 *(continued)*

within that website without relying on the browser's "back" button. Navigation bars, icons, or text links are to be provided on all secondary web pages to return the user to primary web pages.

- The DOP web should be simple and intuitive to navigate. Common images, layout, or icons within a website create a look and feel which over time becomes intuitive to a user. For example, provide a "back to DOP main page" button at the bottom of every page.

Attempt to maintain a shallow link structure. Data should be less than 3 clicks away from your homepage. The goal of the DOP web is to provide users with the information that they need quickly and easily.

- Frames will be permitted as long as adequate navigational aides are provided for the site and it is coded to allow easy and complete exit when the viewer wishes to move to another page. Please refer to Appendix A for a discussion on how to provide this navigation.

3. Hyperlinks

Links should provide the "big picture" of a website. Follow the descriptions of the links down a web page. If they are identified accurately, they should give the user the purpose and scope of the entire website without reading any other text.

- Make links obvious. If a link is important, do not put it inline with text. Put it after the paragraph, ideally with some sort of icon and without extraneous text. Important links should never be buried in text that is likely to be quickly scanned. Inline links should be related information that may be of lesser utility.
- All data links must be in a prominent place on the page, preferably near the top or at least within sight of the user. No user should have to scroll past large graphics or nonessential text to get to data.
- Create sensible links. The link should always be the title or description of the destination. Do not use "here" or "click here" as a link.
- If you use icons, buttons, or banners to link to data, provide text links as well. Web pages containing image maps are required to have text links to the same resources.
- If a link that is "under construction" must be created, do not activate this link until at least minimal page content is provided. It is a waste of time for a user to click to a page whose content is solely text stating "under construction".
- Hyperlink colors may be modified, but it must remain obvious that they are hyperlinks. Avoid setting "new" and "visited" links to the same color. Keep in mind that monitors display colors differently, so review your pages on several monitors before you release them.
- End with a slash (/) when linking to a directory: Some browsers do not provide a directory listing if your URL does not end in a slash.

(continued)

Figure 7.3 *(continued)*

4. Menu Pages

Menu pages are web pages with lists of links. The following guidelines are set to enable a user to find what they are looking for quickly and easily without having to spend a lot of time searching for information within a website.

- Shorten and organize lists of links. Keep lists to manageable sizes so that users are not confronted with many lines of elements to read. If you have more than seven or so elements in a list, consider splitting the list into smaller sections.
- Avoid text blocks. Lists should be stripped of any text that is not necessary to describing the link's function.
- For online help menus giving access to other resources, size the page to fit on 24 lines.

5. Documents and Text

With the ability to put almost any document on the web server comes the responsibility of maintaining the integrity of the data contained within and of maintaining the relevancy of that data. Be professional in your work. Spell check your text, edit carefully, and maintain high standards.

- Shorten or split apart large documents. It's easier to find information in a short document than in a long document. Many users find it burdensome to read more than a few screens of information as a single entity. They often absorb only what is on the first screen and if not interested, will not bother to scroll down. For textual documents, attempt to keep the length to less than 1½ screens.
- An advantage with a long, continuous document is that it is easier for a reader with scroll bars to read in an uninterrupted flow. If a document is fairly long, internal anchors will enable the user to move around the document quickly and easily.
- Normal text colors may be changed to lend cosmetic appeal to pages or to call attention to specific text items.
- Maximum web-page resolution is 800 pixels. If a web page wider than 800 pixels is viewed on an 800-pixel-wide resolution monitor, the layout can become distorted or unreadable. Remember that browser scroll bars take up about 50 pixels.

6. Images and Icons

Avoiding putting images and icons on your web page that have no relevance to the information or that do not add value in terms of navigation, context, or consistency. The following guidelines are to be followed when the use of images and icons on DOP web pages are considered.

- Always provide ALT tags. The ALT tag is used to provide text that is displayed by viewers that disable graphics.
- Using height and width information in the image tag increases the loading speed of your web page, as well as preserves the format of the web page during loading.

(continued)

Figure 7.3 *(continued)*

- Browser clients may not display images the way your client does, or the viewer may have a different monitor resolution. If formatting your page around images, remember that the line breaks are a function of the window width. Add the BR or BR=All directive if you want a break to occur in a specific place.
- Avoid unnecessary use of animated gifs. Many of these cause instability in web browsers, especially if many are used on the same page or in succession.

7. Background Images and Colors

Background images and colors should be used with extreme care. Make sure that the text is readable against background colors and not obstructed by background images. Also, many browsers are set by default to ignore background images or colors while printing. If you have a dark background image with white text, the printer would print the text as white and it would be unreadable to the person who printed it.

- Background images and background colors must contrast well with text.
- Keep in mind that other monitors may display background colors differently than your monitor, so view pages on several systems to assure readability.
- If your data has the possibility of being printed, do not use dark backgrounds with light colored text.

8. Frames Navigation

Frames create independently changeable and (sometimes) scrollable windows that tile together to organize the pages in a website. Each frame holds its own HTML file as content, and the content of each frame can be scrolled or changed independently of the others. In a way, each frame almost becomes its own "minibrowser." Use of frames is discouraged at DOP because they can make web navigation difficult.

9. The FRAMESET Container

- Frames are contained in a structure called a FRAMESET, which takes the place of the BODY container on a frames-formatted web page. A web page composed of frames has no BODY section in its HTML code, and a page with a BODY section cannot use frames.
- Because no BODY container exists, FRAMESET pages can't have background images and background colors associated with them. (They are defined by the BACKGROUND and BGCOLOR attributes of the BODY tag, respectively.)
- The FRAMESET tag has two attributes: ROWS and COLS (columns). <FRAMESET ROWS="value" COLS="value"></FRAMESET>
- A curator can define any reasonable number of ROWS, COLS, or both; something has to be defined for at least one of them. If there is no definition for more than one row or column, browser programs ignore your frames completely. The "value" in the generic FRAMESET line is a comma-separated list of values that can be expressed as pixels, percentages, or relative scale values. Remember that bare numeric values assign an absolute number of pixels to a row or column,

(continued)

Figure 7.3 *(continued)*

values with a % (percent sign) assign a percentage of the total width (for COLS) or height (for ROWS) of the display window, and values with an * assign a proportional amount of the remaining space.

10. The FRAME Tag

- The <FRAME> tag defines a single frame. It must sit inside a FRAMESET container, like this:
- <FRAMESET ROWS="value">
  <FRAME>
  <FRAME>
  </FRAMESET>
- Note that the FRAME tag is not a container, so unlike FRAMESET, it has no matching end tag. An entire frame definition takes place within a single line of HTML code. A curator should have as many FRAME tags as spaces defined for them in the FRAMESET definition. In this example, the FRAMESET established two rows, so you need two FRAME tags. The FRAME tag has six associated attributes: SRC, NAME, MARGINWIDTH, MARGINHEIGHT, SCROLLING, and NORESIZE. Here's a complete generic frame:
  <FRAME SRC="url" NAME="window_name" SCROLLING=yes|no|auto MARGINWIDTH="value" MARGINHEIGHT="value" NORESIZE>
- The most important FRAME attribute is SRC (source). A curator can have a complete frame definition using nothing but the SRC attribute, such as this:
  <FRAME SRC="url">
- SRC defines the URL of the content of a given frame.
- A curator cannot use plain text, headers, graphics, and other elements directly in a FRAME document. All the content must come from HTML files as defined by the SRC attribute of the FRAME tags. If any other content appears on a FRAMESET page, it is displayed, and the entire set of frames is ignored.
- The NAME attribute assigns a name to a frame that can be used to link to the frame, usually from others in the same display. The following example creates a frame named Frame1:
  <FRAME NAME="Frame1">
- The Frame1 frame can be referenced via a hyperlink like this:
  <A HREF="http://www.yoursite.net" TARGET="Frame1">
- Note the TARGET attribute is what references the name of the frame.
- If a curator doesn't create a name for a frame, it simply has no name and can't be targeted. All frame names must begin with an alphanumeric character.
- MARGINWIDTH and MARGINHEIGHT control the width of the frame's margins. They both look like this:
  MARGINWIDTH="value"
- The value is always a number and always represents an absolute value in pixels. For example, the following creates a frame with top and bottom margins 5 pixels

(continued)

wide, and left and right margins 7 pixels wide:
<FRAME MARGINHEIGHT="5" MARGINWIDTH="7">

- MARGINWIDTH and MARGINHEIGHT define a space within the frame within which content does not appear. Border widths are set automatically by the browser, not by HTML code.
- Frames automatically have scroll bars if the content specified for them is too big to fit the frame. Sometimes having scroll bars ruins the aesthetics of a page, so HTML coding is added to control them. That's what the SCROLLING attribute is for. Here's the format:
  <FRAME SCROLLING="yes|no|auto">
- SCROLLING has three valid values: yes, no, and auto. Auto is assumed if no SCROLLING attribute appears in the frame definition. Yes forces the appearance of a scroll bar. No keeps them away at all costs.
- The user can normally resize frames. If a user moves the mouse cursor over a frame border, it turns into a resize gadget that lets them move the border where she wants it. Doing so always changes the look and feel of a frame. Curators can use the NORESIZE attribute to keep users from resizing frames. Here's how:
  ><FRAME NORESIZE>

11. Navigation for a Site with Frames—Targeted Windows
  - TARGET names the browser window to use when jumping to the specified URL. If a window with that name doesn't already exist, the browser opens a new window and calls it by the TARGET name.
  - If a curator doesn't feel like adding TARGETs to every link on your site, an associated new BASE tag names the default.

II. Grammar and Mechanics

A. Abbreviations and Acronyms

- Spell out the abbreviation or acronym on the first use, and follow with the abbreviation in parentheses to prepare readers for your subsequent use of the abbreviation.
- The general trend is away from using periods in abbreviations, unless confusion might result. Thus, we get *DOP* rather than *D.O.P.* and *ASP* rather than *A.S.P.*
- Do not use the ampersand (&) as a replacement for *and.* Use the ampersand only when it is part of an official name of a company, product, or other proper noun.
- Abbreviate only *Ave., Blvd.,* and *St.* and only with a numbered address. Spell them out and capitalize when part of a formal street name without a number.
- Abbreviations and acronyms should be restricted to situations in which they enhance comprehension, that is, when your copy refers repeatedly to a lengthy name or term that has a commonly accepted abbreviation. Be aware that familiarity with most abbreviations and acronyms is context sensitive and field dependent. If you use *DOP* in your copy, will it be clear immediately to all your readers whether you mean *Downtown Orlando Partnership* or *Downtown Orlando Project?*

(continued)

Figure 7.3 *(continued)*

B. Capitalization

1. General

- Official names and proper nouns are capitalized. Common nouns and various shortened forms of official names are not capitalized. Use the full, official name the first time it appears in a document or section of a document.
- Standard style guides, including the *AP Stylebook* and *The Chicago Manual of Style,* require lowercase letters in running text. The lowercase style is becoming the preferred style for external communications, in part because the media observe that style.

2. Job and Position Titles

Capitalize job titles only when they immediately precede the individual's name or when they are named positions or honorary titles (as in the last two examples).

- It's common knowledge that former President Bill Clinton loves to golf.
- The president, George W. Bush, took the oath of office under cloudy skies.
- The president of the United States serves a four-year term of office.
- When former Governor Bob Graham visited the Downtown Orlando Partnership offices, he was impressed.
- When Bob Graham visited the Downtown Orlando Partnership offices, he was a governor.

3. Structures and Places

Capitalize the full official names of buildings and formally designated places in downtown Orlando:

- City Hall
- TD Waterhouse Arena
- Bob Carr Theatre.

4. Do Not Capitalize:

- city of Orlando, the
- state of Florida, the
- university, the (when it stands alone in reference to the University of Central Florida, for example)
- seasons: spring, summer, fall, winter.

C. Punctuation

1. Commas

- Use comma before *and* or *or* in a series.
- Use commas to set off direct quotes; place commas and periods inside right quote marks.
- Commas set off nonrestrictive clauses.
- Separate independent clauses joined by a coordinating conjunction with a comma.

(continued)

Figure 7.3 *(continued)*

2. Dashes

En Dashes

Use en dashes between inclusive numbers and with compound adjectives when one element consists of more than one word. (On the Macintosh, en dashes are created by hitting the option and hyphen keys.)

- You'll find the examples on pages 23–26 of your quarterly report.
- He's taking the earliest Seattle–New York flight.

Em Dashes

Em dashes are used to denote a sudden break in thought that causes an abrupt change in sentence structure. Traditionally, in all uses except most newspapers (and in some display typography, such as headlines), a dash is set without a space on either side. Especially in certain formats and type sizes, adding a space before and after a dash can make the text look "gappy."

- Professor Rose—who had driven over a skunk earlier that morning—gave his lecture on the importance of the olfactory sense in mammals.

3. Hyphens

Most questions about whether to hyphenate or not can be readily answered by consulting your dictionary. DOP follows the example of the *Chicago Manual of Style:* Omit hyphens from adjectival compounds where there is little or no risk of ambiguity.

Common uses for hyphens:

- Abbreviations of UCF campuses: UCF-Orlando
- Compound adjectives such as self-sufficient.

4. Apostrophes

When indicating the possessive for names, use an apostrophe followed by an *s* even when the person's name ends in *s* or another sibilant. A sibilant is a hissing sound. The two traditional exceptions are *Jesus'* and *Moses'*.

- Jones's theorem
- Marx's ideas

5. Colons

Use colons to introduce a series or a list, especially a list preceded by *as follows* or *the following*. Capitalize material after a colon if it constitutes a complete sentence. Use a colon to introduce an explanatory phrase or sentence.

- Conference participants should bring the following items: alarm clock, laptop computer, eye drops, coffee maker, and pillow.
- The implication of the mayor's challenge was clear: Residents must not be overburdened by bureaucracy if they are to enjoy a fulfilling Orlando experience.

6. Semicolons

Use semicolons in lists whose items include commas. Use semicolons to separate closely related clauses.

(continued)

Figure 7.3 *(continued)*

- Officials at the meeting included: Glenda Hood, mayor; Buffie Paulauski, DOP executive director; and Jeb Bush, Florida governor.

7. Ellipses

Use ellipsis points to indicate that material has been omitted from the middle of a quotation. Do not use ellipses at the beginning or end of a quotation even if you start or stop in the middle of the quoted sentence. Ellipses are created with three period characters, with one space on either side of each character:

- . . . not ...
- Mayor Hood's speech began with a request that audience members "leave all video cameras, tape recorders, and still cameras with the staff . . . at the entrance."

8. Periods

Use periods in the following situations:

- at the end of a declarative sentence
- at the end of a quoted passage that also ends a sentence, even if it is not the end of the sentence in the original passage
- with abbreviations at the end of items in a vertical list, such as this one, if some or all of the list items are complete sentences at the end of a vertical list that is punctuated with commas at the end of each item.

9. Quotation Marks

Commas and periods always go inside quotation marks. Colons and semicolons always go outside quotation marks. With question marks and exclamation points, use the following: If the punctuation is part of the quotation, put it inside the quotation marks; if it's not part of the quotation, put it outside.

Use quotation marks

- to indicate the exact words that someone spoke or published
- the first time you refer to a nickname
- the first time you use a term or phrase ironically or sarcastically.

D. Numbers

1. General

- Spell out one to nine. Use numerals for 10 and above.
  She gave five dollars to A Gift For Teaching, which was 8 percent of her annual dividend check.
  When she turned 21, she realized that she'd rather be a city commissioner.
- Spelling out large round numbers is preferred.
  She gave the museum more than a hundred thousand artifacts.
- Use a combination of numerals and words with numbers in the millions and larger.

(continued)

Figure 7.3 *(continued)*

The population increased by 2.3 million.

- Use a comma for numbers with more than three digits.
  Average corporate donations for 1998–99 were $6,038.
- Hyphenate fractions when they are spelled out.
  A four-fifths majority voted in favor of the amendment.

2. Telephone Numbers

Now that all local numbers require use of the area code, do not put the area code in parentheses. Instead, simply use a second hyphen: 407-648-4010.

3. Time

Use numerals with a.m. and p.m. (small caps or lowercase letters) to indicate specific times. Use noon and midnight in place of 12:00 p.m. and 12:00 a.m., respectively, for clarity.

- The seminar will begin at 2:30 p.m.

D. Inclusive Writing

Communicate in a manner that does not exclude particular individuals or groups. Also, avoid becoming trapped in euphemisms. The basic idea is to treat people as equal individuals.

1. Sex and Gender

Avoid the awkward *s/he* and *his/her*. The easiest way to write copy that applies equally to men and women is to use plurals. If the singular must be used, use both pronouns, joined by a conjunction.

- To be successful, employees need to do more than attend training classes regularly; they also need to practice good study skills, take advantage of the company health club, and get sufficient sleep.

2. Disability

When writing about individuals with disabilities, use "person first" language, that is, person who uses a wheelchair. Similarly, blind employees would be preferable to the blind. Do not capitalize blind, deaf, or any other term relating to people with disabilities.

- Special arrangements may be made for employees with hearing, vision, or physical disabilities.

3. Race and Ethnicity

Current practice and preference is to type the names of non-European Americans without hyphens.

- African American (American of African descent)
- Asian American (American of Asian descent)
- European American (American of European descent)
- Hispanic American (American with ancestors from Spain, Mexico, Puerto Rico, Cuba, South and Central America)

(continued)

Figure 7.3 *(continued)*

- Native American

F. Publications, Presentations, and Reports

Titles of books, journals, movies, and TV and radio programs are styled italic with initial caps. Titles of articles, episodes, short stories, book chapters, poems, conference papers, and essays are styled Roman and enclosed in quotation marks. Titles of reports, workshops, conferences, and so on are also set in Roman text with initial caps.

- *O-Town News,* available through the Office of News Services, contains a wealth of facts, stats, and information about the city of Orlando.

As you can surmise from its detailed table of contents, this reference document includes general guidelines about writing and designing web pages as well as more specific design and style conventions for the organization's web pages. Glancing through the table of contents, you'll probably notice many of the traditional style categories, such as "abbreviations and acronyms" and "numbers." We also point you to the section covering the important but often overlooked topic of inclusive language. Missing are specific, more prescriptive guidelines for "spacing," "levels of headings," and other design features; depending on the organization's desire for uniformity across its web pages, the design guidelines might be a little loose. They are rhetorically oriented, however. The first few sets of guidelines, in particular, revolve around the web user's needs and expectations. By teaching rhetorical strategies, this guide performs an additional service to the organization and its staff. Because the design guidelines are flexible enough to apply to most websites, we invite you to consider them alongside the web guidelines we offer later in this chapter.

The guide reproduced in Figure 7.3 is also noteworthy because of its accessibility and, in the second half, specificity. In addition to creating a table of contents, the writers make their guide accessible through informative, clearly distinguishable headings and subheadings, bulleted lists, short blocks of text, and indented examples in a different font. In terms of specificity, we were impressed with the explanation of the frame's navigation system and the examples in the grammar and mechanics section.

Designing Your Document

Document design is not something technical and professional communicators leave until the end of the text production process, but something they create throughout. Because design is crucial to a text's accessibility, clarity, coherence, and persuasiveness, it must be orchestrated carefully. In addition, designing a document cannot be done overnight as it involves the coordination and adjustment of several elements. The more time you give yourselves to experiment with and test your design, the more effective your text will be.

Before we discuss the process of creating document grids, we will review the basic elements you will need to consider in designing your service-learning text.

- columns
- rules, boxes
- header/footer
- spacing, including blank space, indentions, leading, and margins
- font or typeface, including size, type, and style
- format of written text, including line length, justification, and use of lists, callouts, and marginal glosses
- headings and subheadings
- color and shading
- graphics (tables, figures, and logos), including types, amount, and placement

Figure 7.4 illustrates several of these elements at work in the University of Florida Community Campaign (UFCC) newsletter mentioned in the preceding style sheet section. As the figure shows, designing even one page of a newsletter can involve the complex coordination of multiple design elements. Before we arrive at the first heading and the main text, we see a header in italics, a title with a logo and line art, and a horizontal rule to set off the title.

First and foremost, your design decisions should be based on your analysis of the rhetorical situation, including purposes, audiences, and uses. Visual design can have an added impact on your audience's expectations, values, and motivation. Think of how much the design of a website affects your ability to find what you're looking for and your motivation to continue exploring the pages. When planning design, begin by reviewing your answers to the invention questions in Chapter Three. If you've completed a discourse analysis, review this as well. Although we present common design conventions below, every discourse community forms its own version of these conventions.

Ensuring the rhetorical suitability of a text involves not only adapting it to your discourse community and audience, but also accounting for the extrarhetorical constraints within which you have to work. These can include such things as page constraints, which can affect font size, use of blank space, and so on, and cost constraints, which can affect your ability to use color or certain visuals. In short, these considerations will enable you to craft your text's design to seize advantage of and conform to the kairos at hand. Visual design is not an arbitrary or completely subjective element of a technical text, but a thoroughly rhetorical one.

As a larger discourse community distinct from other groups of visual communicators such as advertisers or journalists, technical and professional communicators also tend to follow more general design guidelines, which we review below. We also discuss some of these in Chapter Three and in our discussions of particular genres throughout the text. You will notice that some of these guidelines are based on principles such as concision, clarity, emphasis, and tone, principles that we normally apply to style. As Kostelnick and Roberts explain, these principles are also applicable to visual design. Being visually concise means using

Volume 4 Issue 2, December 2000

UFCC Campaign Update

UFCC Exceeds Goal of $725,000

The first University of Florida Community Campaign was conducted in 1993 – in that first year we raised $440,580 in support of 39 participating health and human services agencies. Since then, the Campaign has grown tremendously, and the number of agencies that receive support from UF employees has increased. The 2000 UFCC has 58 charitable organizations in the Alachua County area that will be helped in their efforts to offer many services including child care, recreational opportunities, counseling, legal aid, environmental protection, disease prevention, and medical assistance.

The following graph shows the growth in dollars received through the UFCC – from $440,580 in 1993 to **$735,290** in 2000.

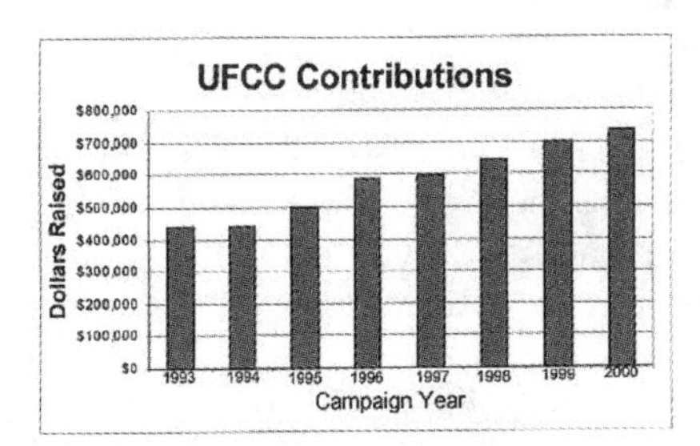

The UFCC offers employees an opportunity to give back to the community; they may contribute through payroll deduction or as a one-time gift.

Approximately 6000 UF employees contributed over $735,000 in support of the 58 agencies participating in the 2000 UFCC. Our employees who give through volunteer work and those who provide financial contributions to these agencies support a more caring, healthier, and safer community for everyone.

Visit our website at
www.ufcc.ufl.edu

Each fall UF employees find pledge cards and brochures listing charitable organizations in their mailboxes or on their desks. And each year thousands of these employees voluntarily take a few minutes to fill out the cards, pledging one dollar or more to help those in need.

Frank Catalanotto, Dean,
UF College of Dentistry
Chair of the 2000 UFCC

This year the University again called on its 12,500 employees to fill out those cards. This is what UFCC is all about – "people helping people" and UF employees reached out to help others in their community.

On behalf of the UFCC Steering Committee, I want to thank each and every one of you who donated and participated in this year's Campaign. The University has come together, and we sincerely appreciate everyone's generosity in reaching out and touching the lives of others.

The coordinators, representatives, and Steering and Planning Committee members who worked on the Campaign are commended for their efforts. Many people in our community will benefit from their combined hard work. We are a community, and in helping others we help ourselves and others that we know.

I have enjoyed being the leader of UF's Campaign this year. Learning about the various agencies firsthand has encouraged me even more to want to serve others.

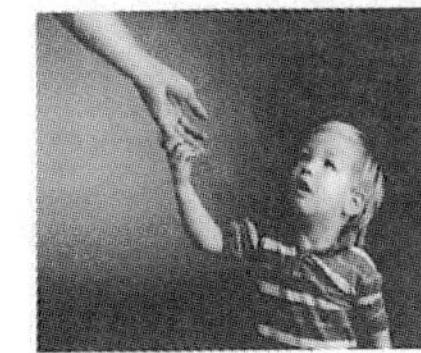

1

Figure 7.4 Newsletter page with multiple design elements

the fewest possible visual elements necessary to create the effect you desire. All visual elements, including blank space, should serve distinct rhetorical purposes. This is why we want you to carefully plan the design of your texts in relation to their rhetorical situations. Visual clarity refers to the understandability and readability of, say, the font or design of a visual aid or use of color. A clearly designed text enables the reader to efficiently make sense of and use the information it presents. Emphasis, which can refer to the prominence and memorableness of the information presented, can be created through color, font size and style, visual aids, and a range of other visual elements. As we have already discussed, sometimes writers need to create kairos by grabbing readers' attention through design. Like stylistic choices, design choices affect the tone and register of a text. Certain fonts, for example, convey a more casual tone than others. One principle that applies across all of the following guidelines is consistency: Always apply design features consistently unless you have a good rhetorical reason not to. Remember that the following guidelines are not hard-and-fast rules but suggestions based on reader-based research and conventions that must be adapted to better fit the demands of the rhetorical situation.

Font

- Use serif fonts for long blocks of text that are meant to be read linearly, and use sans serif fonts for titles, headings, and subheadings and for text that is part of graphics (e.g., titles, captions, labels). Especially on a screen, less bubbly fonts and fonts with thicker lines are easier to read.
- Limit the font types, especially on the same page, so that you don't confuse readers with unintended visual cues. We recommend no more than two font types. When possible, use fonts from the same typeface family (e.g., Arial, Arial Black, Arial Narrow). Experiment to determine which font types are complementary.
- Limit font colors to two, except in web documents, which need different colors for the main text, links, and visited and active links. Consider that readers will probably notice textual elements in color before those in black and white.
- Strategically and sparingly use italics, boldface, and other font styles to emphasize key information. Too many variations in font style can make the text look unprofessional and can overload readers. Italics is especially hard to read on a screen. You've probably noticed that boldfaced text can be more easily discerned with some font types than with others.
- Varying font size is one of the most effective techniques for differentiating levels of headings. Don't overuse this tactic, however, as too much variety can be distracting.
- If you make the main text of a printed text smaller than 11 point, it will likely become difficult to read for many. Certain fonts such as Times New Roman are already smaller than others. Online text should be larger, and text in visual aids accompanying oral presentations should be still larger. The default font size in PowerPoint is large for good reason.

Spacing

- Most readers prefer generous margins. Although reducing standard margins will enable you to include more information on a page or screen, this may well detract from the text's ethos, making it look jumbled and unprofessional. Consider the binding when determining margins; if the left margin will be bound, make it larger than the right margin. In many printed texts, the bottom margin is larger than the top margin.
- To keep the columns of a text visually distinct, allow plenty of space in the gutters between them, especially if the text in the columns is ragged right. You can also use vertical rules to separate such columns visually.
- Keep in mind that text justified on both sides is more difficult to read than text that is only left justified.
- Use blank space to create chunks of information (i.e., rhetorical clusters) and pathways through the text. The spacing should signal to readers which parts of the text go together and should enable them to find and retrieve specific information.
- Use more space between sections than between elements within a section. For example, use more space above a heading than below it.
- When possible, don't place a heading or begin a section at the bottom of a page, breaking it up over multiple pages.
- For double- or one-and-a-half-spaced text, use indents to set off paragraphs. For single-spaced text, indent or skip a line between paragraphs. Indent lists if you want them to stand out further.

Graphics (Tables and Figures)

- Choose graphics that will be familiar to your audience and that best fulfill your rhetorical purpose. Like technical jargon, complex or abstract graphics such as 100 percent bar graphs and schematic diagrams are not usually appropriate for novice audiences. Make sure every graphic has a specific purpose, even if is motivational rather than informational; don't include graphics just as decoration.
- Keep individual graphics as simple and visually concise as possible. For example, don't include unnecessary rules or cells in tables, and crop photographs to eliminate unnecessary details. Don't try to do too much with one graphic.
- Visually set off graphics with extra space and/or by boxing them.
- Integrate graphics into the rest of the text by introducing them, helping readers interpret them, and stating the conclusions you want readers to draw from them. Don't expect graphics to speak for themselves. Although discussing graphics may seem redundant, this kind of redundancy can provide readers multiple ways to make meaning from the text and can reinforce key ideas.
- To prevent the reader from having to flip back and forth, place graphics important to the main text as close as possible to their verbal references and explanations. Try to keep a graphic and its accompanying text on the same page.

- Use specific titles, labels, and captions where appropriate to point out specific features of a graphic and generally make it more informative.
- Emphasize key elements in a graphic by varying the font, using color or shading, or using labels. Do this sparingly, however.

Color

- With cost constraints in mind, limit the use of color, especially for textual elements of printed texts. Emphasize with color selectively, strategically, and consistently.
- Use color to filter information, as in distinguishing different levels of headings or types of visuals, but don't only depend on color to do this.
- To ensure clarity, use high-contrast color schemes, especially in online texts. If using shading, make sure the different shades are easily distinguishable. Use reverse-color blocks (e.g., white text in a black block or yellow text in a dark blue block) sparingly, as they stand out powerfully and can distract from other important information.
- Choose harmonious and complementary colors in a multiple color scheme.
- Consider using different shades of the same color for queuing or showing hierarchical distinctions between types of information. For example, you could use dark green for major headings and a significantly lighter green for subheadings. Most web readers will expect links and visited links to be in different colors, however.
- Think about the connotations of particular colors for the audience, and play on these connotations strategically. For example, use bright red to highlight warnings in a set of instructions, or blues and greens in a text about ocean preservation.

Let's take another look at Figure 7.4 to see which guidelines the designers of the UFCC newsletter employed on the front page. Notice first the basic two-column design of the page. The designers create plenty of space between the columns to make them visually distinct, and the lines of text in the columns are neither too long nor too short to prevent fast reading. The bar graph, photograph, and circled information are set off with extra space, and the graph is boxed to give it distinct borders like the other visuals.

In general, the graphics are visually concise and informative. Notice how the final visual uses an unusual camera angle and shows only the adult's hand so as to focus on the child as one of the main recipients of the fundraising campaign. The simple bar graph might have also been rendered as a line graph. The two most important visuals—the bar graph and photo of the UFCC chair—are strategically placed and fairly well integrated into the text.

Although you can't see this, the designers use color sparingly and strategically. Dark blue, which contrasts well with the white background, is used for the title, headings, photo caption, circled information, and part of the bar graph. All of these but the bar graph are rendered in a noticeably different font type to show emphasis and to help readers filter information. The designers bold the crucial number amount above the bar graph.

The designers employ a highly readable sans serif font for the main text and use enough leading to keep the lines from blurring together; a serif font such as Times New Roman might have been even easier to read. Also, although the second font is elegant, a font with thicker lines would provide better readability. The size of the font, however, may compensate for its thin lines.

Creating Document Grids

In developing your style sheet, you have been recording decisions about your document's design, including its font, spacing, and design of graphics. But your style sheet won't help you visualize the document the same way a **document grid** can. Document grids, which can range from basic thumbnail sketches early in the project to full-size mockups later, are basically blueprints that show what the document will look like and how its parts will fit together. They can help you clarify and add to the design decisions in your style sheet. A **page grid** does the same thing on a smaller scale. Like style sheets, grids are particularly useful for invention and revising/editing.

Thumbnail Sketches

Beginning grids, often called thumbnails, are rough sketches that help writers try out different designs and arrangements before deciding on the most promising ones. Thumbnail sketches are usually just drawn by hand, though you could use a simple draw function in a word processing program such as Word or create them. We recommend that you draw out thumbnail sketches early in the drafting process, as your text's design will affect the number and type of verbal and visual elements you can include. Unlike detailed grids, thumbnail sketches are not precise or all inclusive; they generally only determine elements such as number of columns, placement of graphics, and spacing patterns.

Before you can draw a thumbnail sketch, you will need to take an inventory of the text's major elements including the type and number of visuals. Even if you haven't drafted all of these elements, you should have a good idea of how much space they will take. In addition, you will need to consider the basic form and size of your text. Although you know what genre or general type of text the agency expects you to produce, you may not have been given much guidance about form. A student in an advanced technical communication course agreed to produce a set of safety and cleanup procedures for the kitchen of the restaurant where she worked. After thinking about her text's audience, uses, and context, she changed the form from a regular-size print booklet to a set of large, laminated posters to be tacked to the kitchen walls. Even if you have been given a specific assignment such as producing a brochure for recruiting new volunteers, you'll need to decide on the size of the brochure, the number of folds and panels (most brochures have either three or four panels on each side), and the folding pattern. In this case, the amount of information you need to include will help determine if you need to go from a two-fold, six-panel brochure to a three-fold, eight-panel one.

Figure 7.5 shows some early thumbnail sketches of an American Cancer Society patient information packet and its pages. The writing group decided to use a folder containing a set of information sheets. The folder also allowed them to include business cards (on the outside of the pockets) and other agency brochures that describe select patient services in more detail (on the right side of the folder).

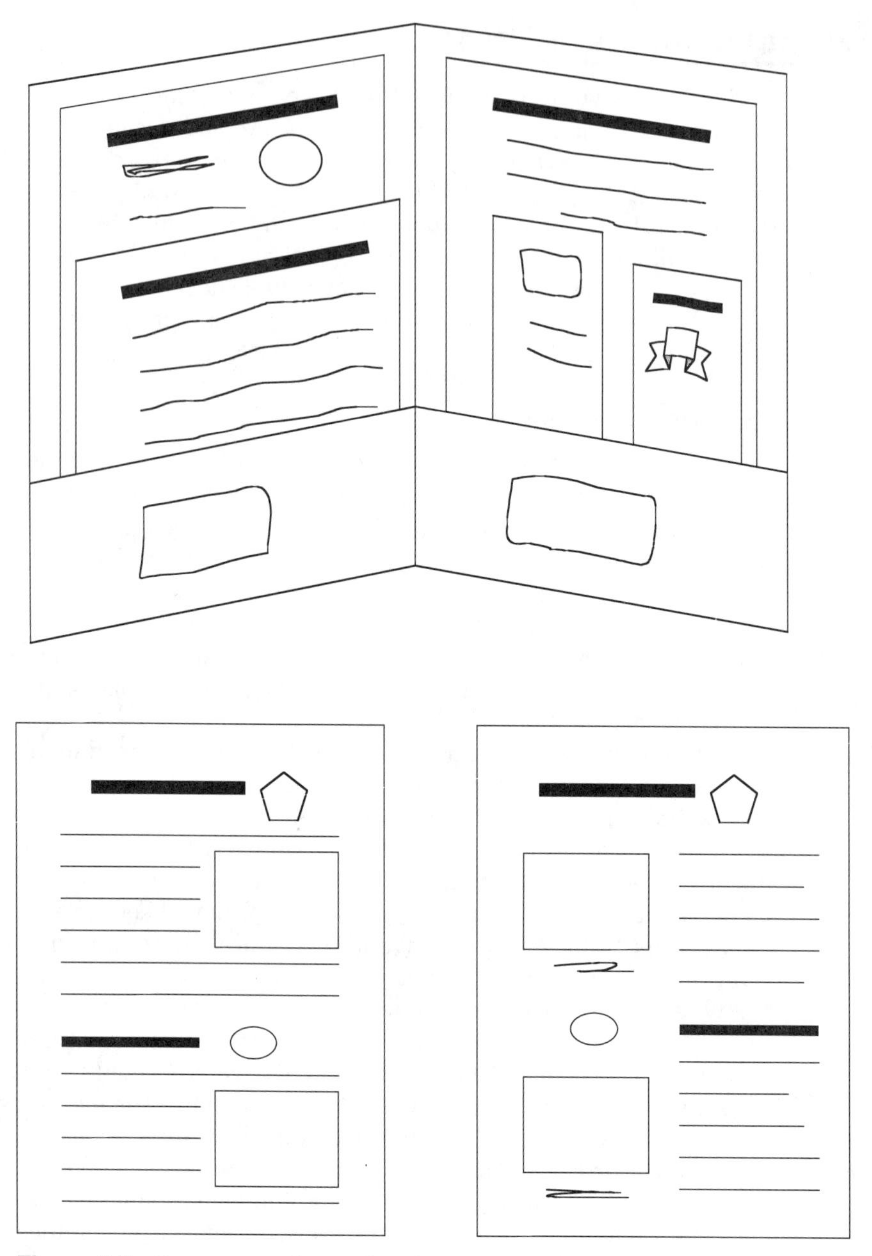

Figure 7.5 Document and page thumbnail sketches of patient information packet

As the first sketch illustrates, the group decided to make the pages differing sizes so that the user could see all of the titles at once after opening the folder. The section requiring the least amount of information—fundraisers—was placed on shorter pages than the two longer sections that overview programs and services. The other two sketches show two versions of the page design for the information sheets. The first uses a one-column design, wrapping the text around the main visuals (photos). The headings and accompanying logos appear near the blocks of text they introduce. In the two-column design of the second version, the visuals, visual captions, and program logos are placed in the left column, and the headings and main text are placed in the right column. The section title and agency logo appear the same way in both versions. We encourage you to experiment with and assess different page designs this way.

As they were planning their document's basic design, the American Cancer Society writing group met with their agency contact person to discuss production and distribution costs. We highly recommend that you do the same before you get too far into the design process. Binding, color, and paper size, type, and finish are just a few of the factors that can affect cost significantly, especially if the organization plans to produce a large number of the texts. Find out your cost and other design-related constraints early on. A student group producing a curriculum guide for students in the First-Year Florida program, a course that introduces new University of Florida students to campus resources and opportunities, was told fairly late in the project that their text must fit on sixteen 6x9-inch pages because it would be produced as part of a student planner. Knowing this earlier would have saved them from much replanning. Another student group produced a beautiful full-color, glossy-finished brochure for their campus's chapter of Phi Alpha Delta, a pre-law fraternity. Although it was beautiful, the group couldn't afford to mass-produce it, and the color visuals didn't translate well into black and white. Deliberating with the agency about cost constraints would have led to a more usable product.

Detailed Grids

Later, after you have drafted the document's sections and selected its graphics, you might create grids that serve as more precise blueprints of the document. For these grids, we recommend that you use a draw program or a desktop publishing program such as PageMaker or Microsoft Publisher. We're advising you to create more detailed grids a little later in the drafting process so that you have more time to study and deliberate about possible designs. In addition, agencies sometimes change their visions of what they want from a document after a project is underway; therefore, you may not want to invest too much time on the details of the design until you know for sure what you are designing. The patient information packet requested by the American Cancer Society started out as a set of brochures.

Following Karen Schriver's recommendation in *Dynamics in Document Design,* you might begin to create a more detailed grid by organizing the elements of the text into rhetorical clusters or combinations of verbal and visual elements that function as a unit and that the reader should encounter together (343). A larger cluster might include a heading, a paragraph of text, and an accompanying graphic; a smaller cluster might include a graphic, its title and caption, and its labels. Forming clusters might help you further determine headings and subheadings, which we advise you to experiment with at this point.

Next, approximate the amount of space each cluster will require, and measure the area you have to work with, dividing it up into a grid with horizontal and vertical lines. You may want to draw the grid lines for margins, columns, and other elements lightly to keep them from interfering visually with the design elements you will place on the grid. Measuring and dividing the area should help you determine if you need more space and how the parts of the clusters might fit together. When plotting the basic grid design, consult your style sheet regarding margin size, spacing, font size, visual design, and other elements. Remember to be consistent with such elements across the text.

Next, design the rhetorical clusters, and plot them into several coherent arrangements within the grid, creating several different paths for your reader to travel through the document. In designing a cluster, consider the hierarchy of textual elements and how they logically and spatially fit together. In creating different grid patterns, it may help first to make several outlines of the text and then to translate these outlines into document grids. We also suggest that you plot transitional elements to connect the clusters. This is the point where you actually draw out the grids, being as precise as possible about the design elements listed above.

As you try to create an accessible design, consider different visual strategies for chunking the elements of a cluster, such as positioning them close to one another and separating them from other clusters through spacing, headings, and perhaps rules. Markel relates chunking to queuing and filtering (see 340–41). Queuing involves visually signifying hierarchical distinctions among elements of a text or cluster, as in making first-level headings larger than second-level ones. Filtering involves creating patterns among similar elements across a text. For example, all headings of the same level should be in the same font size, type, and case; all key terms should be emphasized using the same font style (e.g., italics, boldface); similar sections of text should be aligned and formatted the same way (e.g., in boxes, in lists); similar graphics should follow the same design.

Using multiple columns may give you more options for clustering, queuing, and filtering and may enable you to be more economical with your space. You can make major headings stand out more, for instance, by beginning them in the margin before the text. Larger elements, such as boxed text and graphics, can be extended across multiple columns.

Figure 7.6 shows a more detailed grid of a brochure produced by professional writing students for a writing internship course. The brochure's design made it important to plan clusters of information that would fit on the panels. As the headings in the figure show, the group divided the information up into seven major clusters: the title and other cover elements, elements of the internship course, examples of past internship projects, student comments about the course, requirements, goals, and application information. The clusters are chunked by headings, by extra space between them, and, in most cases, by the panels themselves.

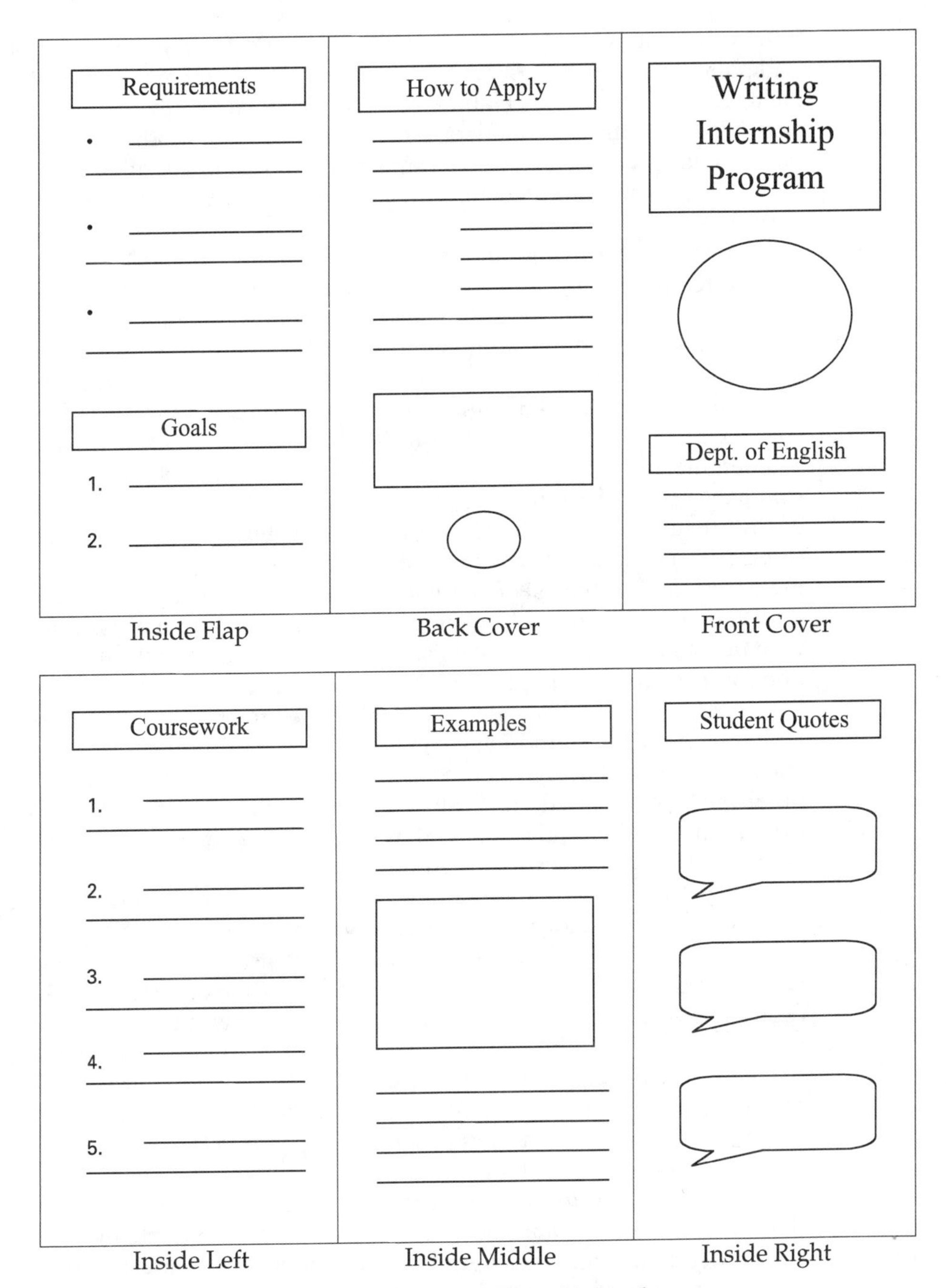

Figure 7.6 First document grid for writing internship brochure

After you have formed several versions of a more detailed document grid, evaluate the strengths and weaknesses of each version, focusing on its coherence or logical flow from the audience's perspective. At this point you should seek feedback from your instructor, agency contact person, and perhaps others about which grid pattern seems the most promising. When seeking others' comparative assessment, also ask for feedback about some of the details of the design such as font type, types of lists (i.e., numbers, bullets, or checklists), leading and justification, color scheme, and the precise placement and design of specific graphics. Experiment with combinations of these elements just as you experiment with grid patterns.

Figure 7.7 shows how the grid for the writing internship brochure was revised after comparing arrangements and experimenting with design features. You'll notice that some of the rhetorical clusters were adjusted and rearranged. The group shortened the requirements because most of the details were included in the coursework section. They also expanded goals to include reader-based benefits and expanded student quotations, and they added a short section on possible sponsors. Given that the inside flap is one of the first panels a reader will encounter, the group decided to move a more persuasive section, including student comments, there. They reinforced these quotations with a list of benefits directly behind them in the inside middle column. Although the requirements section could have been placed closer to coursework, the designers thought it was more important to place the list of possible sponsors close to the examples.

When revising for design, the group replaced a couple of the numbered lists with bulleted ones because establishing a hierarchy among the items was not important. They did use a numbered list to create a checklist under "how to apply." The circular logo was moved to the front of the brochure to signify the university affiliation right away, and the general course description on the front was shortened and boxed. Finally, the designers selected a graphic for the front from a number of options, used a script-like font for the title to better set it off and to conform to the graphic, and changed the font of the headings from the more elegant Garamond to the more readable Arial.

After deciding on a grid pattern and more specific design features, you are ready to record these decisions in your style sheet and to create a mockup grid of the document. If possible, make this a full-size grid. If you're producing a long document containing many pages following the same design, it may only be necessary to create a mockup of one or two pages. This was the case for the students producing the curriculum guide. All sixteen pages conformed to the same design parameters. One way to extend the usefulness of the final grid you create is to save it electronically as a template. This will ensure consistency when you put the document together and make it easier to add pages to the document later on (see the style sheet section in this chapter).

We don't mean to imply that the design process is over at this point. Undoubtedly, you will continue to obtain feedback and to change some of your design decisions until you submit the final project to your class and the agency. If you approach design as a long-term process, you will finish with a more effective document and have only minor editing and proofing to do toward the end of the project. Groups that wait until the final days of the project to work on document design end up with inconsistent, unprofessional, and ineffective documents.

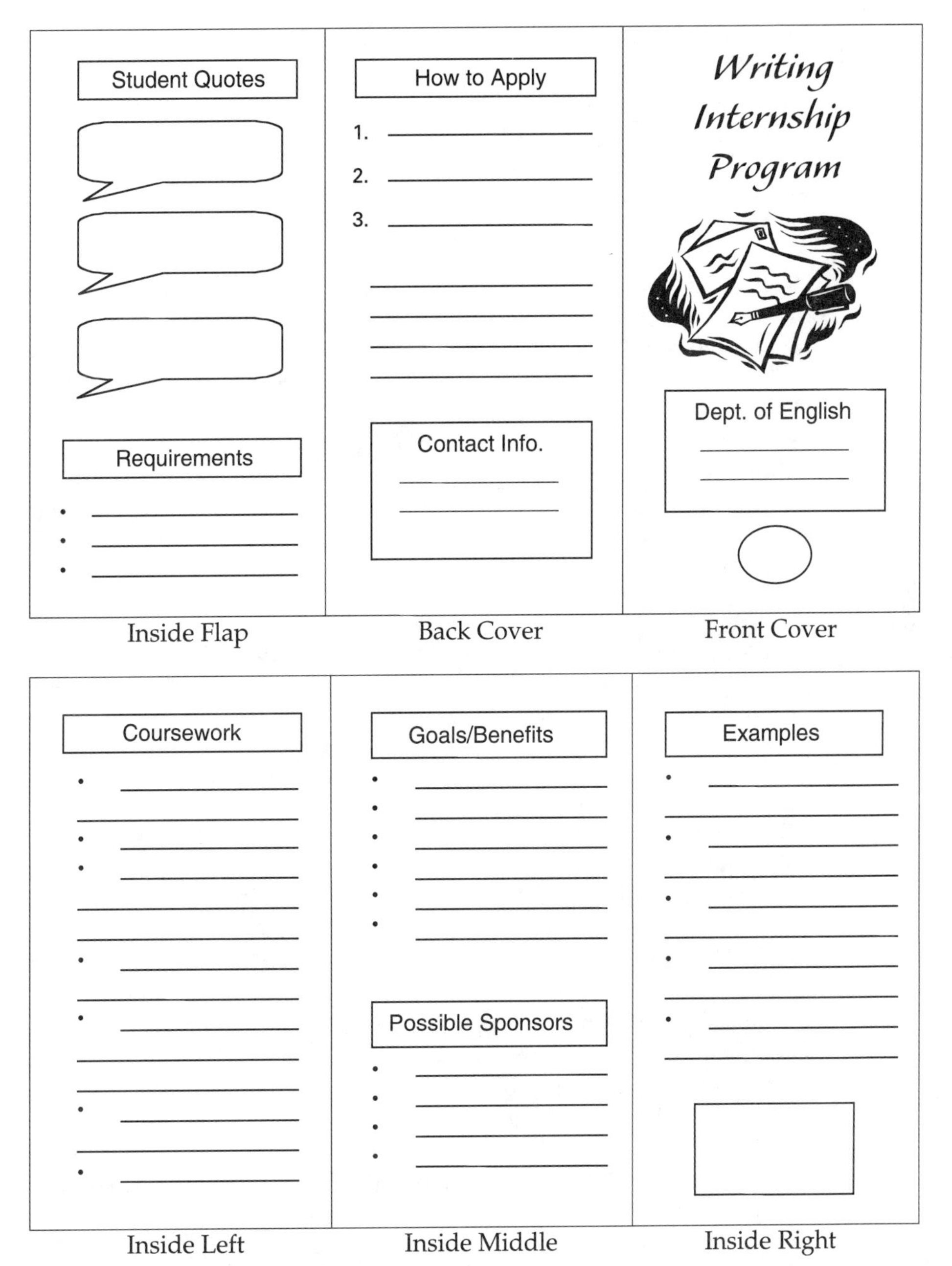

Figure 7.7 Revised document grid for writing internship brochure

To summarize, here are the basic steps of forming more detailed document grids (adapted from Schriver 342–56).

1. Organize the text's elements into rhetorical clusters, including headings where appropriate (a rhetorical cluster is a group of interrelated verbal and

visual elements that function as a unit; in a set of instructions, for example, a subheading, description of a step, and visual showing the step could be considered a rhetorical cluster).
2. Approximate the amount of space each cluster will need.
3. Measure the area with which you have to work, keeping margins and other constraints in mind.
4. Determine the overall design of the grid, including the number of columns, spacing, and design of graphics.
5. Plot the clusters into several different arrangements, paying careful attention to design elements within and across clusters.
6. Assess the different arrangements for their cohesion from the audience's perspective.
7. Experiment with the details of the design such as font type and color scheme.
8. After deciding on a pattern and design details, make a full-size mockup of the document.
9. Record your decisions on your style sheet, and perhaps save them in an electronic template.

Creating grids for web pages and websites involves some additional considerations because of their more dynamic structure and the different user expectations for this medium. Before planning the design of the individual pages, you'll need to map the interrelationships of the pages and their links. Some websites have clearer hierarchies of pages; diagrams of such sites might look more like organizational charts or inverted-tree diagrams. A more associative site involving converging links might require a more cluster-like diagram. Figures 7.8 and 7.9 show diagrams of two different versions—the first more hierarchical—of a writing internship website the students who produced the brochures might have created. The boxes are the pages or "nodes" of the network, the solid lines are the primary links, and the dotted lines are the associative links or links between two pages in different parts of the network. David and Jean Farkas explain various network structures and show their corresponding node-link diagrams in chapters six and seven of *Principles of Web Design*.

In comparing the two figures, note the clearer levels or layers of pages in the first one: first the homepage, then the three second-level pages, which in turn split off into third level pages. Such a structure may give users a clearer sense of the site's structure and hierarchy of content but may also require them to go through several links from the homepage to arrive at specific needed information. An agency worker might not reach the page that explains the benefits of sponsoring an internship, for example, or an interested student would have to go through two layers of links to find examples of other internship projects. The version in Figure 7.9, though still somewhat hierarchical, better facilitates multidirectional movement through the site. Notice the associative links among the second-level pages, for instance, and the converging primary links to the application form, a crucial page for an applying student.

Because your homepage is the conceptual and navigational center of the site, its design is the most important. Here you establish the site's main design features, content areas, and navigational pathways. Like the front panel of a

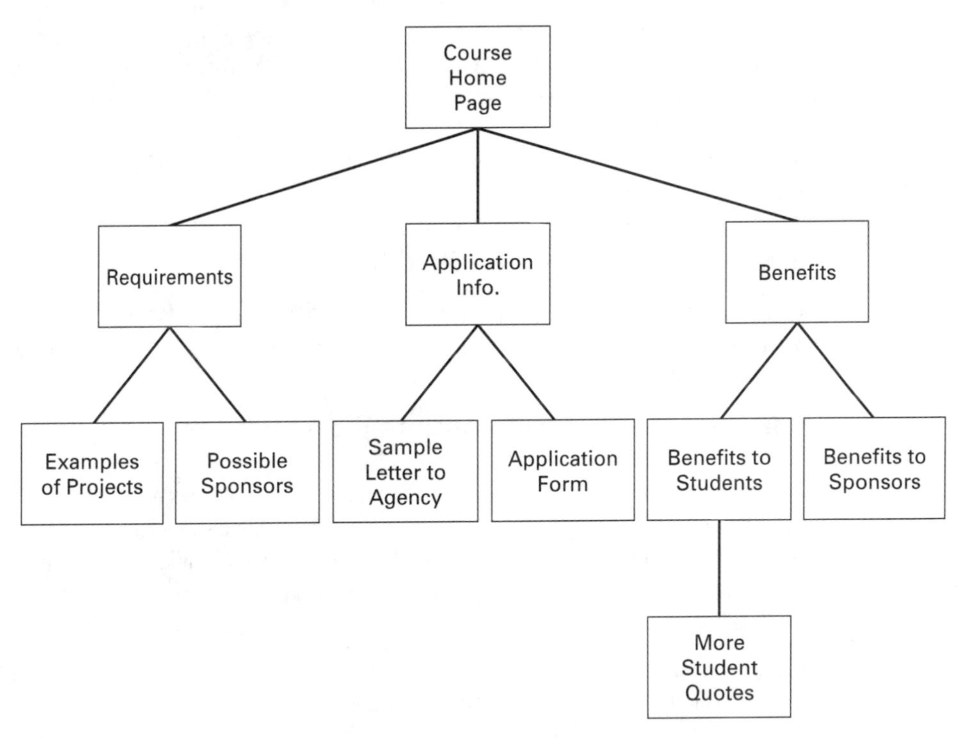

Figure 7.8 Strictly hierarchical site map of writing internship website

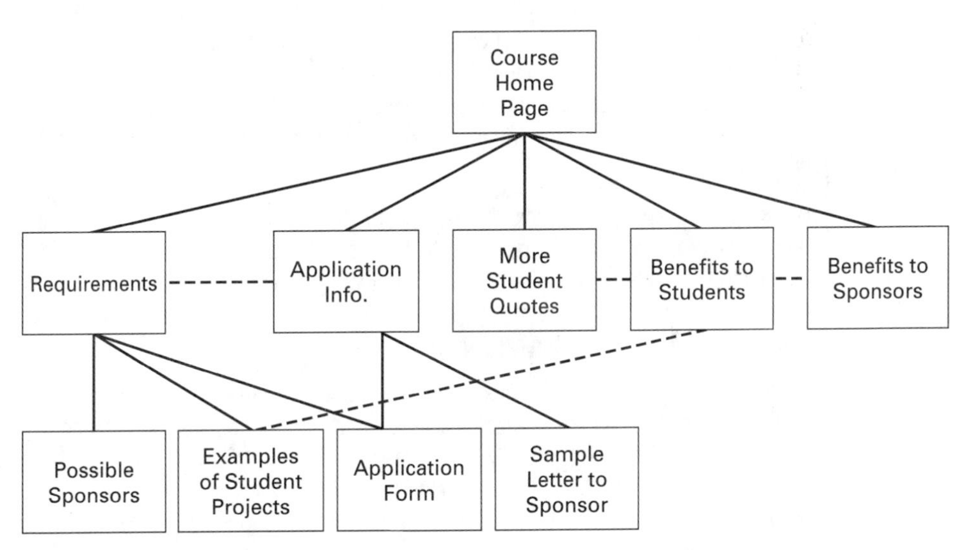

Figure 7.9 More associative site map of writing internship website

brochure, the homepage must also be especially inviting and motivational. The other pages will probably follow a standard design that includes some of the same features as the homepage; therefore, you may only need to make a thumbnail sketch of one of these. In creating the page design, remember that web texts are generally less linear and verbal than print texts. Most web texts use more columns, graphics, and color. As a result, their grids can be even more complicated.

OTHER VOICES 7.B Karen Peters

Karen M. Peters is the manager of instructional design at the Center for Education Technology Services of the Pennsylvania State University and has worked on service-learning projects with technical writing students.

Using A Real-World Approach to Workshop Texts Throughout the Semester

Feedback is a vital component of the learning process. To create successful documents, students need ongoing input that helps to improve their work and refine their writing styles. Sometimes it is difficult for an instructor to provide thorough formative feedback throughout the semester because of time constraints or lack of knowledge about individual students' projects. For that reason, it's important for students to have consistent contact with their agency contact persons.

Each contact person needs to be aware of the objectives of the course and to place these goals at the center of the project. The instructor, students, and contact person should work out a communication/feedback timeline that fits everyone's schedule. The communication and feedback can be virtual at first—students can e-mail their projects to their contact persons. The contact person should plan to visit the class or meet formally with the students at the agency office about halfway through the project. At this first workshop meeting, students should have questions prepared for the contact person based on their initial virtual feedback. This type of activity mimics the real workplace but also can serve as an assessment of student learning. If the session takes place during a class, the instructor can use this opportunity to listen to student questions and get ideas for a follow-up class discussion.

Based on the feedback from this first workshop, students should continue with their projects and be able to e-mail the contact person with questions. Guidelines can be collaboratively set as to how often this process is repeated before the final presentation and project completion. The students and contact person can work together to develop tools for assessing the project's progress, such as style sheets and workshop guides.

I recommend that all contact persons attend the final project presentation session. This is important to the learning process; it gives students the benefit of hearing feedback on their own work and also hearing feedback on the work of their peers. This broadens the students' perspectives and experience. Ideally, I think that the final meeting should be student-centered. Students should prepare presentations for their contractors and be prepared to answer questions from their classmates regarding their projects and their writing processes. This is also a good opportunity for contact persons to point out the similarities and differences between a service-learning project and writing processes in the "real world."

Other Common Service-Learning Genres: Conventions and Resources

This section provides some helpful guidelines for developing several additional genres on which you may be working for your service-learning project—instructions, websites, brochures, and newsletters—and refers you to additional resources for studying them. You may remember from Chapter Three that we approach genres not as fixed formats, but as forms of social communication shared by discourse communities. Along this same line, we define conventions as common reader expectations. Most readers of instructions, for example, expect visual aids that show them how to perform the steps. These more flexible definitions suggest that your invention involves more than following the general genre guidelines below or the general design guidelines earlier in the chapter; you should also adapt these guidelines to the specific characteristics and constraints of your rhetorical situation and to the more specific discourse conventions followed by your organization and readers. In addition to performing a thorough rhetorical analysis, studying sample documents from your agency is still your best strategy (see the discourse analysis section). A group of students writing a grant proposal for a local AIDS project needed to study other proposals written by the organization and related health-care organizations to learn specific arrangement and stylistic conventions that their textbook didn't cover. If your organization has a style guide and/or templates, you'll want to pay particular attention to these, although even templates are usually adapted to the situation.

Approach the guidelines below, then, not as strict rules but as starting points for producing particular genres. You'll probably want to supplement your rhetorical and discourse analyses by consulting additional sources that focus on certain genres and their conventions (see the lists at the end of the chapter). If you were creating online documentation, for instance, *Writing Software Documentation* by Thomas T. Barker would be an invaluable resource. Finally, you might want to look at the book's appendixes—the sample projects written by some of our students—to get ideas about design, arrangement, and other elements.

Although the guidelines below are general, they are based on rhetorical concerns, primarily on how readers will *use* the text. You'll notice that some of the guidelines also adhere to such design principles as clarity, concision, and emphasis that we discussed earlier. In most cases, adapting the guidelines below will enable you to better fulfill your users' expectations and needs.

Instructions

All of us have had to struggle through a set of inadequate instructions for using a device, performing a process, or following a procedure. To avoid creating this kind of frustration for your reader, think carefully about how she or he will use the document. Indeed, think of yourself as the user's advocate whose main goals are to help the user feel confident and comfortable performing the task. Unlike manuals, online documentation, and other texts that are used primarily for reference, instructions are followed step by step to complete a specific task. Online instructions often take the form of a tutorial. Think about the difference between online documentation, which provides you with a databank of information and help about various topics and processes, and an online tutorial, which takes you step by step through performing a process. Whether you are designing instructions for using a piece of office equipment, performing a volunteer activity, or using a software program, consider the following components and user-based strategies.

Components

- An *introduction* that explains the document's purpose, audience, and scope and provides a conceptual overview of the task or process at hand. Here you might motivate readers to use the instructions and forecast the remaining parts.
- A *list of items* needed to complete the process. This could include the information needed to complete a form, the tools needed to repair a machine, or the applications needed to run a program. If the instructions show the reader how to use a device, you might provide a visual of the device and its major parts.
- *Safety information.* Most instructions provide overall safety information in the introduction but then place specific cautions and warnings by the steps to which they apply.
- *Step-by-step directions,* usually presented in a numbered list and in active, imperative style. This is the main part of the instructions, where you guide the user chronologically through the steps of the process. Be sure to clarify what each action will produce. Perhaps a particular prompt should pop up on the screen or a light on the copier should appear.
- *Useful visuals.* This is a critical part of instructions, especially for those that guide a reader through using a machine or tool. As Anderson points out, visuals can show users where things are, how to perform steps, and what will result from performing steps (545). Action-view drawings can be especially useful, as they can enable you to more easily focus on a particular action. Visuals should be concise and easy to read, should include all parts that are referred to in the document, and should represent accurately what the reader will see when she or he is performing the actions. Visuals should be placed as close as possible to the words that explain them.
- *Troubleshooting advice.* Many instructions place user-centered troubleshooting tips at the end of the instructions, sometimes in table form. You might also embed specific troubleshooting advice (e.g., how the user should proceed if the step doesn't produce the right result) in the step-by-step section to keep users from having to move to the end of the document in the middle of a step.

Design and writing strategies

- Perform a *task analysis* from the user's point of view and determine *user-centered, manageable steps.* A task analysis breaks the process down into every step and substep a user must take when performing the process from preparation to completion. It also asks questions about what the user will need, the user's circumstances, what might go wrong, and how much time the process will take (Lay et al. 445). After listing every step and substep, determine (based on the user's familiarity and level of expertise) which ones the user can intuit, which ones can be collapsed, and which ones need to be explicit and explained in detail. You would undoubtedly give different instructions for changing a tire to someone who doesn't know what a jack is than to someone who understands the equipment needed and who has previously observed the process.

- *Make the step-by-step directions highly accessible.* Accessibility is especially important given that users might be moving back and forth from the instructions to the performance of the task. To make it easier for readers to find their place in the text, use a numbered list, visually chunk the steps (through blank space), and give each step a subheading that stands out and summarizes it. You also might emphasize key words visually within each step. While visually separating each step, cluster related steps under larger headings to give the user a better sense of how the steps relate to each other.
- Use *action-oriented language* in the headings and steps themselves. As you know from experience as a user, most steps are written in active, imperative voice. Keep in mind that the user will be reading your words and attempting to translate them into actions. Given a choice between "Tab A should be placed into Slot C" and "Place Tab A into Slot C," most readers would prefer the latter.
- *Distinguish the types of information in the command.* Use variations in font or type style to identify direct commands and other information, such as explanations and troubleshooting advice. Maintain a consistent order as often as possible, making an effort to lead each entry with an action.
- Place safety information *before* the steps to which it relates so that the user doesn't arrive at the information too late. *Visually distinguish* among types of safety information. The words *note* and *important* are used to alert readers to potential damage to equipment or critical phases of a process; the terms *danger, warning,* and *caution* connote various levels of possible injury to the user. *Visually emphasize* all safety information, perhaps through color, by boxing the information, or by using small icons. In addition to the ethical and professional obligation to accommodate users, instructions for possibly dangerous processes must also meet legal requirements for ensuring safe use and even anticipating possible misuses. We recommend that you avoid producing these types of instructions unless you are very familiar with the subject matter.
- *Test with actual users.* In Chapter Nine we'll introduce you to some important processes for evaluating your documents, including usability testing. Though it's important to obtain reader feedback on every text we produce, this is especially critical for instructions. When you are very familiar with a process, it's often hard to anticipate what your reader, presumably someone who knows little about it, will need to know or will have difficulty understanding. Those of us who have changed a tire, for example, may not

recognize all of the detail and help needed by someone who has never changed a tire before and who knows little about the tools required for the process. You might present your task analysis and actual instructions to a few prospective users to see where you need to clarify, elaborate, and increase the accessibility.

Resources

Cunningham, Donald H., and Gerald Cohen. *Creating Technical Manuals: A Step-by-Step Approach to Writing User-Friendly Instructions.* New York: McGraw-Hill, 1994.

Helyar, Pamela S. "Products Liability: Meeting Legal Standards for Adequate Instructions." *Journal of Technical Writing and Communication* 22.2 (1992): 125–148.

Kemnitz, Charles. "How to Write Effective Hazard Alert Messages." *Technical Communication* 38.1 (1991): 68–73.

Redish, Janice C., and David C. Schell. "Writing and Testing Instructions for Usability." *Technical Writing: Theory and Practice.* Eds. Bertie E. Fearing and Keats W. Sparrow. New York: Modern Language Association, 1989. 63–71.

Special Issue on Instructions. *Journal of Technical Writing and Communication* 26.4 (1996).

Websites

Many of our students' service-learning projects in recent years have involved creating or updating websites for nonprofit and campus organizations. Although you may not have much experience designing websites, you have probably viewed a lot of them and therefore may still have some experience on which to draw as the user's advocate. Because of the open nature of the web, the audiences of websites can be difficult to predict. This does not mean that an audience analysis is not possible, however; it just means that you'll need to consider strategies to connect your targeted readership to your website (see below). Most organizational websites are not intended just for members but are designed also to inform others about the organization and even recruit them. The instant access to other websites and the exploratory browsing of many users also means that your website will be competing with many others for your audience's attention; thus, your website should be quite inviting. Keep in mind the mostly nonsequential ways in which you read websites and the qualities of websites that make you want to keeping browsing and that make it easier to do this.

Because web design is such a rapidly growing area, design possibilities and conventions are constantly changing. Below you'll find some basic suggestions that nevertheless could be viewed as generic conventions for this changing medium. If you take on a project like this, you'll definitely want to consult additional resources. We recommend that you start with the 25 guidelines in Appendix A of *Principles of Web Design.* As you will see, the first three guidelines emphasize fundamental rhetorical concerns, namely your purpose, audience, and theme or "core message that connects your website to your audience" (Farkas and Farkas 334).

General considerations and strategies

- *Choose site-creation tools.* Web pages are created through a code called Hypertext Markup Language (HTML). If you are already proficient in using HTML, you may choose to code the pages yourself. If not, you can use one of many

WYSIWYG (What You See Is What You Get) programs that function like word processing programs, allowing you to use relatively simple menu commands to design your page. These range from something like Netscape Composer, a component of the free Netscape Communicator, to more sophisticated programs like Dreamweaver. If you use one of these programs, we suggest that you still at least look at the basic structure of your document's HTML code by selecting an option that allows you to view or edit the "source" of the page. Learning the basic grammar of the language may eventually enable you to go in and tweak the code to make minor changes in your document. Get a general sense of what you can reasonably learn and do with each creation tool before selecting one. In addition to assessing your skills, consider the organization's access to the tools and the staff's experience with using them.

- *Begin with the rhetorical situation.* Some web creation programs provide templates of websites that you can follow or adapt. Instead of starting your design invention with such templates, however, start by considering the elements of the rhetorical situation with your agency contact person. This may be a useful exercise for the contact person as well; some agencies feel a kind of general pressure to have a presence on the web but don't have a clear sense of the site's purpose. Have one or more detailed conversations with your contact person and possibly other staff members to determine exactly what the site should include, how it should look, how it might be used, and so on.
- *Plan, plan, plan.* Websites, which are networks of interconnected pages, present complex design challenges. Besides coming up with an appealing main page and basic design template, you will have to plan the content of each branching page and the interrelationships among the pages. We highly recommend creating and experimenting with different diagrams of the page and site design (see the end of the document grid section).
- Follow *copyright and privacy* requirements. Don't take material from other websites and put it in yours unless you have permission from their authors to do so. Exceptions to this include clip art and other graphics from sites that explicitly make their content available for free. Keep in mind privacy considerations if you use pictures on the site. Get permission from clients and employees to use their images in the page. Appendix B of *Principles of Web Design* overviews relevant copyright law.
- Help the agency register their site with major search engines. Current instructions for quick and efficient registration are available online through most engines. Explain this process to your contact person, and obtain permission before registering the site yourself. In addition, consider ways to connect your site to other websites that your target audience is likely to visit. This, too, is part of attending to the distribution of your text.
- Design a site that someone on the staff has the software and skills to *maintain.* A good way to supplement such a project is to write a short set of site-maintenance instructions. Don't promise updates in the content or other things that may be hard to deliver. Despite their best intentions, staff members may not be able to keep up with commitments to update statistics or to keep current news items on the site.

- *Finally, test the site.* Ask potential users to attempt to find information on the site, perhaps even using a kind of online scavenger hunt approach. See what steps are difficult for them, what they find frustrating, and what elements take too long to load. Include users with low-end browsers and equipment.

Design strategies

- *Prioritize readers' needs and uses over flashiness.* We have noticed that many students who are creating websites for the first time find the experience so enjoyable and challenging that they have a hard time focusing on the site's audiences, purposes, and uses. If one flashing neon sign is good, some students rationalize, then ten of them are great. Think about the websites that you actually visit to find important information; the most useful ones are usually quite simple. We suggest that you focus less on creating pages full of flashy effects than on creating pages that are accessible, readable, easy to navigate, and logically connected. Bells and whistles such as busy backgrounds, gratuitous images, and sound files often distract readers from important information, complicate navigation, and decrease clarity. Save such elements for a fun personal page rather than an agency website.
- *Create simple documents that are easy to print.* If your audience members will be accessing the page(s) to find information, some will want to print them. Most of us have had bad experiences with documents that were cut off or otherwise distorted in the printing process. We suggest that you arrange information so that it will fit easily onto a standard page, allowing generous margins.
- *Build in document flexibility to accommodate different users.* Accessibility for people with disabilities is being increasingly recognized as an important concern for web designers. For example, federal laws require that any online materials that are necessary for completion of a course must be available in a text-only version that is "readable" by accommodation software of several types. One resource that can help you evaluate your website's accessibility to people with disabilities is http://www.cast.org/bobby. If many people are accessing your website, it's safe to assume that it will be viewed on a range of monitor sizes, operating systems, and browsers. All of these variables can alter the appearance of your site significantly. Design tables based on a percentage of the window rather than on a number of pixels to make them more easily adaptable to a range of situations. Preview the document on a smaller screen size. Keep download time to a minimum, recognizing that some users may have computers with slower modems and/or processors than you do. Avoid including large video files or complex animations that may make accessing the site frustrating. If you need to design a site with numerous visuals that might be slow to load, create two versions of the document, including one that's primarily text based.
- *Present information in manageable chunks.* To minimize the user's need to scroll horizontally and vertically, make the lines short enough to be viewed on a smaller screen size, and limit the amount of text you present on each page. It's better to develop several pages with shorter text segments than to make readers keep scrolling down or across.
- *Create a user-centered navigation system.* Users should be able to move easily around the site without having to use lots of browser functions or reenter a

URL. Include buttons that allow readers to return to the main page, jump up and down on a longer page, and go back to a previous page. To help users move quickly across the major pages, include a table of contents (in hyperlinks). Most sites place this navigational aid in a table or frame at the top or left side of every page. For more complex sites containing a great deal of information, you may want to create multiple tables of content, each one using different keywords, to give users multiple avenues for finding specific information. Users of a complex site might also benefit from a site map—that is, a map revealing how the website's network of pages is structured—linked from the homepage. Make sure your headings are reader centered and content specific.

- *Make the site visually coherent.* Design consistency and coherence is even more important for websites than for printed texts, because online users depend more on visual cues and can move in and out of the site more quickly. Make sure you place headings, the table of contents, and other navigational aids in similar places in all the pages of your site so that users will know where to look. Users should be able to tell when they are still within your network of pages and when they have followed a link out of the site. Some websites use a different color for external links and/or warn readers about which links will take them out of the site. You can create a common feel to your site not only by repeating the basic design structure, but also by consistently using the agency icon, colors, and other identifying information.

Resources

Cubbison, Laurie. "A Heuristic for Defining the Purpose of a Client's WWW Site." *Business Communication Quarterly* 60.4 (1997): 95–98.

Farkas, David K., and Jean B. Farkas. *Principles of Web Design.* New York: Longman, 2002.

Lynch, Patrick J., and Sarah Horton. *Web Style Guide: Basic Design Principles for Creating Web Sites.* New Haven: Yale UP, 1999.

Niederst, Jennifer. *Web Design in a Nutshell: A Desktop Quick Reference.* New York:O'Reilly & Associates, 1998.

Special Issue on Web Design. *Technical Communication* 47.3 (2000).

Tovey, Janice. "Organizing Features of Hypertext: Some Rhetorical and Practical Elements." *Journal of Business and Technical Communication* 12 (1998): 371–380.

Brochures

Despite the growth of the Internet and new media, brochures remain one of the most common types of documents our students produce in service-learning projects. They can be easily mailed to a potential client or sponsor, handed to a client on site, or left in a public space to be taken by interested passersby. When brochures are placed in a kiosk or a display stand, they usually compete with other brochures for attention. This is why the design of the cover panel is especially important. Most brochures are intended to publicize the organization and/or its programs to people outside of or new to the organization. They may be distributed to potential volunteers, potential donors, or people in the community who might benefit from the organization's programs. Most readers first skim or scan brochures and then decide whether to search for more specific information.

Because of the physical construction of brochures, this scanning is more linear than that of web readers.

Because this is such a familiar genre, some students underestimate the kinds of challenges it can raise. The example in the earlier document grids section might have opened your eyes to some of these. The strategies below also apply to related documents such as fact sheets.

General considerations

- *Keep in mind the agency's schedule for using the document.* Some organizations overhaul their publicity materials every year or two; others use the same text for many years. If your agency plans to use the text for a long time to come, avoid including information that's likely to become obsolete, such as detailed schedules, fees, or even staff members' names if there is high turnover.
- *Consider other concerns such as mailing and photocopying.* If the agency is likely to mail the brochures, design the outer panel with the agency's return address in one corner and blank space for a mailing address in the center. Even if the agency has funds to create a slick multicolor version for the first run of the brochure, design the document so it can be photocopied easily in black and white in the future. To accommodate this possibility, either use black-and-white compatible images or create two different versions of the text.
- *Create a template.* As we mentioned earlier, one of the most important contributions you can make to your agency is producing or revising document templates. Save your brochure layout in a word processing or document design program as a template, and give a copy of it to the agency on disk.

Design strategies

- *Design the cover of the brochure carefully.* Choose a descriptive title for the document so that readers will know what they are getting when they pick it up. Place the title at the top of the front panel so that it can be seen if the brochures are placed in a kiosk or in a folder. Instead of using generic-looking clip art on the cover, use something that underscores the brochure's purpose and the agency's mission. If the agency has a logo, include it along with the title.
- *Consider users' reading processes as you plan the document's arrangement.* Whether you are using a three-panel or four-panel design, an accordion-style or inside-flap fold, consider the order in which a typical reader will likely encounter the brochure's panels. As you're developing your document grids, practice opening the document as a reader might. Be sure to group information in rhetorical clusters and to arrange these clusters in a logical, reader-based order.
- *Use visuals strategically throughout the document to motivate and inform readers.* This kind of document is often used to introduce unfamiliar readers to the agency and to encourage them to find out more and become involved. Readers often scan brochures before deciding whether to read the details. Visuals such as photographs, drawings, and text boxes can help draw readers in and make them want to read on. Don't include visuals just to jazz up the text,

though; make sure that they also have distinct rhetorical purposes, as with a photo showing volunteers in action or a text box including testimony about the benefits of membership. We recommend that you place visuals strategically throughout the document and balance verbal and visual elements. Don't include so much text that it is hard to absorb or so many visuals that the brochure has little substance. Place visuals that supplement specific text chunks close to those chunks, but also know that you can place visuals across columns/panels to make the layout more varied and appealing.

- *Use blank space generously and strategically.* Most readers prefer generous margins. Allow plenty of space between columns (in the gutters) to keep the columns from bleeding into each other. Include adequate space around visuals as well so that they don't looked crammed into the brochure. This can be especially tricky when wrapping text around a visual. For readability's sake, we suggest that you only left-justify your text, leaving the right side ragged. Given your space constraints, decide whether or not you want to hyphenate words at the ends of lines.

Resources

Addison, Jim. "Brochures: A Teaching Rhetoric." *Technical Writing Teacher* 10.1 (1982): 21–24.

King, Janice. "Brochures and Data Sheets." *Technical Communication* 40.3 (1993): 550–552.

Ryan, Charlton. "Producing Brochures in the Technical Writing Classroom." *Society for Technical Communication Page.* <http://www.stc.org/proceedings/ConfProceed/1993/PDFs/Pg205207.pdf>.

Newsletters

Like brochures, newsletters generally enable both linear and nonlinear reading, dividing the text up into loosely connected rhetorical clusters that often include visual elements such as clip art. Readers of newsletters are selective about what they read and depend on headings or headlines to help them decide what to pursue. Readers often receive newsletters in busy workplace settings and aren't likely to give the texts a high priority. This is one reason why newsletters must be motivational. Unlike many brochures but like fact sheets and annual reports, newsletters are directed mainly to members of an organization and can therefore include detailed information and terms familiar to this internal audience. While fact sheets are usually straightforward, informative texts intended for readers interested in particular issues, newsletters are more journalistic texts intended to motivate and mobilize a broader range of readers about multiple topics. Newsletters also serve to maintain an organization's sense of its own identity. Newsletters and individual news stories often employ journalistic techniques that are less applicable to other types of professional texts.

- *Identify the publication's purpose.* Most newsletters are founded on basic mission statements. Their goal might be to provide information, promote events, build goodwill, inspire and mobilize readers, or build community. Work with your contact person to understand the newsletter's purpose, and keep it in mind as you create documents and design the text.

- *Keep stories brief and fast moving.* Remember that newsletter readers expect a small amount of information on a range of topics. Because they are reading to learn the facts, include such information as the name, purpose, location, time, and date of an upcoming event. List a contact person, and include information about event costs and registration deadlines.
- *Create compelling headlines.* Make these catchy but highly informative so that they both pique interest and enhance scannability. Consider using subheadings to break up lengthy stories. Also consider using jumplines (e.g., "continued on p. 3") to indicate where readers should go to continue a story.
- *Write in a journalistic style.* Although you'll want to vary your sentence structure to make the style more interesting, stick mainly with direct constructions (e.g., subject–verb–object) that can be processed quickly in a busy setting. To grab readers' attention and keep them involved, lead with action or a provocative detail, and then get to the point of the story quickly.
- *Design the appearance for motivation and scannability.*
- *Use a variety of column, headline, and visual sizes* to avoid a static and unengaging page. Consider using an uneven number of columns, such as one single-width column and two double-width columns.
- *Avoid clutter.* Remember that whether they realize it or not, your readers are comparing your text to the professionally produced magazines and newspapers they see every day. Notice how the visual concision of most of these texts makes them easier to read.
- *Use images to draw readers into stories and reinforce messages.* Photographs and clip art often have motivational functions. More technical figures such as pie graphs or bar graphs often serve more informative purposes. Carefully crop photographs and provide captions for them.
- *Print the newsletter on white, ivory, or light gray paper,* using dark ink for better contrast and readability. If you use color to motivate, reinforce meaning, or reinforce the identity of the organization, keep the color scheme fairly simple. Simplicity will keep the production cost down as well as aid clarity.
- *Put the most important items in the upper part of the far left and far right columns to conform to typical reading patterns.*
- *Create or, if possible, adjust the newsletter masthead to meet audience needs.* Be sure this section of the document captures the document's mission and includes the date, issue number, sponsor's name, and contact information.
- *Keep in mind postal guidelines as you design the text.* Some regulations will limit the amount of space you must leave blank on one panel to facilitate mailing.

Resources

Beach, Mark, and Elaine Floyd. *The Newsletter Source Book.* 2nd ed. Cincinnati:Writers Digest Books, 1998.

Blyler, Nancy Roundy. "Rhetorical Theory and Newsletter Theory." *Journal of Technical Writing and Communication* 20.2 (1990): 139–149.

Fanson, Barbara. *Producing a First-Class Newsletter: A Guide to Planning, Writing, Editing, Designing, Photography, Production and Printing.* Edmonton, ON: Self Counsel Press, 1994.

Lewis, Janet L., et al. "Newsletters: Who, Me?" *Society for Technical Communication Page.* <http://www.stc.org/proceedings/ConfProceed/2000/PDFs/00070.pdf>.

Keep in mind that when you write and design texts on behalf of an agency, you are taking its ethos or credibility into your hands. Pay close and careful attention to details of both style and visual design, deliberating about these concerns throughout the writing process. Above all else, the writing and design process must begin with and constantly return to your analysis of the rhetorical situation and organizational discourse community. Like writing, design is a thoroughly rhetorical concern.

Activities

1. Based on examples in other chapters and the criteria we've offered here, create a writing workshop guide for your main service-learning document, focusing on design issues. Ask peers in another group to use it to assess your documents.
2. Find out if the organization you plan on working with has a style guide or uses a professional style manual such as the *AP Style Manual.* If so, look through the table of contents and some of the examples to get a sense of the kinds of style issues covered. If not, continue studying the organization's texts, making a list of the style conventions and patterns they follow.
3. Obtain a copy of a style sheet used by a professional editor. You may be able to find such a document online, or you may need to contact an editor and request a copy. Make notes about how the style sheet is arranged and the kinds of guidelines it provides.
4. Interview your contact person and another member of your agency about how they define the characteristics of their discourse community. Be sure to ask open-ended rather than closed (yes or no) or leading questions. Write a one-page memo summarizing your findings.
5. Draw thumbnail sketches of four or five of the sample texts you collected, including two or three from outside the agency. Make a list of the layout patterns and design elements that seem effective from each one.
6. Take the texts you used for your discourse analysis and evaluate their designs, using the guidelines for font, spacing, graphics, and color that we provide. Where do they conform and where do they depart from these guidelines? If they depart, do they seem to have good rhetorical reasons for doing so? Write a short report in memo form explaining these differences.
7. The article "Is This Ethical? A Survey of Opinion on Principles and Practices of Document Design" by Sam Dragga (see citation below) addresses ethical issues related to the design of technical documents. It begins with a survey that was administered to technical writing students and professionals across the country. Take the survey, and then consider the study results described in the article. With classmates, discuss issues this article raises and apply some of them to your own process of creating documents.

Works Cited and General Design Resources

The Associated Press Stylebook and Libel Manual. Reading, MA: Addison-Wesley, 1998.

Dragga, Sam. "Is This Ethical? A Survey of Opinion on Principles and Practices of Document Design." *Technical Communication* 43 (1996): 255–265.

Dragga, Sam, and Gwendolyn Gong. *Editing: The Design of Rhetoric.* Amityville, NY: Baywood, 1989.

Farkas, David K., and Jean B. Farkas. *Principles of Web Design.* New York: Longman, 2002.

Hilligoss, Susan. *Visual Communication: A Writer's Guide.* New York: Longman, 2000.

Jones, Scott L. "A Guide to Using Color Effectively in Business Communication." *Business Communication Quarterly* 60.2 (1997): 76–88.

Kostelnik, Charles, and David D. Roberts. *Designing Visual Language: Strategies for Professional Communicators.* Boston: Allyn & Bacon, 1998.

Kumpf, Eric P. "Visual Metadiscourse: Designing the Considerate Text." *Technical Communication Quarterly* 9.4 (2000): 401–424.

Porter, James E. "Intertextuality and the Discourse Community." *Rhetoric Review* 5 (1986): 34–47.

Schriver, Karen A. *Dynamics of Document Design.* New York: Wiley, 1997.

White, Jan V. *Editing by Design: A Guide to Effective Word-and-Picture Communication for Editors and Designers.* New York: Bowker, 1982.

Chapter 8

Assessing Your Progress

As a member of a technical or professional field, you will likely spend a great deal of time in your career reading and writing documents. In previous chapters we have discussed several kinds of texts you might create in a service-learning project, but, excluding correspondence, the most common type of document you can expect to encounter in most workplaces is the report. Reports can range from simple one-page memos to lengthy, complex presentations of findings or recommendations. The basic rhetorical principles we've explored when presenting other writing tasks apply to writing reports, too. To produce an effective one, you must identify your readers and consider their purpose in reading and the use they'll make of the document. You must design your document to meet their expectations of form and to answer their questions.

In this chapter we'll focus on the **progress report.** We'll explain what a progress report is and identify several components of it. We'll include sample student progress reports and identify the strategies the writers used to accomplish their goals. We'll ask you to use this assignment to examine your ideas about professional ethics and the ways in which your values may shape the work you do as a technical or business professional. Finally, we'll discuss how this assignment can help you focus on completing and evaluating your final project.

Progress Report—Rhetorical Situation

Purposes

As its name indicates, the progress report updates readers about your project's progress, letting them know where you are and how things are going. A type of intermediary report, the progress report is written between the project proposal and the final project. In a job situation a progress report might come at any point in the process. Many projects require updates at regular intervals—daily, weekly, monthly, or quarterly—depending on the overall time frame. These regularly scheduled updates are called **status reports,** and keep the involved parties aware of developments. For your class project, you might write several informal updates to your instructor, perhaps focusing on a different aspect of your project in each one. You may also present one or more oral progress updates, which share many features with their written counterparts. For more on this kind of assignment, see Chapter Ten, "Presenting Your Project."

The progress report assignment is a kind of reality check for you and your agency contact person and your instructor. It will help you to assess your accomplishments and refine your plan of action for completing the final stages of the process. It's also your opportunity to request additional information or assistance with the project. Writing a thoughtful and thorough progress report can give you a sense of the tasks that lay ahead. It can also smooth your path to successful project completion.

Audiences

Most progress reports take the form of a letter or memorandum. As you know, letters are sent to parties outside of your organization, and memoranda are used for internal correspondence. Depending on the scope and type of the project, a progress report may have more than one audience. For your class project, you may write two versions of the assignment—perhaps a memo to your instructor and a letter to your agency representative. If you are an employee or a regular volunteer at the agency, you may choose to write a memo for both. You may also include other members of your class in the audience, as they may be able to learn from your experiences or give you helpful advice about dealing with challenges.

Although both audiences will want to know how your project is going, they likely will have slightly different expectations for your report. Your instructor may want to know how your progress corresponds to class expectations and major due dates, as well as how your group is working with each other and the agency. Your contact person will be concerned mainly with the actual product you are developing and how it is proceeding. Writing related but slightly different progress reports to these two audiences will also enable you to be more frank with your instructor about such obstacles as problems acquiring information from the agency and communication challenges with your contact person. If you write two reports you can ask candidly for problem-solving advice from your instructor while diplomatically communicating your needs to the contact person. You may also need to explain to your instructor workplace jargon with which your contact person is familiar. Writing a separate report can simplify this problem.

When you produce these documents in the workplace, you may write with your client in mind as the primary audience. In most cases, though, you'll have secondary audiences of your supervisors and/or employees. Your document will likely be used as a quickly read update for most of these parties and then filed for future use as a reference. Because of these layered audiences, uses, and purposes, you will want to strike a balance in the document in terms of the level of detail you use when presenting such information as background and project specifics. The challenge is to provide enough information to remind your audiences of the important issues you're addressing without making the text so lengthy that it is a burden to read. You don't want to leave your reader out of the loop on important information or imply that she or he is not aware of the history of the project. Many progress reports are not lengthy, but their content is important. Emphasize concision as you write this document. Present the information in a straightforward way with relevant visuals and a clear conclusion.

Progress Report Parts

You may recall that in Chapter Five we discussed the format of the proposal as a guideline for future documents as well as a roadmap for your work activities. Because the proposal is the contract upon which you and your supervisors agreed, it makes sense to base your report roughly around the parts of the project that your proposal outlines. Your audiences should already be familiar with these parts; thus, this consistency will help your readers quickly make comparisons and understand your update. Naturally, this document will not be as long or as detailed as the proposal. You will rely on the readers' familiarity with the project for the most part, recapping the big picture and updating the reader on the status of the pieces, especially those that are subject to change.

Introduction

Like the introductory summary of your proposal, the opening of your progress report is your opportunity to set the context for the document with a purpose statement, forecasting sentence, and other elements. If your subject line is specific and informative enough, it might convey the report's purpose as well. The subject line and/or purpose statement should indicate the report's type and its specific topic because your readers may be tracking several projects simultaneously.

Perhaps the most important part of the introduction is the statement of status, which specifically tells the audience where you are in the larger scheme of the project. Because this is part of the "bottom line" readers will be seeking, you should make this statement fairly early in the document. First, though, you'll need to give your readers a framework for positioning your status, perhaps by reminding them of the project's major parts in a sentence or two (see your proposal).

In most cases you should provide an updated timeline—updated from the proposal, that is—in the introductory section to help readers visualize your status and progress in the larger framework of the project. Whatever form this chart takes (we advocate a Gantt chart in Chapter Five), it should clarify visually the activities that have and have not been completed. Readers should be able to determine how the project is progressing by simply glancing at this chart. Alternatively, progress report writers sometimes place an updated timeline toward the end of the body section of the report as a way to summarize their progress and possibly to set up requests for schedule changes.

In addition to describing how much work you've completed on your project, you'll also want to use the progress report to describe how well you've done it.

You can include evaluations of your collaboration, your texts, and your learning experience. You also can take this opportunity to preview your assessment of your final project. In doing so, you might review your major accomplishments, which you will describe later in more detail. Finally, the introduction might forecast the remaining report contents briefly.

Team B.A.R.K., the student writers of the sample progress report in Figure 8.1, reinforce their subject line with a purpose statement in the first sentence. Next, they recap the project's purpose and major tasks, providing a framework for positioning their status. Although they might have arrived at a statement of status even sooner, perhaps in the first paragraph, they do so in the second paragraph in which they also summarize their accomplishments and look ahead to future work. They could also wait and provide this summary toward the end of the report. The writers could have also assessed their accomplishments as they were summarizing them. They do, at least, anticipate the success of their final product at the beginning of paragraph three, in which they also forecast the report's main parts.

Body

The body of the report tells your readers what you've accomplished and how the project is going in specific terms. Although you mostly want to emphasize your ongoing successes, this is also the section where you discuss glitches you have faced and your plans for addressing them, propose modifications to your plan of action, and request assistance from your readers. Don't include fluff or speak in generalities in this section, but point your audience to concrete details. Use language that evokes clear images, not hazy impressions.

You can arrange the main part of your report in several different ways: around time blocks, around tasks, or around objectives.

- *Time blocks.* Some projects lend themselves to time-based updates and are therefore primarily organized around time blocks. These might include projects that are on an especially tight schedule or, alternatively, projects that span a long time period. Some readers will want a thorough chronological breakdown that includes what you have accomplished, what you are working on, and what you have left to do. Other readers will mainly want to know what you've accomplished in the most recent phase of the project. In the workplace, your client or employer may ask for updates on your progress at regular intervals, for instance. This format might also be used if you are working on a project for which you bill a client based on the number of hours you spend on tasks. This kind of report might coincide with a billing statement and give the client or sponsor a sense of how her or his money is being spent. Using a chronological arrangement might make it easier to discuss obstacles you have faced and how you have responded to them.
- *Specific tasks.* You may also arrange your report around major tasks. If you do this, the tasks you discuss should correspond with those you detail in your proposal, unless of course you have modified the project since then. This kind of arrangement might make it easier for readers to relate your progress description to the updated timeline, although it could also become fairly complicated

Team B.A.R.K.
University of Florida
ENC 3250: Professional Communication
kroland@ufl.edu

March 25, 2001

Audrey Holt, Volunteer Coordinator
Alachua County Humane Society
2029 NW 6th Street
Gainesville, FL 32609

Subject: Progress Report on Volunteer Training Manual

Dear Audrey,

This report will formally update you about Team B.A.R.K.'s progress on the Volunteer Training Manual we are producing for the Humane Society. Our work on this manual, a text that will serve as both an instructional and reference text for new volunteers, includes researching and gathering materials, designing the layout, drafting the sections, soliciting feedback from our ENC 3250 course and from you, and making the final revisions. Throughout this process we must keep in mind the agency's cost constraints and the manual's expandability.

Team B.A.R.K. has already gathered most of the information, established a standard document design and style, and settled on an arrangement of the overall text and each major section. We are currently drafting the sections.

We are confident that we will complete a successful, well-organized, and attractive manual to serve the Humane Society and its volunteers. The remainder of this proposal will further explain our accomplishments, describe our current and future work, and update our project timeline.

Progress

Task 1: Gather information for manual

Completed:
With the help of your staff, Team B.A.R.K. has gathered information about the agency's goals and mission statement, operational information (hours of operation, services, etc.), and policies on issues such as euthanasia.

(continued)

Figure 8.1 Team B.A.R.K.'s progress report

Figure 8.1 *(continued)*

Ongoing:
We are in the process of gathering and synthesizing materials on volunteer requirements and a few of the procedures such as how to sign-in, how to handle grievances, and what to do if injured.

Task 2: Determine manual's design
Completed:
Our first goal was to create an overall design for the manual. It will be a 8.5 x 11" spiral-bound booklet approximately fifteen pages long. We have also designed the pages for most of the sections, standardizing the font, spacing, and use of visuals. This template should help you expand the manual in the future.

Ongoing:
As we are still working with you to determine what kinds of information to include in the volunteer job description pages, we haven't yet committed to a page design for this part of the manual. It will generally follow our standard design template.

Task 3: Select and design visuals for manual
Completed:
To break up the text and give it a friendly tone appropriate for its audience, we plan to enhance the written explanations with visuals wherever appropriate. These include clip art to accompany the headings, call outs for volunteer quotations, and photographs of volunteers in action, especially for the volunteer job descriptions. Our main concern is selecting visuals that actually serve a purpose and that will photocopy well in color or black and white, because you want to produce the document with your newly acquired copy machine.

Ongoing:
We are still combing the Humane Society's archives for a few final pictures of volunteers in action. We will probably take a few photographs ourselves. We are also researching a mascot that will point out important sections and add humor throughout the manual. We are trying not to use the typical dog or cat, but to choose something more unique.

Task 4: Arranging and drafting the manual's sections
Completed
Team B.A.R.K. has outlined the manual's parts and received some preliminary feedback about our arrangement from our instructor. We have grouped alike information to create a cohesive flow. We have also drafted the manual's title page, opening, and introduction.

Ongoing
The Humane Society will deliver a list of volunteer requirements and other related information early this week, which will enable us to draft those sections. We are also in the process of drafting the agency's welcome letter and a frequently asked questions section.

(continued)

Figure 8.1 *(continued)*

Updated Timeline

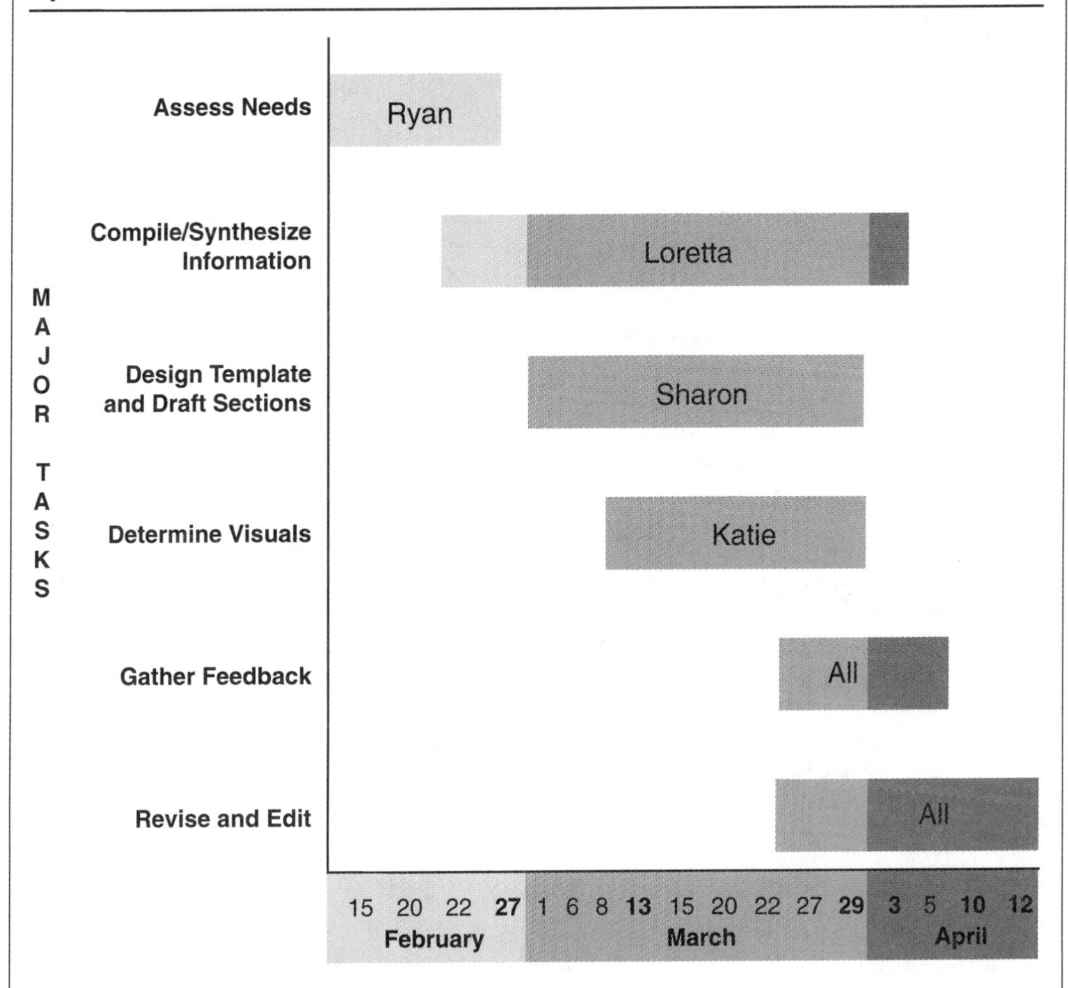

The Gantt chart above shows the updated parts of our project and their general time frames. As you can see, we are only slightly behind our original timeline, mostly due to waiting on needed information. We still plan to finish the entire project on schedule. Future planning will be devoted to combining all of the sections in a coherent manner, getting feedback on and polishing the design and style, and selecting an affordable printing process.

In conclusion, we expect our writing project to result in an informative and effective volunteer manual. We look forward to completing and polishing the manual after receiving additional formal feedback from you. Thank you for taking such an active role in facilitating our progress so far.

Sincerely,

Team B.A.R.K.

if you were working on several phases at once. Notice in Figure 8.1 how Team B.A.R.K. provides an updated project timeline toward the end of the body to visually summarize their progress by task. Organizing the report primarily around tasks might also make it easier for you to ask for assistance with a particular part of the project.

- *Major Objectives.* In this kind of arrangement you might address more big-picture concerns in the document sections. This would be appropriate if your readers want a more general update about your progress toward your proposal objectives or if you need to significantly revise your objectives or tasks. For example, a student who agreed to create an event-planning guide for a magnet school found herself adjusting the shape of the project as she progressed. Because the contact person wasn't entirely sure what she wanted from the guide at the proposal stage, the student had to reassess her work at the progress report phase based on general goals rather than individual tasks. Rather than individual tasks, she used such categories as *needs assessment* and *revised project expectations* to describe her work to date.

Actually, many progress reports use some combination of these models, as in the sample report in Figure 8.1. Notice how the primary headings of the body section revolve around the major tasks or phases of the project and how the subheadings under each heading revolve around chronology—past accomplishments and present work. This group does a nice job of making their organization explicit and information easily accessible. If their reader wanted to know how they were doing on a particular phase of the project, she or he could find this information quickly. Follow Team B.A.R.K.'s lead by using informative headings and subheadings in this section.

What to Emphasize in the Body

Whichever pattern of organization you use, you'll want to emphasize the following types of information in the body of the report.

- *Successes.* One of the most important rhetorical purposes of this document is to assure your agency contact person and your instructor that your project is progressing successfully. This assignment should help you to realize how much you have accomplished. Some students find that the assignment motivates them to complete tasks on which they may be lagging. When you discuss successes you've had along the way, mention specifics. Don't just assure everyone that things are going well and that you'll definitely finish on time. Identify particular accomplishments. Mention a surprising contact you've made. Quantify the number of drafts you've produced, or describe a concept from your field that you've seen in action. In their progress report, students working with the Central Florida Zoological Park project discussed how their specific collaboration strategies enabled them to produce two additional texts beyond what they had committed to in their proposal.
- *Logistical glitches you've faced and your plans for getting past them.* Sometimes even the best-planned projects will be complicated by such logistical prob-

lems as lack of access to information and materials or difficulty in scheduling group work time. It's important to present this kind of problem in an honest way, taking responsibility for whatever aspects of the situation you could have avoided. It's also a smart rhetorical move to explain how you have addressed and rectified the problem or to at least offer a plan for doing so. Describing your proactive response can function as a rhetorical "buffer" to the "bad news" of your problems. For example, when four students working on a research proposal for a nutrition program had collected all the data for their report, they realized that none of them had access to a software program that would allow them to compile the data in a useful way. Though they had planned to use some software available on campus, when they came to the point of actually inputting the data, they realized that they, as students, could not gain access to the software. They explained this problem in their progress report but also described two possible courses of action for escaping from the situation: They offered to do some extra training on campus to qualify to use the software; they also offered to help their sponsoring agency to acquire software for use on this project and on others in the future. They laid out the costs and benefits of each plan and asked their contact person to advise them on the best course of action.

- *Specific assistance you need from your reader.* Many readers treat progress reports as reference documents and don't necessarily feel a burden to respond or to act on them. If you need particular assistance from your contact person or instructor, make a clear and direct request. Mention a reasonable date by which you need the assistance, and make it easy for the reader to provide the information or help to you. Offer to drop by and pick up needed materials at the reader's convenience, for example. Two students designing fundraising documents for a community mentoring program were unable to complete their task until their contact person helped them locate or take pictures for inclusion in their projects. They had to highlight their need for this assistance in their progress report. They identified a logical deadline for receiving the materials and explained the revisions they planned to make in their documents in the event that they didn't receive them. They offered a Plan B, but their clear request helped them avoid having to enact it.
- *Modifications to your plan of action.* When you write progress reports as a professional, you'll find that your clients, supervisors, and other audiences will want updates about the budget and schedule. Changes in these elements will clearly affect the outcome and delivery of the final product; therefore, you may need to ask for your reader's permission to enact them. In addition to explaining why the changes are needed, you might also emphasize what they will help you accomplish, thereby putting a positive spin on the request. Scheduling is a particular concern in a class project. The term is, after all, a finite period of time. In one class, two students were working on a brochure about stress management for a local teen outreach program. Part of their commitment to the project included researching the time and costs of producing the brochures. After a great deal of research with local printing shops, they found that they needed to revise their plan of action to complete the project

within the schedule and budget to which they had agreed in their proposal. They made a plan to simplify the brochure, deliver it to the printer earlier than originally planned, and incorporate fewer expensive production elements such as colors and pictures. They carefully worded their progress report to explain the need for these changes and were able to avoid an unpleasant ending for the project.

- *Appendixes.* If you actually want to show your readers part of the research you've completed, documents you've designed, or other "products" of your work, you may also want to enhance your descriptions above with appendixes. Be sure to refer to the appendixes in the report. If you decide to do this, choose wisely, and include only material that is critical to the project. The purpose of the progress report is to update the reader; you don't want to burden her or him with a great deal more work and reading material. The most important appendixes to include are items your reader has requested.

Conclusion

This section of the progress report, often consisting of just a paragraph or two, is your final opportunity to highlight your main accomplishments, assess your progress, and instill confidence in your readers about your project. This is not the place to rehash your problems, although you might formally request permission to adjust your schedule here.

Most progress reports also end by referring to the impending successful completion of the project and expressing appreciation for the audience's assistance. Some progress reports reiterate the final project delivery date here. Although you are providing a service to the agency, your contact person and/or members of her or his staff also are likely to be working hard to make this project a useful educational experience for you. Be sure to acknowledge assistance they have given you and thank them in advance for other help you expect. If your report includes a request for more assistance from them, you will likely garner more cooperation if you express appreciation for whatever effort they have made. Team B.A.R.K. tersely but effectively concludes the progress report in Figure 8.1 with both of these rhetorical moves. In pointing to the project's completion, the writers also remind their contact person of their upcoming workshop with her.

Tone

Supervisory audiences and clients sometimes use progress reports to decide whether or not to support or intervene in a project. In any case, most progress reports discuss situations (e.g., encountering an obstacle) that could be interpreted in multiple ways. For these reasons and others, tone is a crucial factor in the report's efficacy. Keep the following guidelines in mind as you draft and revise your text.

- *Maintain a tone that is consistent with your proposal.* Make sure your progress report doesn't sound much more formal or much more casual than your pro-

posal. Don't use new or different terms for critical concepts or p
project. Look for buzzwords or jargon that you might have ac
working on the project and with which your reader might not be fam
for example, you have come to refer to an agency by an acronym, be sure to explain that term briefly so that you are inviting your reader into the project rather than shutting her or him out.

- *If your report is a team document, write in a united tone.* Though work distribution needs may require that one person be the primary author of the first draft of the report, all members need to contribute to the writing process. In melding these different contributions, make sure you convey the same attitude about your project's success, your handling of problems, your need for assistance, and so on.
- *Don't write the report like a sales pitch.* Though you may indeed be working to garner favor and support as you write this text, don't present it in a tone that is superficial or exaggerated. Avoid such empty adjectives as *great* or *excellent.* Avoid a pattern of overstatement by deleting such words as *perfectly* or *very.* Your responsibility in this report is to present an accurate and useful assessment of the status of the project. Don't use this task to bargain for a high grade. At the same time, you want to sound confident in your self-assessment, which means avoiding such phrases as *we feel* and *we believe.*
- *Don't complain or place blame.* If you're having a tough time with a project, it may be tempting to use the progress report assignment as an opportunity to make someone or something else accountable for delays or other problems. Avoid this. Transferring responsibility for the project won't make you look better. Depending on who is reading your text, this kind of move could make you seem immature, irresponsible, or unfair. Even if some action or inaction on the part of your contact person or instructor has hindered your progress, it won't help your situation to complain about it. Describe the sequence of events that brought you to your current situation in objective terms, perhaps using passive voice, and request specific and reasonable kinds of assistance with an eye toward the future.

Style Focus: Active Style

We've already discussed the importance of concision and coherence to technical and professional communication. Given that this type of communication is often designed to report on and help people perform particular tasks, clarity of action is another of its preferred stylistic attributes. By *active style,* we mean more than just using mostly active voice. Creating an active style involves viewing the subjects and verbs of your sentences as actors and precise actions. These actors don't always have to be people, although they often are.

As Williams points out, creating an active style can make your sentences clearer and easier to follow, an important consideration for most workplace texts. Readers better understand sentences that are structured around basic actor–action patterns. Using strong action verbs, like using precise nouns, will also make your messages more informative. Perhaps most importantly, writing in an active style can give your prose more rhetorical force. Imagine a set of instructions directing readers in passive voice sentences such as "The length of the door should now be measured." Rhetorical force is especially important in more overtly persuasive texts such as proposals and recruitment materials.

As we suggested earlier, the progress report is both informative and persuasive. Active style can further both of these general aims. It can help you explain your past and current work more clearly and more precisely, and it can help you create an industrious and accomplished ethos.

In this section we will take you through four basic techniques for creating a more active style, using the progress report as a sample context. As in earlier chapters, we give examples of how to implement these techniques in a sample document. Figure 8.2 shows how Team B.A.R.K. revised some of their sentences to create a more active style.

Convert Passive Voice Constructions to Active, Where Appropriate

Because your main goal is to tell your readers how you are progressing, it only makes sense to position yourselves as actors in many of the sentences. To avoid beginning too many sentences the same way, though, you might alternate using *we* and your group name and look for ways to begin sentences with dependent clauses or with nonhuman actors.

Many of you probably already are adept at recognizing and converting passive voice sentences. It helps to look for a *to be* verb followed by a preposition. Sometimes the preposition is implied, as in the example "design elements were established [by us]." In essence, a passive voice construction takes the actor and makes it the (often implied) object of a preposition.

Let's go through a couple of examples in Team B.A.R.K.'s report. Look at how the writers have revised the last sentence in the second paragraph in Figure 8.2. The formerly passive voice sentence—"The sections of the manual are now being drafted."—leaves the reader with the vague sense that the drafting is happening somehow, somewhere, by someone. The writers revise the sentence by inserting themselves as the actors, making it more concise in the process. Later in the report, under Task 4, Team B.A.R.K. revises this even more awkward description of their arrangement work: "The manual's parts have been outlined and preliminary feedback about our arrangement has been received." Outlined and received by whom? The instructor? Agency personnel? The writers needed to clarify the actor and seize the opportunity to reinforce their accomplishments regarding this task.

Of course, not all passive voice sentences are inappropriate, even in a progress report. You can use passive voice sentences strategically to create a given–new information chain, express action with no clear actor, emphasize information other than the actor, and even avoid attributing action to an actor. In another sample progress report, a group of student writers used the following sentence to explain their revised timeline: "All completed work has been shaded in the timeline." Because it is clear from this sentence who the actors are, and because it is not important to emphasize this particular action as an accomplishment, passive voice seems appropriate here.

Team B.A.R.K.
University of Florida
ENC 3250: Professional Communication
rroland@ufl.edu

March 25, 2001

Audrey Holt, Volunteer Coordinator
Alachua County Humane Society
2029 NW 6th Street
Gainesville, FL 32609

Subject: Progress Report on Volunteer Training Manual

Dear Audrey,

This ~~is a progress~~ report ~~on the work being done to create~~ will formally update you about Team B.A.R.K.'s progress on the Volunteer Training Manual we are producing for the Humane Society. Our work on this manual, a text that will serve as both an instructional and reference text for new volunteers, includes researching and gathering materials, designing the layout, drafting the sections, soliciting feedback from our ENC 3250 course and from you, and making the final revisions. Throughout this process we must keep in mind the agency's cost constraints and the manual's expandability.

Team B.A.R.K. has already gathered most of the information, established a standard document design and style, and settled on an arrangement of the overall text and each major section. ~~The~~We are currently drafting the sections of the manual ~~are now being drafted.~~

We ~~feel~~ are confident ~~in the completion of~~ that we will complete a successful, well-organized, and attractive manual to serve the Humane Society and its volunteers. The remainder of this proposal will further explain our accomplishments, describe our current and future work, and update our project timeline.

Progress

Task 1: Gather information for manual

Completed:

With the help of your staff, Team B.A.R.K. has gathered information about the agency's goals and mission statement, operational information (hours of operation, services, etc.), and agency's policies on issues such as euthanasia.

(continued)

Figure 8.2 Revised progress report

Figure 8.2 *(continued)*

Ongoing:

We are in the process of gathering and synthesizing materials on volunteer requirements and a few of the procedures such as how to sign-in, how to handle grievances, and what to do if injured.

Task 2: Determine manual's design

Completed:

Our first goal was to create an overall design for the manual. It will be a 8.5 x 11" spiral-bound booklet approximately fifteen pages long. We have also designed the pages for most of the sections, standardizing the font, spacing, and use of visuals. This template should help ~~your expansion of~~ you expand the manual in the future.

Ongoing:

As we are still working with you to determine what kinds of information to include in the volunteer job description pages, we haven't yet committed to a page design for this part of the manual. It will generally follow our standard design template.

Task 3: Select and design visuals for manual

Completed:

To break up the text and give it a friendly tone appropriate for its audience, we plan to ~~put visuals with~~ enhance written explanations with visuals wherever appropriate. These include clip art to accompany the headings, call outs for volunteer quotations, and photographs of volunteers in action, especially for the volunteer job descriptions. Our main concern is selecting visuals that actually serve a purpose and that will photocopy well in color or black and white, because you want to produce the document with your newly acquired copy machine.

Ongoing:

We are still combing the Humane Society's archives for a few final pictures of volunteers in action. We will probably take a few photographs ourselves. We are also researching a mascot that will point out important sections and add humor throughout the manual. We are trying not to use the typical dog or cat, but to choose something more unique.

Task 4: Arranging and drafting the manual's sections

Completed

~~The manual's parts have been~~ Team B.A.R.K. has outlined the manual's parts and received preliminary feedback about our arrangement ~~has been received~~ from our instructor. We have grouped alike information to create a cohesive flow. We have also drafted the manual's title page, opening, and introduction.

Ongoing

The Humane Society will deliver a list of volunteer requirements and other related information early this week, which will enable us to draft those sections. We are also in the process of drafting the agency's welcome letter and a frequently asked questions section.

(continued)

Figure 8.2 *(continued)*

Updated Timeline

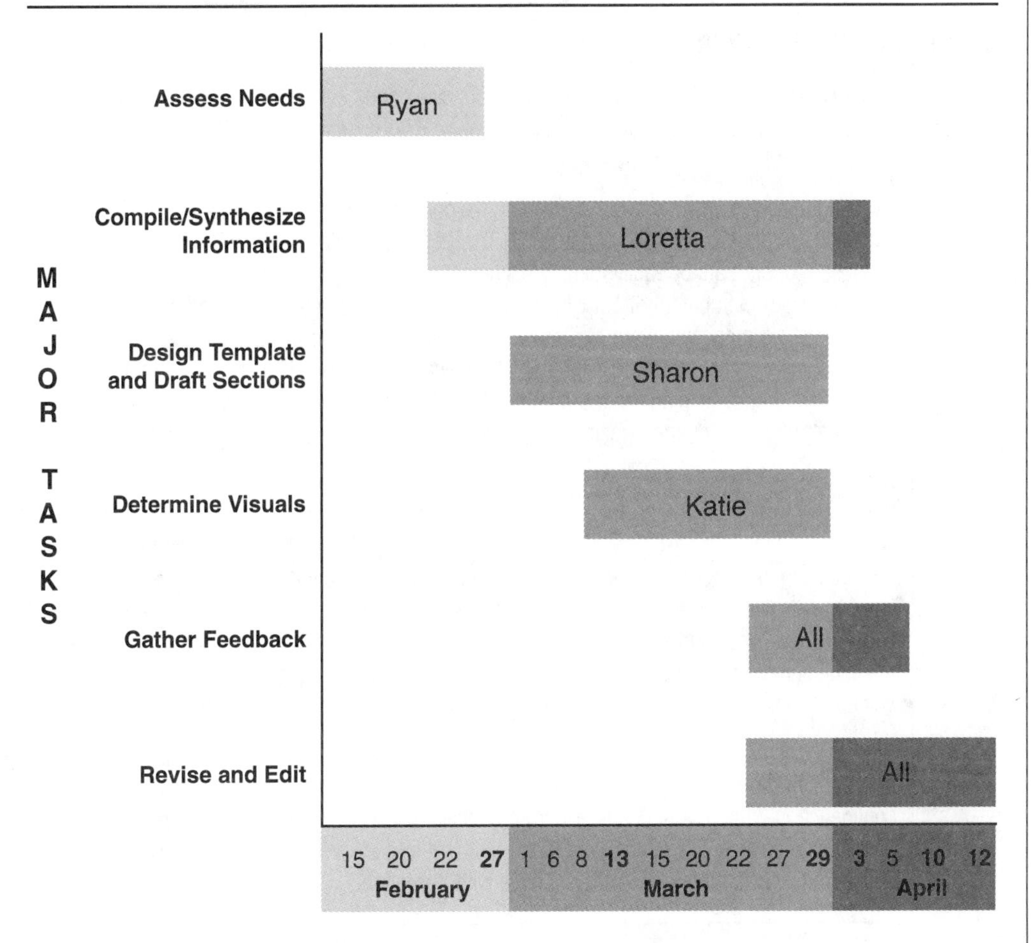

The Gantt chart above shows the updated parts of our project and their general timeframes. As you can see, we are only slightly behind our original timeline, mostly due to waiting on needed information. We still plan to finish the entire project on schedule. Future planning will be devoted to combining all of the sections in a coherent manner, getting feedback on and polishing the design and style, and selecting an affordable printing process.

In conclusion, we expect our writing project to result in an informative and effective Volunteer Manual. We look forward to completing and polishing the manual after receiving additional formal feedback from you. Thank you for taking such an active role in facilitating our progress so far.

Sincerely,

Team B.A.R.K.

OTHER VOICES 8.A Kristen Kennedy

Kristen Kennedy, Ph.D., is senior editor of Technology and Learning Magazine. *In the past, she has taught service-learning professional writing courses at University of Rhode Island and Wake Forest University.*

Overusing passive voice is often a mark of inexperience—with subject matter, writing style, and audience. In rhetorical terms, inappropriate use of passive voice can compromise a writer's ethos because readers cannot easily identify sentence agents. Moreover, passive voice tells me that a writer is lost in the content about which he or she writes and lacks command of material. My students' working proposals were riddled with passive voice not only because they thought passive voice sounded more "professional," but also because they weren't entirely confident in writing about their agency or plan for the community. In the latter case, passive voice indicated uncertainty and unclearly defined goals. The unwieldy nature of the project expressed itself in passive voice.

Student writers should keep in mind that even experienced writers struggle with awkward and confusing lapses into passive voice. For example, when I changed careers from academic writing and teaching to work as an editor for a trade publication, my confidence in my new role was shaky. Not surprisingly, I had a difficult time editing other writers for their overuse of passive voice because I wasn't yet comfortable editing non-academic writing. Also, I wasn't always the expert on the subject matter I was editing. With time, and as I grew more comfortable with this new idiom, I gained confidence.

Of course, confidence grows with experience, but as a student you may have to talk yourself into confidence in your writing. How? Do what most professional writers do: fake it. One way of faking confidence in writing is making deliberate choices to use active voice when you're feeling pretty passive and unsure. The course and the project will give students the experience that builds confidence. In the meantime, remember to write with a swagger.

Earlier we advised you to avoid attributing blame to your supervisors or collaborators when describing problems in your project. Passive voice can be a useful strategy for accomplishing this through sentence structure. Stating "the brochure materials have not been received," for instance, can help you avoid pointing a finger toward the contact person who hasn't sent them.

Replace Weak Verbs with Stronger, More Precise Ones

Converting active voice sentences to passive voice ones is one way to eliminate *to be* verbs. As we discuss in the section on writing résumés in Chapter Four, though, you should be on the lookout for other instances of weak and vague words as well. In choosing your verbs, think of them as the actions of characters, and determine the specific actions you want to convey.

Chapter Four shows how expletive constructions can make sentences wordy as well as passive. The opening sentence of Figure 8.2 is another example of such a construction: "This is a progress report on the work being done to create the Volunteer Training Manual for the Humane Society." In their revision of this sentence, Team B.A.R.K. replaces the construction with a more precise actor—the *report*—and action—*update.*

The writers of the sample report revise another weak verb in the first sentence under Task 3. Instead of "put visuals with written explanations wherever appropriate," they write, "enhance written explanations with visuals wherever appropriate." This change replaces the verbal phrase *put visuals with* with the single, more precise verb *enhance.*

Change Nominalizations to Verbs and Make Actors the Subjects of Those Verbs

Nominalizations are usually longish nouns that have been converted from their more common verbal forms. Because they convert verbs into nouns and often require additional words, nominalizations can make a sentence more passive and unnecessarily long. In many cases they can be converted back into verbs, in the process trimming extra words from the sentence and establishing an actor for the more precise action.

Figure 8.2 includes examples of nominalizations. The first sentence of the third paragraph originally read "We feel confident in the completion of a successful, well-organized, and attractive manual to serve the Humane Society and its volunteers." Here the word *completion* usually takes a verbal form of *to complete.* The students probably used the nominalization to sound more formal, but they ended up stripping the sentence of rhetorical force. In changing *completion* into the verb *complete,* the writers create a more proactive sentence. In another sentence under Task 2, the writers of Team B.A.R.K. revise the nominalization *expansion* into the verb *expand,* similarly inserting an actor in the sentence and ridding it of the unnecessary preposition *of.*

Get to the Main Verb Fairly Quickly in the Sentence

One easy technique that Williams recommends is counting the number of words it takes you to arrive at the main verb of the sentence. If you are suspending the reader too long before reaching the action, you might consider adjusting the sentence structure. Consider this action-suspending sentence, also written in passive voice: "At the time of this progress report, Task 1 and Task 2 are completed." Here we don't arrive at the action until the very end of the sentence and even then the actor is unclear. By revising the sentence to "We have now completed Task 1 and Task 2," we foreground the action better and earlier.

If Things Become Complicated

Most of the guidelines we've offered above pertain to projects that have gone basically according to plan. One of the reasons why service-learning approaches are challenging and rewarding is that sometimes things become complicated.

Agency budgets, leaders, and priorities change. Schedules don't operate on a semester or quarter system, and agencies are often facing more serious concerns than our projects. If your project has hit a barrier, the progress report is your opportunity to find a creative and productive way to convert the difficulty you face into something positive for yourself and the agency with which you're working.

For example, one student in a service-learning course was helping a local veteran's advocacy group to write a grant proposal to gain funding for a homeless shelter for vets. When he started the project, it looked like a fairly straightforward process. The agency knew which grant it wanted to secure and had all of the appropriate forms. Bart's job was to sift through some information the staff had compiled and put together a document that explained the group's objectives, history, and qualifications. Unfortunately, the agency representative did not have a clear sense of what would be involved in opening the shelter when he made the decision to pursue the project. After Bart had spent several weeks working on the documents, he learned that the project would be abandoned. In Figures 8.3 and 8.4, you'll see two informal progress reports he wrote to his instructor. As you can see, one is dated late in October and the other in mid-November.

To: Dr. B
From: Bart D. Leahy
Date: October 29, 2000
Re: Meeting With the VVA

As you requested, I am providing you with a progress report on my grant proposal to create a shelter for displaced veterans for the Vietnam Veterans of America. This has been a fun project, as it is one I truly believe in. Just to recap, here are some of the milestones I've passed along the way.

Based on your and my classmates' input, I have written several drafts of the initial proposal prior to and after the original due date, the most recent being October 16. I have found that successive drafts have refined my thinking and writing about the project. The Audience Analysis and Organizational History assignments resulted in yet another draft of my grant proposal. The redrafting has been feeding off of itself, as I try to get a single message across: "Give these people money!"

Because this seems to be the proper forum for airing reservations and concerns, I can share a couple with you about this project. I looked through the 12 pages of federal forms that I copied from a book from HUD and realized that the standard proposal format you've provided us to follow might not fit the government's forms. Writing for the government seems to involve a lot of bullet points and filling out forms. "Persuasion" consists of facts and background. I think the most my writing can add to the mixture is clarity and

(continued)

Figure 8.3 Informal progress memo that discusses an ethical dilemma

Figure 8.3 *(continued)*

organization. Pathos is not exactly a bureaucrat's line of work. I sense that there will be more work for me after the semester is over as I try to adapt the writing I have done to the appropriate forms.

I am concerned about the timing of my assignment. I am trying to complete assignments, but I am at the mercy (for lack of a better word) of my contact person and how much information he is willing or able to give me. Mostly, this has been deadline pressure. What usually ends up happening is that I receive an assignment, forward my questions to Dick Everson, and then wait. Part of the fault is mine, as I sometimes ask for information two days beforehand. Another problem is that Dick has not yet assembled much of the specific information I have requested. Despite repeated requests over the last month, I have still not received specific details about the shelter's budget. Although Dick might be very committed to this project, it is apparent to me that he and the VVA have not worked through all the details yet.

The good news is that Dick has been attending grant-writing classes to clarify the questions he needs to answer. Also, I plan to visit a VVA meeting on Tuesday, November 14, to obtain more specific details about the project, get a better feel for the membership, and also acquire the information I have not been able to get from my e-mail correspondences.

The ethical dilemma I'm facing is minor, but probably obvious to you. I am trying to satisfy your requirements for a complete presentation while at the same time not harass my clients. I would like to help VVA after the class is finished, and I'm trying not to come across as pushy. This is a matter of personal style with me as well as a matter of weighing consequences. Although I have absolute deadlines for the class, VVA does not have a definite deadline or even a starting date for their grant request. In this case, I believe the client's needs come before my academic needs, so I will continue my low-key approach. However, if you absolutely require a budget from me, I will at least get an estimate from Dick to complete the assignment. If VVA had a more pressing deadline, I would undoubtedly be more persistent in my efforts. I hope you understand my decision. I am willing to consult with you about any suggestions you might have for speeding up the process.

This is a good class, and I am glad that I have had the opportunity to interface with "the real world." Hopefully, this experience will also bolster my resume and make me more employable. I look forward to the completion of this project and the results it brings.

In the first report, Bart describes some challenges that he faces as a student—concern about how deadlines for the project match up with deadlines for the course and about whether or not the work he's doing on the project will meet the course requirements. In the second report, he is expressing a different kind of concern. At this point, his project is in total flux. All he really knows for certain is

To: Dr. B
From: Bart D. Leahy
Date: November 15, 2000
Re: Meeting With the VVA

Well, as you have said at least once, working in the real world is always an eye-opening experience. Having said that, here is how my eyes were opened at the VVA meeting last night.

Dick Everson (my primary contact) and Rich Hall, another VVA member, went to a symposium for homeless support groups recently. At that convention, they learned that VVA Chapter 811, while motivated to do this work, does not have the experience or wherewithal to start a homeless shelter as originally described. In fact, the people at the conference convinced them that the best thing for VVA to start would be a "homeless veteran watchdog group." In addition, Dick and Rich were advised that the best way to get started as a "watchdog" for homeless vets would be to partner with another homeless agency (or coalition of agencies) in the Central Florida area. This, obviously, has major implications for the project I was originally proposing. Put bluntly, that project has been scrapped or put off until the far future.

What Dick was proposing as a start-up to the actual homeless shelter would amount to a freelance watchdog group. In an effort to make the watchdog process less adversarial, the VVA would start out by helping local shelters in the following ways:

- Taking a census of homeless vets
- Finding out if the needs of the vets in existing homeless shelters are being met
- Helping the local shelters prove that they are helping veterans.

I suggested an additional strategy for VVA to pursue, once they are more established in their watchdog role. I thought it might be worthwhile for them to start to perform some of the actions they plan to do at their own shelter, such as:

- Performing a needs assessment for homeless vets in the shelters
- Referring them to the counselors they need
- Providing basic needs that might not be available at the shelter where they are living—soap, clothing, food

Proven experience in the field would give VVA members the right background they need to run their own homeless shelter eventually, as they originally wanted. In the meantime, I'd be happy to help. I have already volunteered to assist with a research effort on this new project. Hopefully that will fill some of the final project requirements.

(continued)

Figure 8.4 Progress memo that updates instructor about dilemma

Figure 8.4 *(continued)*

I also intend to make a fuller proposal to VVA about how they can proceed on this new venture. Judging by their reaction to some of my comments, I would guess that this latest venture has also not been 100 percent thought through yet.

Any suggestions you would have about how to help VVA on this level would be appreciated. I will, of course, continue my research.

that the document he planned to produce is no longer needed. Despite an urge to panic, he chooses to offer some ideas about possible options in this report and reflects on what this derailment has taught him. Bart's situation is unusual with almost any kind of project, though it's more likely to happen when students are producing financial documents such as grant proposals. But this kind of problem does occasionally happen on a smaller scale in all kinds of projects. The key to emerging successfully from a difficult situation like this is focusing on your primary objective of developing your writing skills. Take such challenges as opportunities to use your rhetorical abilities in new ways. Be ready to implement a backup writing plan with your instructor's help if your project becomes as messy as this one did. Remember that your primary course goal is to improve as a technical or professional writer.

Facing Ethical Challenges

You may have noticed that Bart refers to an ethical dilemma he faced in his project in the first draft of his progress report above. He was in a complicated situation and had to make a decision about how to proceed. In Chapter Three we referred to three categories of ethical dilemmas that technical and professional communicators might face—those dealing with obligations, ideals, and consequences. Bart's problem was primarily a question of obligation. Specifically, he had to determine how to balance his separate obligations as a student—to meet deadlines, complete specific assignments, and so on—with his obligations as a consultant—to complete the tasks he committed to despite scheduling and other kinds of barriers. He wasn't sure how to be honest about this problem with both parties, so he addressed it in his progress report to obtain guidance from his instructor.

We consider the process of reflection that Bart and other students experienced to be a critical part of the progress report assignment. We suggest that you take some time at this point in the process to consider the kinds of choices you have had to make throughout the semester. When we introduce the progress report in our own classes and advise students that we want them to reflect on an ethical dilemma that they have faced over the course of the project, some are surprised. They're doing community work for a nonprofit agency, a campus group, a local school, or a business with a service commitment. What could be ethically complex about that? Some students believe that such groups are safely tucked outside the rat race of commerce and are unlikely to face overt ethical quandaries like we

OTHER VOICES 8.B Erica Olmstead and Amber Feldman Goldberg

Erica Olmstead is a graduate student in technical writing at the University of Central Florida and works as a technical writer for the United States Census Bureau in Washington, D.C. Amber Feldman Goldberg is also pursuing a master's degree in the UCF technical writing program and works as a graduate assistant and a consultant at the University Writing Center.

We thought we had a great project idea for our proposal writing class. While working with the Women's Studies program we heard about a plan to create an archive of letters, diaries, magazines, and other materials from women of Central Florida throughout the area's history. The planners needed help in preparing a grant, raising awareness about the project, and soliciting contributions to the archive. We felt that this was a good idea for a writing project.

We started our project with the idea that the organization was ready to submit a grant proposal but soon discovered that this was not the case. When we met and talked with the faculty and staff members involved with the project, we realized that there was more to do than would be possible in a semester. We felt as though we were in over our heads. Suddenly people were relying on us to make decisions that went beyond our writing project. We were becoming aware of the complex politics surrounding the archive idea. The group was not yet organized well enough to provide even the basic information that was needed to apply for a grant. For example, there were conflicting stories about whether the archive would be housed on campus or elsewhere in the community. This information was essential; we could not write a grant proposal that did not include a description of the location in which the materials would be stored.

Despite these setbacks, we didn't want to give up on the project. By the time we got to the progress report phase, we had to decide how to refine the project to fit our course requirements and time frame. We consulted with our writing professor (Melody) and with the director of Women's Studies to determine our role. We decided that we wanted to give the group a recommendations report and help them get to a point where they would be ready for the grant application process. We became personally involved in the project. We interviewed other members of the group and offered to help with the grant submission after the semester ended.

During the semester we wrote as much of the grant as we could. Then we gathered the limited information that was available and wrote a report that contained research we had done on other grant opportunities so they could continue with the project in the future. We included suggestions for ways the group could meet the national archive standards. We gave them detailed instructions for completing the grant proposal that we started which wasn't due until three months after the semester ended. Finally, we created a PowerPoint presentation that the group could use to promote the archive and get community support for the project.

By not giving up on our project and by redefining our focus, we contributed more to the group than we had expected to. We felt that writing a grant proposal that they would be unable to actually submit would have been a waste of our time and theirs, so the shift in our focus worked out for all of us. We made our project something that was manageable for what we needed to do for our class while also directing the group towards a complete grant application. We were able to use the process of writing a progress report to help us think through our goals and solidify our plan of action.

might imagine in business or politics. But your work in this kind of setting can definitely present such challenges. You may find yourself dealing with questions about how much information to reveal on a particular topic, how to deal with complex budgeting concerns, or how to negotiate your own role in the process. As you write your report, reflect together on any decisions with which you've had to struggle in these areas. Below are some general categories of ethical dilemmas that student writers often face.

- *Honesty.* In our experience, most students' ethical concerns relate to the general concept of honesty. Sometimes students are asked to write documents that contain information they don't consider to be accurate or truthful. One group of students working for a university medical research program found themselves in this difficult position when they had to prepare a budget for inclusion in a report. Because one member was an employee of the office, the group knew that some of the information their contact person asked them to include in the report was not entirely accurate. They had to struggle with how to handle that situation, especially when their contact person told them that this kind of minor dishonesty was a standard practice in the field. He assured them that the ultimate bottom line would not be affected by the slight modification and insisted that this move was critical to the success of the project.

  This kind of budget dilemma is not uncommon. This is not to say that nonprofit agencies and the other kinds of groups with which students typically work on these projects are generally dishonest or that their financial practices are always so complex. The point here is that if you are working with an agency budget, you may feel uncomfortable with some financial practices. This may stem from your own idealistic images of how nonprofit or campus organizations work; it may be a genuine case of inappropriate action; often it's something in between. If you are in this kind of situation, you'll need to be prepared to present your concerns to your contact person to learn as much as you can about how financial matters work and to complete your project successfully. Involve your instructor in this process if you need to.
- *Privacy.* As we mentioned in Chapter Five, privacy matters sometimes come up when students are working with documents in a nonprofit or government agency. Your project may involve working with clients' private records or with sensitive agency information. Although it's best to avoid such situations altogether, sometimes it is impossible to do so. When one student was producing a brochure for a facility that provided housing and education services to new and expectant teenage mothers, she often worked with her contact person at the agency office. She learned about individual girls' experiences and had to make decisions about how much of that information to relate in her documents. Although she believed that the details of some of the stories might be compelling enough to encourage her readers to donate money and services to the facility, she had to balance that goal with concern for the girls' rights to their dignity and privacy. Struggling with this question as she wrote her progress report helped this student to make a decision with which she felt comfortable.

- *Labor issues.* In Chapters One and Five we described the kinds of work that are appropriate for a service-learning writing project. They include such activities as researching, drafting, designing, and editing documents. Occasionally, an agency contact person will misunderstand these parameters and ask students to do other kinds of work. The progress report is a good place to clarify the responsibilities you accepted in your proposal. Earlier we explained the situation that one student who was working with a charter school faced. Though she had agreed to put together an event-management guide for the school, her contact person sometimes pressured her to take on other kinds of duties related to planning a particular event. Because the agency representative did not understand the student's role completely and seemed to confuse it with an internship, she expected the student to do more than was appropriate given the parameters of her class project. By writing an explicit progress report, Wendy was able to clarify these concerns. Though she took on some of the other duties as a volunteer after the class ended because she was genuinely committed to the cause, it was critical that she separate class responsibilities from volunteer ones.
- *Conflicting values.* Though you may work hard at the beginning of your project to choose an agency whose priorities match your own, at some point in the process you may feel that your contact person or even the organization in general does not have the same ideals that you do. One student working on a project for a nature conservation group expected to be writing documents about endangered species and habitat preservation when he agreed to work on the agency's fall newsletter as one of his service-learning projects. He was disappointed and disillusioned when he realized that the primary purpose of the newsletter was not to encourage people to value nature more, but rather to encourage them to donate money. He was frustrated when he was expected to write long columns praising individual donors for their generosity or describing fundraising parties. When he expressed this frustration in his progress report, his contact person reminded him that the readership for the newsletter was already convinced of the value of the natural world. The function of this particular document was to remind these readers of the importance of continuing to give funds to support the agency's other efforts. When they finished their discussion, Bryan felt better about the newsletter. The agency contact person appreciated his enthusiasm and encouraged him to write at least one piece for the publication that emphasized his environmental interests.

As you prepare to write your progress report, work with your group to think about these concerns and to examine your service-learning experience. Consider addressing these issues in your report to find a solution to current or future problems. You may want to employ some of the project management strategies we discussed in Chapter Six. Be sure to talk with your instructor, put your concerns on the table, and work together to develop a plan to deal with these problems as they arise. Keep in mind that issues that may seem quite serious to you may be simple misunderstandings. You may not have the experience or background needed to understand certain policies or procedures. In any case, avoid being judgmental in your written report and in your personal interactions with your contact person. Approach your concerns as an opportunity for learning and reflection. At the same time, don't agree to things that require you to compromise your ethical standards. If you reach an impasse on an ethical issue, ask your instructor for advice on negotiating a way through your difficulty. This kind of reflection gives you a chance to stop and think about the kind of professional you want to be in the future and the kinds of standards you want to uphold in your work.

An Ethical Dilemma and a Rhetorical Challenge

Another reason why we stress ethics as a frame for thinking about your progress report is that students' ethical quandaries can sometimes lead to meaningful changes in an agency culture. For example, a group of University of Arizona students were working on a writing project for a nonprofit agency in Tucson. The three students, who were majoring in social work, engineering, and business, were producing some fundraising documents to help support a program called the Lifebook Project. This project provides children who have been in the foster care system for two or more years with elaborate personal scrapbooks or *lifebooks.* Community volunteers work with family members, teachers, counselors, and others to compile letters, pictures, report cards, and other documents for the children to keep as records of their lives. Many kids in this system don't have the baby books, photo albums, or oral family histories that others take for granted. The project helps these children to develop a better sense of who they are and where they've been through their lives, which helps to build their confidence and self-esteem. The books are also useful to adoptive parents who want to know about a child's background to develop strategies for helping her or him to grow as a person. One of the most important pieces of this project is the narrative, a life story that the volunteer writes with the help of the partners listed above. This story presents even the toughest parts of the child's life—abuse, abandonment, or the death or arrest of a parent—in a straightforward way. The writers reiterate that the child is not to blame for these events and that she or he has the potential for a good future.

These big three-ring notebooks, full of artwork, stickers, stencils, stories, and artifacts, mean a great deal to the children who receive them. Because they believed in this project, the three students decided to create several documents to solicit funding for its support. It had initially been funded by a major national grant for children's programs, but like many projects, was only funded through this source for the first few years. As the end of the grant period loomed, the agency needed help from the students to acquire financial support for this important project.

The students got off to a good start on their work, learning a great deal about the history of the program, the kinds of work that went into each lifebook, and the tremendous effect these books had on their recipients. They also learned that although the project was supposed to be primarily a volunteer effort, one employee, the project coordinator, was actually doing most of the research, writing, and compiling that went into the books. Though her official responsibility was to provide training and resources for community volunteers, she had slowly begun to do most of the other work as well. This didn't seem to be a major problem initially; as long as books were being made and distributed, it seemed, it hardly mattered who created them. But as they researched funding organizations around the nation that might be willing to support the project, the students found that very few organizations whose interests and funding patterns matched their needs were willing to provide money for staff salaries. This is not uncommon: Many grantors and donors will provide funding only for such things as equipment, materials, training resources, and other similar items. This forces agencies to have a basic solid foundation for staffing and overhead expenses and helps funders to narrow their giving to agencies that are viable. It created a real problem for the students, however, because the funding agencies they were targeting would not support the project unless it incorporated legitimate involvement from community members. They came to understand the reason for these regulations and to agree with the principles behind them. Many grantors prefer matching funding—if the community puts in so many resources in the form of hours worked or facilities donated, and so on, the grantor will "match" those contributions with its financial gift. This ensures that the program it is supporting has a value to the community and has ongoing community support that will secure its continued value and existence beyond the funding period. In a sense, it provides a kind of insurance for their investment. In the case of the Lifebook Project, the lack of community involvement was a serious weakness.

The students found themselves in an ethical dilemma. Should they acknowledge this problem in the grant proposal and the funding solicitation letters they were writing? They could simply represent the project in an ideal way—not accurately reflecting the work that the agency representative was doing. After all, that was the official version of the program, and they were not employees. They could also be honest about the situation and try their best to find a few possible funding sources that did not prohibit funding of salaries. They quickly ruled out the first option, deciding together that it would be dishonest to misrepresent the situation. For a short time they considered going with the second option. They researched funding sources and tried to identify some that would be willing to support salaries. They soon realized that this decision would narrow their options too much, leaving them with only one or two possible sources. That limitation would sharply curtail their chances of success.

Finally, after deliberation together and with Melody, they chose to tackle this challenge in their progress report. In the document they updated their successes and the information they had acquired about funding agencies that would be appropriate for the program. They also mentioned that to qualify for a wide range of funds the program would have to be reorganized so that the project coordinator was doing less of the actual book production work and was spending her time training, supervising, and assisting volunteers instead. They presented the written text to their agency contact person in a face-to-face conference at the office. They explained their findings and suggested that the program might be better served by reimagining staff roles. Though they went into the meeting with some apprehensions about the complexity of their status as students, they approached the session in a professional way and supported their position with carefully formulated evidence. The agency director saw their point and found the evidence they presented compelling. Ultimately, she made a decision to recalibrate the project coordinator's daily work with her original job description. The program was strengthened as a result—more books were produced, more volunteers were involved, and more children and adoptive parents were served. In this case the progress report process was not only an opportunity for students to reflect and report on their accomplishments and challenges, but also an opportunity for them to shape the agency and, by extension, their community in a meaningful way.

Writing Workshop Guide for Progress Report

1. Skim the report quickly as if you were the instructor or agency contact person. Make notes about your initial reaction. Does the report immediately give you a clear understanding of where the project stands and when it will be completed? Does it give you a clear sense of what kind of response the writer expects from you? Note these items on a separate piece of paper so that you can compare them with your response after reading the entire report carefully.
2. Read the report introduction. Does this section give you a sense of the bottom line on project progress? Does it allow you to ascertain quickly whether or not the writers are on schedule and if they are facing any major problems? Does it forecast the contents of the report? Suggest revisions that would enhance clarity in these areas.
3. Describe the document's arrangement. Is the report shaped around time blocks, specific tasks, or major objectives? Assess the appropriateness of this selection. Suggest revisions that might make the text easier to understand and navigate.
4. Turn to the report's timeline. Is the chart easy to interpret? Does it give you an immediate sense of the writers' accomplishments and remaining work? Do the writers use effective shading and other design features to convey information?
5. Consider the sections in which the writers discuss their tasks in detail. Does this analysis section give you a clear sense of the kinds of challenges the writers have faced and how they have dealt with them? Put a star by specific details that give you a clear image of the group's activities.
6. In reading the report, can you easily ascertain what has gone well with the project? Place an exclamation mark next to successes summarized by the writers. Offer suggestions for highlighting others that may seem underplayed or hard for the writers to recognize.

7. Apply the following checklist to the report, noting each component with the abbreviation listed next to it.
   a. statement of status: ss
   b. plan modifications: pm
   c. glitches: g
   d. requests for assistance: ra
   e. expressions of appreciation: ea

   Mention components that are overlooked or need more development. Offer comments on the arrangement of the components, suggesting reordering as needed.
8. Does the report include a reference to an ethical dilemma or another major barrier to progress completion? If so, analyze the way in which the writers present this problem. Do they present a clear and blame-free description of the situation? Do they describe a solution—either one they've already enacted or one they plan to implement? Do they articulate the ethical or practical bases for their decisions and present a convincing argument about their appropriateness? Does their plan of action seem focused on future solutions rather than previous errors? How could they revise this section to make their position, priorities, and plan clearer?
9. Does the report include a clear conclusion that recaps accomplishments and forecasts remaining activities and project milestones? Does the report ending seem abrupt or too lengthy?

Evaluating Group Members and Group Dynamics

In addition to assessing your project for your instructor and your contact person in the progress report, you'll want a "reality check" about your group's interpersonal dynamics and each member's contribution at this stage. If you took to heart our advice in Chapter Six, this self-assessment is something you have been doing informally all along. Formalizing this through a self/group evaluation form or a special self/group assessment meeting, however, may help you be more candid about problem group members and more pointed about what these members and the group should do to ensure a smooth final phase of the project.

Start by brainstorming about the requirements for a good project. What do all group members need to do to ensure success in terms of both process and product? These criteria can include the following:

- Share group goals.
- Follow group guidelines.
- Share leadership responsibilities.
- Meet group deadlines.
- Attend group meetings.
- Participate in group meetings and decision making.

- Do fair share of writing or revising.
- Volunteer for unexpected minor tasks.
- Answer group correspondence.
- Communicate openly about problems.
- Ask for feedback.
- Provide useful feedback.
- Listen to other members, and show respect for their ideas.
- Motivate other group members.
- Facilitate other group members' participation.
- Handle the project professionally.

Figure 8.5 shows a self/group evaluation form developed by some of Blake's students; it's based on the criteria they saw as most important for their projects.

Although the sample form states that the students' ratings will be confidential, it also suggests that they should discuss product and process-related problems openly with fellow group members. Your group may decide to develop its own evaluation form, one that elicits honest assessments of group members without quantitative ratings. Whether or not you use a form, sharing your perceptions of how the group is working may open your groupmates' eyes to shortcomings of which they weren't even aware. For example, one group member may be struggling because she finds it difficult to compose her part of the project alone, without immediate input from her peers. Another group member might be taking on too much work because he volunteers for extra tasks before anyone else has a chance to do so. Sharing your assessment also gives you the opportunity to collaboratively develop strategies for rectifying and avoiding future problems. Although you want to be forthcoming in such a meeting, you should also be respectful and constructive. Most people don't respond well to solely negative criticism or to relentless finger-pointing. Whenever possible, approach both problems and solutions as group concerns while still making clear each person's responsibilities. Part of being a good leader is being able to motivate others to raise their standards and rise to the challenge. Ask yourself how you might help a struggling group member get back on track without taking over his tasks.

Developing a self/group evaluation form may come in handy later when you evaluate your project in its final stages. Although you might have to add or change a criterion or two, you may be able to simply change the tense from present to past for some. In Chapter Nine, "Evaluating Your Project," we discuss how you can use your self-assessments to develop part of your more formal evaluation report to your instructor.

Activities

1. Brainstorm individually, listing all the concerns you have about your project. Consider all of your worries, frustrations, and complaints. Also note positive feelings—things that excited you and that you have learned. Put your notes aside for a couple of days and then review them again, identifying concerns that need to be addressed with your groupmates, with your instructor, and

Self/Group Evaluation

Use the table below to rate each member of your group, including yourself, in the following categories on a scale of 1–10, 10 being the highest. Then total each person's score. Below the table or on the back, explain those ratings that need it, summarize the main problems of your group, and propose possible solutions to those problems.

Your ratings will be confidential, though I encourage you to discuss task- and process-oriented problems with your group so that you can together propose solutions.

Your name: Date:

Criteria	Group Members			
Meets group deadlines				
Attends and comes prepared for group meetings				
Actively contributes to meetings and group decision making				
Does fair share of writing and revising				
Submits high-quality drafts following style sheet and other group guidelines				
Listens to other group members; openly communicates with them				
Respects other group members' ideas; facilitates their participation				
Actively solicits feedback; constructively provides feedback to other members				
Total				

Figure 8.5 Self/group evaluation

with your contact person. Give each of these matters some thought before presenting them to the group as a whole in a meeting.

2. As a class, form five small groups. Assign each group to tackle one of the scenarios below. Read through the dilemma assigned to you. Identify the major

ethical issues connected with the dilemma. Work together to draft your would-be progress report; then share your draft and a summary of the concerns you addressed in your discussion of the situation with the class as a whole.

a. You are working on a top-secret, high-tech development project that is going to change the world as we know it. You have joined forces with a major corporation to fabricate your product. This corporation has publicly announced its forthcoming new techno-whatzit. One of your employees has, despite his excellent record and signed loyalty oath, leaked the information to a competitor who will now begin to manufacture the product at about the same time your company will. Sure—you're going to sue the employee for breaking his loyalty oath, but that takes time. What are you going to tell the CEO of the company with which you're working? You have a progress report due next week.

b. Your project is going very well. You are accomplishing exactly what you set out to do for the group with which you're working. You do have one problem, though: Your star engineer, the person on whose reputation your organization's reputation hinges, has recently been jailed for selling drugs. Your company can still do the job as contracted, but it'll have to be done without this person, about whom the CEO of this company asks regularly. How will you handle this in your progress report?

c. You have completed half of your project. The work is going well; you've accomplished what you said you would by this point. You've used 75 percent of the funds allocated for the project, however. Your proposal suggested that you should still have half of the money remaining at this point. You are sure that with some creative maneuvering that will not jeopardize your project's outcome, you can manage to stay within the budget. How do you handle this in your progress report?

d. Your proposal was honest and accurate, based on the information provided in the RFP and all of your discussions with the contracting company's vice president of new projects, the company president's son. Now that you are deep into the project, however, and the contracts are signed, supplies are bought, and so on, you find out that Sonny, the VP, misrepresented a major aspect of the project. This is going to put you over budget and behind schedule. How will you handle this situation? You have to write a progress report to the president.

e. Your project has been going very well. This morning, however, one of your assistants spilled coffee on the disk of data given to you by the company for which you are working. You didn't panic initially because your company motto is "Always Back It Up." When you asked the assistant where she had put the back-up, she said that in this one case she hadn't followed this policy. You know for a fact that your contract company has a copy of the data. You can't possibly complete the project without it. How will you deal with this issue? Will you use the stationery with the company motto on it?

3. Write a one- to two-page narrative about an ethical dilemma you have faced outside of this class, preferably one from a workplace. Analyze your dilemma,

considering its root(s). Was this a question of dishonesty, a breach of privacy, a labor concern, a clash in basic values, or another kind of problem? Work with a small group of classmates to discuss strategies you used to deal with that dilemma and others you might use if you confronted a similar situation in the future.

4. Write a metanalytical memo about your collaboration process. Describe the specific strategies you've used to facilitate your group work, keeping in mind some of the terms and concepts from Chapter Six. Identify tactics that worked well and others that were less successful. Working with your groupmates, develop a list of tips for collaboration that your instructor can share with future classes. Consider posting these tips to the website for this book or for your course.
5. Research your field's professional code of ethics. Respond to it, adding your own slant and augmenting it with additional concerns you might have.

Chapter

Evaluating Your Project

The final two chapters of the book will help you bring your project to a productive close by providing some strategies for evaluating your documents and presenting them formally to audiences. So many times we downplay these two important phases as we rush to complete a work-related project or a writing task due at the end of the term. The following evaluation and presentation activities will enable you to assess what you've learned in the project, ensure that you create the most polished, audience-centered project possible, and articulate your accomplishments (and challenges) to others.

Perhaps you recall the list of components required for a successful service-learning project that we discussed in Chapter One. There we explained that (1) service-learning relates directly to course goals; (2) service-learning addresses a need in the community; (3) service-learning involves developing reciprocal relationships among the college or university and the communities in which it is embedded; and (4) service-learning involves critical reflection on the student's part. In this chapter we'll describe some ways to formalize these goals through the evaluation of your project. Your instructor's evaluation can help you better relate your project to course goals, and assessment by agency members and prospective audience members can help you determine how to make your text more helpful and effective. Also importantly, your deliberation with fellow group members and classmates can help you think through the less obvious effects your texts might enable (e.g., the fostering of certain types of community relationships) and can help you judge your project against shared ethical criteria.

Document evaluation is a critical part of technical and professional communication. In our everyday lives, most of us have read such documents as instructions for assembling a bookshelf or handouts describing campus parking policies that are, at best, difficult to understand. Most of us have also experienced the frustration of creating documents that are designed to instruct or inform—perhaps driving directions to your home or a formal letter of complaint about a company's service—and finding that they do not accomplish the goals for which they were created. Sometimes, no matter how clear an idea is to us, our readers don't get the message we have in mind. Perhaps it's because we are using a vocabulary that is too technical. It may be that we've drawn too heavily on our own unique experiences and left our readers out of the loop. As we explained in Chapters Two and Three, technical and professional communication are inherently audience centered. It makes sense, then, that technical and professional communicators often invite audience members into the process of document creation.

It is not as though the process of assessing your work critically is new for you at this point. Through ongoing workshops, meetings and correspondence with contact persons, teacher/student conferences, and informal assessments, you've likely already done a great deal of evaluating by the time you reach the final stages of the project. You probably have a fairly clear sense of your documents' strengths and weaknesses. As you move toward completing your texts, though, we invite you to participate in some formal evaluation activities that will help you to learn more about the documents you've produced and the effects they are likely to have on audiences.

In the following pages we'll describe usability testing, a kind of document analysis that you may use to identify your documents' strengths and weaknesses. This process allows you to present your texts to *real* readers and document users and to evaluate how well their use of the texts matches your design objectives. We will expand our previous discussions of editing your texts with some detailed strategies for making a final series of editing passes in a structured, systematic way. We'll describe ways in which you can elicit and incorporate feedback from your agency contact person productively. We'll expand our discussion of the use of style sheets, consider ways in which you can make the final editing phase a collaborative process, and explain how you can put your findings throughout these processes into an important workplace document, the evaluation report.

Obtaining Feedback from Members of Your Targeted Audience

Later we'll go over the procedure for workshopping your texts with and negotiating your feedback from your supervisors (especially those at the agency) and your peers. Now we want to discuss getting feedback from perhaps the most rhetorically important audience—the targeted users of your texts. They are the people who will ultimately determine whether or not your texts achieve their purposes.

You may have ready access to these users; for example, you may be producing a text to be used primarily by members of the organization or by a campus audience with whom you are familiar or of which you are a member. In these cases you probably won't have much difficulty acquiring feedback from prospective users. This is one advantage of doing a project with a campus organization—more than likely your audience will be students like yourselves. Even if you could be considered a member of the targeted audience, you should still solicit feedback from prospective readers outside of your group. Because you are researching the topic and otherwise working for the organization on the writing project, you aren't in the same position as your audience. When a group of advanced professional communication students took on the task of creating an online tutorial for a web-authoring tool, they drew on the members of the group who fit their audience's profile—student lab users who were new to creating web pages. As group members learned more about the web-authoring tool to guide their audience through the process of using it, however, they became more expert than their audience and therefore needed to go outside their group for this audience's perspective.

You may be producing texts for an external group of users in the community. In some cases, such audiences are very willing to provide feedback, as with potential pet adopters who are targeted by a Humane Society website. In other cases, your audience might be harder to pinpoint or to access, as with cancer physicians and patients or users who want to remain confidential. If you are addressing a community audience, ask your contact person to identify some appropriate potential users with whom you can work during this process. Also ask her or him to let you know about any confidentiality concerns you should keep in mind. Finding representative readers/users is crucial to obtaining rhetorically sound responses. In certain cases, you may have limited or no access to members of the prospective audience, however, and may need to ask classmates or others to serve as testees. In Other Voices 9.A, Karla Kitalong provides suggestions for handling the testing this way. Another option in such an instance or even when you can find prospective users is to design a user testing guide for the organization to use later down the road, perhaps when they are making the final adjustments before distributing the texts or even after the final texts reach the hands of the audience. A group producing a volunteer training manual for an organization, for example, might want to leave the organization with directions for conducting user tests during the next volunteer training period, or a group producing instructions for clients in a clinical setting could develop a guide that enabled clinical workers to administer user tests on clients in the actual setting. Such testing might help the organization identify glitches with the documents that your regular workshopping didn't pick up.

It's probably a good idea to garner your audience's feedback at several stages of your project, one such stage being when you are analyzing this audience and determining its needs. For example, the group of students who were creating the web tutorial wisely asked other novice lab users what web page elements (e.g., pictures, links, color, frames) they would want to learn to create. One of the most important stages at which to get prospective readers' comments is after you have a complete, usable draft (including design elements). At this stage, they can offer you more specific diagnoses of which parts work well and which need improvement. You can collect such feedback formally or informally, through structured usability testing (which we describe next) or simply by giving your text to prospective users and asking them what they think. Either way, you should encourage the users to be picky and to express their opinions freely. Remember how cautious your classmates were when you had your first in-class writing workshop? To avoid this pattern in interactions with your targeted audience members, make your inquiries specific. Your group will want to decide about what, in particular, you want feedback and, thus, how to direct the users' critique. As we discuss below, asking your readers specific questions about your text, either orally or in writing, is a good way to ensure helpful responses. You can even show them two different versions of a text and ask them which one they prefer and why.

Conducting a Usability Test

Usability testing is a structured way to gather feedback from your targeted audience. It is used commonly in the design of texts such as instructions and documentation, particularly in the computer industry. The efficacy of user documentation for new software, for example, is often tested along with the design of the software itself. In fact, many companies, such as IBM and Microsoft, employ user-testing specialists or technical writers who perform usability tests as part of their jobs. Conducting usability testing usually saves such companies money by helping them iron out the glitches in their texts, making them more appealing and satisfying to customers.

Usability tests can be helpful assessment tools for other kinds of texts as well, including primarily informative ones such as websites and primarily persuasive ones such as fundraising packets. If you have the time and uncomplicated access to members of your targeted audience, we recommend that you conduct a small-scale usability test. This experience will not only improve your service-learning text, but also be a valuable addition to your list of qualifications as a technical writer; candidates for technical writing positions who have conducted and can talk about usability tests should have an advantage over their competitors.

The basic technique of user testing is to have members of your prospective audience—preferably at least five or six, although even a couple could provide helpful feedback—read and use your document without any help or interference from you. You want them to use the text in the most realistic setting possible. The basic purpose is to determine how well your text achieves its purpose and how usable it is. More specifically, you want to find out, from the user's perspective, which elements of your text are effective and which could be more user-friendly. For example, where do readers need more explanation, become confused by visual elements, or have a hard time following the arrangement? To make these determinations, you want to gather as much information about the testees' understanding and use of your text as possible.

When collecting users' responses, remember not to help testees or otherwise interfere with their use of your text. With this in mind, you can gather this feedback in a number of ways, including the following.

- Observe users in action with the text, noting where they become confused, are turned off, or have problems finding information;
- Tape-record or videotape users in action, asking them to voice their thoughts as they go along; this is called a **think-aloud protocol;**
- Interview or administer a short questionnaire to testees after they use the text(s), perhaps asking them to comment on particular features or asking them to explain why they had problems at certain points;
- Create a user-testing guide that asks students to answer a few questions while they perform the process.

Some companies and universities house sophisticated usability testing labs that can videotape and audiotape users, save a log of moves users make on computers, and gather their responses in other ways. You don't need to do all of this to generate plenty of useful feedback, however: Most of our students conducting usability tests have simply taken observation notes and developed a short set of questions. Whatever techniques you use, explain to the testees any necessary background information, the purpose of the usability testing, what to expect from the testing, and approximately how long the testing will last. Put all of this in writing as part of a testing guide.

You might have more than one group member take observation notes, so that you're not relying on only what one person notices. Also, you might divide your paper into two columns; write down where users seem to be having problems in one column and what you think might be causing the problems in the other. If you can't determine the causes from observing, you can always ask the testees about them after the test. Many users will need some guidance about what to do and what kinds of feedback to provide. If you use such a testing guide, limit the number of questions you ask; we recommend no more than five or six. Make questions for a guide and/or questionnaire open-ended rather than closed or leading questions. Questions that invite yes/no answers—such as "Does the introduction give you all of the background information you need?"—are unlikely to generate helpful feedback. Leading questions, such as "Do the pictures motivate you to read the brochure?" are also unlikely to generate frank, helpful answers. We are not saying that all of your questions should be short-answer questions or that all of them should relate to specific elements of your document. You might consider including a rating question, such as "On scale of one to ten, ten being the highest, how would you rate the design of the drawings in the step-by-step instructions? Explain your rating." You might also consider combining specific, pointed questions with more general ones because the latter type can give testees more opportunity to comment on what they see as important. General questions such as "What aspects of this website make you want to read it? What elements turn you off?" let the reader decide what to focus on.

Consider the post-test questionnaire produced by the students creating the web-authoring tutorial. This questionnaire includes the following combination of general and specific open-ended questions.

1. On a scale of one to ten, ten being the most helpful, how would you rate the task of replicating a specific web page as a way to learn web page authoring? Explain.
2. In the sections that gave you difficulty, what about the tutorial caused these problems (e.g., word choice, arrangement, amount of information)?
3. Which images were the most helpful? Which were less helpful or not necessary? Where should an image be added?
4. How did the framed links ease or complicate your navigation through the tutorial?
5. How did taking this tutorial enhance or inhibit your confidence and knowledge in using an HTML editor?

Notice how the questions are neither too taxing nor so simple that they could be answered by yes or no. Our only major critique is that question two might be too difficult to answer in retrospect rather than during use.

Another useful questioning strategy is to have the testees answer questions before, during, and after the testing. Before and after questions enable you to measure directly how your document affected users and how well it answered their questions and lived up to their expectations. Asking testees what questions they have about the campus film festival before they view a website about it, for example, will help them determine more easily how well the website answered their questions. Giving testees questions to think about as they actually use the document can also help them record and remember their impressions and suggestions. The questions below are taken from a test guide developed by student authors of a brochure for a campus organization. Note how one set of before, during, and after questions asks about the brochure's persuasiveness, showing how you can ask about motivation as well as ease of use and comprehension.

Before

1. What kind of information do you expect to learn about Phi Alpha Delta?
2. What might convince you to join Phi Alpha Delta?

During

3. Draw wavy lines under confusing sentences or sentences you have to reread.
4. Point out where the style seems too friendly or too formal.
5. What about the design turns you off?

After

6. What questions do you still have? What information could be easier to find?
7. What about the brochure makes you want to join the organization?

Before you finalize your list of questions, you might run them by your instructor and/or exchange them for those of another group in the class. You could even try them on a single audience member to make sure they'll generate useful responses. If you use a written questionnaire, be sure to state at the top how much time it should take and to leave plenty of space after each question for the testee's response.

OTHER VOICES 9.A Karla Kitalong

Karla Kitalong, Ph.D., is an assistant professor of technical communication in the English department of the University of Central Florida. She has written several articles and successful grants related to usability testing.

I have overseen technical communication projects involving "real" organizations for several years, and one of the recurring *touchy* issues that I've seen has been what I call an *expectation gap*, the often considerable discrepancy between how students visualize the experience of working with a contact person and what the contact person expects of the interaction.

(continued)

(continued)

One common problem scenario involves how best to present information. When students and contact persons reached an impasse in the project development process, and I was called in to offer mediation, I would find myself facing what might seem to be a brick wall; in one case, in particular, a student I'll call Jana was creating an information packet for a private, non-profit social service organization. The packet was to be sent to potential donors who might wish to contribute to the organization. The agency representative, David, had explained to the student all of the constraints of the project, including cost and time, and had provided rough drafts of the content, so Jana thought she understood both the opportunities and the limitations of the project.

By the designated deadline, Jana had created several alternative designs for the donor information packet. However, David rejected each of them, preferring that the student emulate a design that he had picked up as a sample at a conference several years before, but that he had neglected to present at the outset of the project. Jana judged the sponsor's sample design to be outdated and banal. Ultimately, she lost her enthusiasm for the project because she felt that the sponsor was not seriously considering her ideas. She ended up creating a packet based on the sponsor's design, but her heart wasn't in it, and she lost the opportunity to extend her technical and business skills.

Now, students' so-called "real world" experience is often limited; in this case, Jana lacked both negotiating skills and confidence in the knowledge she had gained in her coursework. She was unable to persuade David to seriously consider her design ideas.

To their credit, however, students—including Jana—often know the capabilities of relevant software, or have a vested interest in learning it. In addition, by the time they get to a senior project course, they should be experienced in writing and document design. Finally, they have usually been exposed to innovative or effective new ways to present material that clients wouldn't necessarily be able to visualize or evaluate. On the other hand, sponsors can be expected to know the audience's characteristics, needs, and preferences, and to understand the context in which a document will be used. And, after all, their company's or organization's name will be on the product. In the best-case scenario, student and organization strengths complement each other to result in a more effective project. In the worst-case scenario, students and sponsors come to blows—at least figuratively—over how to proceed.

After Jana's unsatisfying experience, I decided that students needed an authoritative way to talk to sponsors, so I instituted a required mini-usability test just after mid-semester, as the students were completing the first useful drafts of their projects. The students were required to write a task-oriented test scenario in which they specified the audience members' attributes and the proposed context for the document's use. Then, the other members of the class assumed the personas outlined in the scenario to give the students desired feedback on their project drafts. The system was not ideal; for example, when students were working on highly technical projects, their classmates were unable to provide good feedback. But the usability test did give students some external validation for their decisions, which they could take to their sponsors when necessary. In addition, the unit gave students some basic familiarity with usability testing procedures, which served them well in their first jobs.

Some good sources for usability testing information can be found on the Society for Technical Communication's Usability SIG website (http://stc.org/pics/usability/), the Jakob Nielsen's site (http://www.useit.com), and at the Usability Professionals' Association website (http://www.upassoc.org/).

The User-Test Report

After you conduct user testing or gather less formal feedback from your targeted audience, you'll want to, as a group, synthesize and interpret this feedback. Perhaps one group member took observation notes and two others administered interviews or questionnaires. Your group won't be able to capitalize fully on your testees' responses unless you analyze them together. You might notice patterns across the responses, or you might have contradictory responses that you need to sort out (perhaps with the help of your instructor). In some cases, your testees might point out a problem with your document but not explain the cause or suggest a solution. All of this will be easier to process as a group. Next, you'll also want to form a revision plan collectively, based on your testees' feedback. This involves determining the changes you want to make and who will make them.

Your instructor may ask you to describe the results of your user testing and your revision plan based on those results in a short **user-test report,** probably in memo form. Such a report would not only update your instructor about this part of your project, but also would give your group a formal mechanism for accomplishing the tasks we describe in the previous paragraph (as well as a record of your work).

Your user-test report should include the following parts:

- an informative subject line
- a purpose statement that identifies the rhetorical purpose of the report and perhaps forecasts what it will cover
- an overview of the testing methods, including the who, when, where, and how of the testing
- a summary of the test results (e.g., your testees' responses, your observations, etc.) and any necessary discussion of these
- a summary of the changes you will make based on the test results (this could be considered the conclusion and recommendation section of your report)
- appendixes with the interview questions/questionnaire, if you used one, and, if your instructor so requests, the actual notes and responses from the testing.

Most of the emphasis should be on your test results and planned changes. The methods section should be fairly brief—maybe one or two paragraphs—rather than a play-by-play narrative of what happened.

Figure 9.1 shows the user-test report written by the previously mentioned students who designed the Phi Alpha Delta brochure. After stating the report's purpose and describing the test's methods, the writers arrange their findings according to the specific questions they address, an arrangement that helps their reader (Blake) avoid flipping back and forth from the results to the questions in the attached questionnaire. Note how much helpful information they generated from administering their usability test to just six people, something they did simultaneously in less than an hour. This group might have also asked the first two questions during the research and the audience analysis phases earlier in the project.

MEMORANDUM
TO: Professor Blake Scott
FROM: Kathy, Alan, Bob, and Sam
DATE: November 22, 1999
SUBJECT: Usability Testing of Phi Alpha Delta Informative Brochure

This memo presents the methods and results of our usability test as well as the revisions we plan to make based on the results.

Methods

We conducted the usability test to discover how our audience members—undergraduate students interested in law school—would react to our brochure about Phi Alpha Delta (PAD). We wanted to get their assessment of the brochure's information, persuasive appeal, and design so that we would have a better idea what and how to revise.

To select appropriate testees, we found a number of undergraduate prelaw students in our classes who had heard about Phi Alpha Delta but didn't know much about it. Six students agreed to meet us at the Student Union on November 20 at 1:00 p.m. to read our brochure and answer the questions on our guide. We decided on this site because it is a place where many students learn about clubs and organizations and because our audience will probably read our brochure in such a busy atmosphere. The tests took about an hour in all, as a few of the students were a little late. Most students took around 15 minutes.

We had them answer the following questions before, during, and after reading the brochure:

Before
1) What kind of information do you expect to learn about Phi Alpha Delta?
2) What might convince you to join Phil Alpha Delta?

During
3) Draw wavy lines under confusing or less-than-effective sentences.
4) Point out where the style or design seems too friendly or too formal.
5) What about the brochure turns you off?

After
6) What questions do you still have? What information could be easier to find?
7) What about the brochure makes you want to join the organization?

We found most of the feedback the testees gave us to be valuable; their answers seemed honest and thoughtful. Especially helpful were the patterns we discovered across their answers. For example, three testees mentioned that they wondered about the dues for joining the organization, information we had left out to avoid turning readers off.

(continued)

Figure 9.1 User-test report

Figure 9.1 *(continued)*

Results

We've summarized the six testees' answers under the questions they responded to.

Desired Information about PAD

What PAD is and does; when it meets; who its members and officers are; who can join; how to join; the costs and obligations of joining; the benefits of joining; law school admission/LSAT information; mock trial information; contact information

Persuaders to Join

Networking opportunities; meeting law school officers; gaining knowledge about law schools and the LSAT; meeting people with similar interests; other benefits

Confusing/Ineffective Sentences

"I've felt PAD is like a family from the start." (cheesy)
"PAD . . . is a professional service organization promoting professional achievement within the legal profession." (repetitious)

Parts Too Friendly or Formal

The white background with the crest seems too formal; the font, especially of headings, too formal; parts of the last section seem too informal

Turn-offs

The contact information seems buried on the front panel; the color scheme has too many shades of the same color

Missing or Inaccessible Information

Dues/other costs missing; officer information missing; contact information could be more accessible

Persuasive Information

Networking and educational opportunities; social aspects; list of member benefits; professional appearance of brochure, including glossy paper and use of color

Revisions

We plan to make the following revisions based on the input we received from the testing. Most of these have to do with improving the accessibility of some information, adding other information to meet readers' expectations, and improving the design to create a more inviting ethos.

1) Revise two problem sentences to avoid repetition and create a uniform tone that is friendly but professional.

(continued)

Figure 9.1 *(continued)*

2) Move contact information to the back panel of the brochure (where readers will more likely look for it) and put it in a larger font and box.
3) Tell readers that officer information is listed on the website and give website address with contact info.
4) Add dues information, but describe as affordable and mention only after benefits.
5) Change font of headings from Garamond to Arial.
6) Stick with organization's color for crest, but get rid of light pink. Change headings to dark purple for better coherence.
7) Add more motivational visuals, such as members at a meeting with a guest speaker or at a mock trial.

The usability testing will greatly contribute to the overall success of our final project, as the input from six different sources pointed out ethos, accessibility, and information problems that had gone unnoticed.

Because you produced your service-learning texts not just for their eventual users, but also for your course and sponsoring agency, your evaluation should be extended to these two groups as well. This part of your evaluation process should be easier than user testing because you already have practice with in-class writing workshops and because your supervisors and peers have been keeping track of your project throughout the semester.

Coordinating Workshop Sessions with Contact Persons and Classmates

In addition to soliciting feedback from members of your eventual community audience (or at least people reading from their perspective), you'll want to workshop your texts with supervisors and classmates. Your agency supervisors and instructor are, after all, part of your audience and, in a sense, partners in your collaboration. If you've already gathered feedback from community readers, you can use this to shape the next phase of your assessment.

In Chapter Two we discussed Michael Polanyi's model of focal and subsidiary knowledge. To review, focal knowledge is gained by analyzing individual pieces of a process or product, and subsidiary knowledge involves understanding how the pieces relate to one another. In planning your workshops, try to gather feedback from both perspectives. Using the writing workshop guides in this book as examples, formulate questions that ask reviewers to comment on big-picture concerns such as relevancy to one's audience, overall accessibility, coherence of design, and tone, as well as more focal concerns such as the design of certain visuals, particular word choices, and the sections of the document. We've found it useful to begin simply by having the reviewer read/use the document and give an initial assessment, perhaps underlining strong areas and drawing wavy lines under confusing or weak ones.

You may already have a good sense of the areas of the document that need improvement. You may choose to present several alternative versions of the text or its component to your workshop participants, and ask them to select the one that they view as most effective. Use your workshops to focus on a set of specific concerns, and give your participants a set of clear requests. On the same sheet, you might write out brief descriptions of the text's rhetorical situation (i.e., audiences, purposes, and uses) so that your reviewers can make more rhetorically informed comments.

If you are meeting with your contact person for a formal review session, we suggest that you send your materials and your questions to her or him a few days before you plan to sit down for your conversation. This will enable you to more efficiently utilize your time together and to possibly allow her or him to gather input from other staff members. If you are not able to do this, or if she or he does not have time to review the materials in advance, have a plan to keep the meeting moving at a productive pace. Prepare specific questions about your document, and use post-its to mark elements that require special attention. Be sure that each participant has a copy (with numbered pages) so that everyone can follow along and contribute. Bring a copy of any previous correspondence (e.g., a progress report) that might be relevant to the meeting.

Don't expect your contact person to offer as much feedback as your instructor, especially if you're working with a business; your contact person is not a writing teacher and is probably working on numerous other projects as well. You're more likely to receive short directives than coaching advice or detailed comments, and these directives won't always be in writing. Although many contact people have very specific criteria for what the text should look like, others do not and, therefore, need to be prompted with questions (not unlike the ones you developed for the user testing, perhaps).

In our experience, although most contact people offer short directives, such as "change the color scheme and add this logo" or "replace that term with this one," these directives usually make rhetorical sense and are often fairly easy to implement. Just listen carefully, take good notes, and perhaps review your notes with the contact person to ensure that you understood her or him correctly. If the directives come across as curt or overly negative, try not to take it personally and remember that you and your reviewers are working toward the same ultimate goal. Your contact person may suggest a revision that contradicts other feedback you've received from your test participants, classmates, or instructor. Discuss such a gap with your instructor, keeping in mind that your contact person probably knows the material and understands the audience best.

You may also receive unexpected, more substantial suggestions that take the project in a slightly different direction or require too much time to implement. This probably won't happen if you've stayed in close contact with your agency representative during the course of the project, but occasionally a representative will think of a new consideration or change her or his mind about something in the final stages. Often, a supervisor less familiar with the project's progression becomes involved in the evaluation at this point. You may be told, for example, to add a new set of pages to a website or to incorporate a different set of images into a recruiting packet. If something like this happens, don't panic or lose your cool. Instead, carefully listen to the suggestions, diplomatically explain your course deadline for the assignment, and offer to do what you can. You might be able to develop a compromise or an easier way of creating the same effect. Ask your instructor to help you formulate a reasonable revision plan. However your workshop goes and whatever feedback you receive, you'll want to follow up on your meeting with a note or email of thanks for the contact person's time and help.

Once your group has had an opportunity to absorb the information from the meeting, work together to synthesize this feedback with the other feedback you've gathered and to create a **revision plan** or mini-proposal for your final round of revision. You remember from Chapter Five that proposals include solutions—in this case, the changes you plan to make—and management plans for carrying out the solutions. Your revision plan should include a detailed, prioritized list of your planned revisions for the overall document and for each major part or element. It should also clarify which members are responsible for which tasks and how your group will collaborate. You could even put this in a checklist format. Involving everyone in formulating the revision plan will help ensure that each group member understands her or his responsibilities. Your instructor will probably want to see your revision plan. We suggest also giving a copy to your agency contacts in a follow-up meeting or along with a thank-you message to assure them that you're incorporating their suggestions.

At this point you are familiar with the process of peer editing and probably have found good strategies for working with your classmates to improve your texts. If time permits, you may want to have peer review sessions both before and after your workshop with your contact person(s). You can ask your classmates to evaluate the changes you've made and even to give feedback on your revision plan. If you have time for a workshop immediately before your project deadline, you may also want to ask your peers to participate in some of the tasks we discuss in the next sections, including copyediting and proofreading.

Copyediting and Proofreading

At the end of Chapter Six and in Chapter Seven, we discuss approaching revision and editing as multidimensional, time-consuming endeavors for your group but ones that can be aided by creating a detailed style sheet. Here, in this chapter about evaluating your work, we want to expand our discussion of the final rounds of revision and editing. In *Technical Editing,* Carolyn Rude discusses the differences between comprehensive editing (what we've been calling revision), copyediting, and proofreading (13–14, 176). **Comprehensive editing,** which you have likely been doing all along, involves assessing how well the document accomplishes its purposes and accommodates its audiences. In other words, it assesses the rhetorical appropriateness of the document.

Copyediting, which can also be a long-term process but usually occurs after a full draft has been created, more narrowly involves editing for completeness and

OTHER VOICES 9.B Chere Peguesse

Chere Peguesse, PhD, is an assistant professor of English at Valdosta State University in Valdosta, Georgia. She directs the South Georgia Writing Project.

Working with Outside Reviewers

In 1999, I was teaching an evening business writing course in which I assigned students to work in teams to create documents for local campus groups, nonprofit agencies, or businesses. One team decided to design an organ donation brochure for the Wellness Center located in our university recreation facility. The three students were young women in the same sorority. They had just completed a Family Studies course in which they made a presentation on organ donation and were therefore familiar with the research. They already had ideas about the design of the brochure and chose this project because it would be easy. They decided who would be responsible for certain tasks and set to work with no problems. During the first class session in the computer lab, they quickly produced the document, complete with colors and images. A requirement of the project was to obtain an outside reviewer to provide feedback and evaluation, and it was this outside review that provided the deepened learning experience that kept this group revising their document for the following seven weeks.

To evaluate their work, the team chose the director of the Wellness Center who had a Ph.D. in nutrition. Her critique of their brochure upset the team members not because it was harsh, but because they had to reconsider the assumption that their document was done. For example, the caption on the front of the brochure read "Be A Hero: Save a Life" in bold red and black letters. Below this was an icon of a stylized pumping heart. The director pointed out that there are people who choose not to donate their organs for religious or other valid reasons. The caption implied that those people couldn't be considered heroic and the wording might offend others because it sounded preachy. The team members brought the letter to me and asked me if they could switch outside reviewers. When I asked them why, they said that the director's evaluation was unfair. Again I asked why. They realized that "because she didn't like it and we would have to do it over" wasn't a valid answer and I didn't have to say a word. In fact, I got excited that they wisely picked someone who took a real and critical interest in their fine work. I was sure that the critique would help them improve the document so that it satisfied the director's needs.

This is part of what the project is all about: putting aside your own agenda and listening carefully to a contact person's needs and then attempting to fulfill those needs. These students recognized the shift in emphasis only because they were confronted with the complexities of accurately identifying and satisfying several audiences. They revised their brochure four more times, submitting it each time to the director, and the project did indeed take up the rest of the term. The director, in her generosity and genuine interest in the project, reviewed and responded to all four versions of their brochure though the project required only a one-page evaluation from her to be included in the final report.

The director's critique also made these students rethink their own positions regarding organ donation. Two of the women realized that they themselves would not want to donate their organs even though they supported the idea in general. In redesigning the brochure, they realized that they were not a part of its audience. It became an ethical dilemma for them to be persuading others to do something they could not bring themselves to do. The issue remained unresolved by the end of the project, but the fact that it made them aware of the implications of their design cannot be overvalued in my mind.

accuracy, correctness, and consistency. Style sheets are designed to ensure just these qualities, and are therefore quite useful as references during copyediting. Copyediting for verbal and visual consistency is especially important if your group used the divide-and-conquer method of collaboration to produce parts of the text. Even when all of you are following a common style sheet and using a common template, it is all too easy to create inconsistencies in elements like font size, spacing, abbreviations, use of commas, and patterns in sentence structure. Writing workshop guides usually address comprehensive and copyediting issues. According to Rude, print-based copyediting is still standard practice, especially for heavy edits, although more and more technical editing is being done online (86–87). Many editors, including ourselves, work with both hard and electronic copy. Online editing is especially common for electronic texts that are meant to be read/used online. One advantage of editing hard copy is the ability to use standard, easily recognizable mark-up symbols. We've also found it more difficult to catch small design and stylistic errors on the screen. On the other hand, online editing can cut down on some tedious tasks with such functions as "find and replace" and "document comparison." Some editors make changes directly to soft copy, perhaps saving the file under a new name or tracking changes. Others embed changes and queries in another layer of the document.

Proofreading, to which many people reduce editing, happens at the end of the text production process after copyediting changes and other finishing touches have been made. In fact, as Rude explains, one purpose of proofreading is to verify that copyediting changes were made correctly. Proofreading also serves as one last chance to catch any minor errors, including those that might have been an inadvertent effect of the editing process. If the text is an electronic one, we suggest that you proofread it at least once in electronic form as it will appear to readers.

As we mention in earlier chapters, all group members should participate in all three types of editing. We recommend that you collectively discuss the kinds of things for which you will look, the mistakes you find, and the changes you plan to make so that group members don't make unnecessary changes or introduce mistakes.

Because copyediting involves checking a number of elements, and because you likely will have much copyediting to do in a short amount of time, your group might also discuss ways to split up the editing, at least for the first round. After the first round, you might switch tasks to check each other's work or go over each person's copyediting as a group. If each group is assigned multiple tasks, or if you decide against this divide-and-conquer method, try not to copyedit for too many types of things at once. Any professional copyeditor would tell you that careful editing involves multiple editing passes. After all, we can only concentrate on so much at once. At the very least, edit verbal and visual elements separately.

If you split up part of the editing tasks, it's crucial that your group have a system for marking, tracking, sharing, and verifying changes. When we began to edit the first few chapters of this book, we didn't yet have such a system and, as a result, we had a difficult time distinguishing between drafts, deciphering each other's changes and queries, and combining both sets of changes to the same file.

As we discuss in Chapter Six, basic word processing programs such as Microsoft Word can equip you with a range of tools for collaborative electronic editing. First, you might develop a file-naming system that indicates the draft and the editor. If you save using the "versions" function (under the "File" pull-down menu), you can indicate the date/time, editor, and any comments about the draft and your editing. Second, decide how you want to mark and record the editing changes you make. Although the "tracking changes" function can be useful, especially because it enables the writer to reject or accept the changes, the marked-out elements can become unwieldy, especially if substantial changes are made. Making editing changes directly, saving them under a new file name, and then using the "document comparison" function enables you to show changes and preserve a clean, uncluttered copy. For questions or larger comments about the draft, you can use the "comment" function (under the "Insert" pull-down menu) to insert hidden text indicated by highlighting; this functions somewhat like a post-it note for hard copy. Third, develop a system for transmitting edited files to each other. After the person in charge of stylistic appropriateness is finished, for example, she might send the document to another group member to be edited for grammar and mechanics, who might then send it to another group member to be edited for visual consistency. Finally, you might verify changes collaboratively by comparing the final edited copy to the initial one; all of each editor's changes should show up.

If you took our advice in Chapter Seven, you have updated your style sheet and grouped the guidelines into larger categories such as *design elements, grammar and mechanics,* and *illustrations.* Such an arrangement might suggest a way to divide up the copyediting work. The following list shows one way to split up the editing passes.

- completeness and accuracy
- stylistic appropriateness
- grammar and mechanics (this is where you edit for correctness)
- font
- spacing
- visuals/graphics.

The first two focuses involve more than just copyediting, but also editing in light of the audience and other rhetorical concerns. All but the first on the list involve editing for consistency—perhaps the main job of copyediting. However you break up and assign the copyediting tasks, do so strategically, considering who in the group has the best command of grammar and mechanics, who has the best visual eye, who knows the most about the subject matter, and so on. Also, because each group member will probably have multiple focuses, try to assign similar focuses (e.g., font and spacing) to the same person.

After you split up the copyediting tasks, have each person brainstorm about the kinds of things she or he will check. Start with the guidelines on your style sheet. Here's a partial list of questions you might ask about the categories listed above.

1. Completeness and accuracy—Is the information correct from the agency's standpoint? Where does the information need to be more developed for the audience? Where is the presentation of the information ethically questionable?

2. Stylistic appropriateness—Where could the style be more concise and active? Where could the style be more cohesive? Where could the most important ideas be emphasized better? Where could the tone be more appropriate for the audience and purposes?
3. Grammar and mechanics—Are the spelling and punctuation correct? Are the sentences grammatically correct? Where could sentence structure, words, and punctuation better clarify meaning? Are spelling, abbreviations, capitalization, and punctuation consistent?
4. Font—Are font types, sizes, styles, and colors consistent within the main text, visuals, and each level of heading?
5. Spacing—Are margins, gutters (between columns), indentions, leading, spacing before and after headings, and spacing between words consistent?
6. Visuals—Are sizes, titles, captions, labels, surrounding space, and the general design consistent? Do they follow conventional designs and ethical standards?

If you decide to edit hard copies, we suggest using standard copymarking symbols known by everyone in the group; this will eliminate confusion over changes and notations and will give you useful practice. See the detailed charts of copymarking symbols for words/letters, punctuation, and spacing/position in Chapter Four of Rude's *Technical Editing.* You'll probably only need a handful of these symbols. If such symbols are new to group members, you might even include them (with examples) as part of your style sheet.

When copyediting or proofreading, be slow and methodical. This should not be a quick and easy job, but a painstaking one (sorry). Try different tricks to keep your eyes from glossing over mistakes. For editing visual elements, it might help to break the page up into chunks or to edit backwards from bottom to top and right to left. When editing for spelling, try not to read words and phrases as units, but as groups of individual letters. When editing for grammar and punctuation, we recommend reading for the meaning of the sentences. You are more likely to find many grammatical mistakes, such as a misplaced modifier, if you are following how the parts of the sentence relate to one another. As Rude explains, "Editing for style is editing for meaning" (252).

Because proofreading is not as extensive as copyediting, we recommend that each person in the group proofreads the entire document; the more eyes that check the text the better. Also, consider having others proofread the document as well; outsiders will bring fresh points of view. As with copyediting, you might come together as a group to synthesize and enter proofreading changes to ensure that you are all working with the same final copy.

Evaluation Report— Genre and Rhetorical Situation

Reports come in many sizes and have many purposes, from a two-page lab report that explains the methods and findings of a short procedure to a several-hundred-page environmental impact study of how building a new highway would affect the ecology of the surrounding area. We've already covered the short trip and user-test reports and the medium-length progress report. Now we turn to an example of a longer, more formal report.

Along with a proposal and one or more progress reports, many long-term projects require a **completion report.** The results and conclusions of **empirical research projects,** for example, are usually shared with others in formal reports or journal articles. Professionals hired to solve problems or provide services for an organization usually give their clients final reports that document their work. Even a company's **annual report** could be considered a completion report about its activities and accomplishments during the past year. As these examples suggest, completion reports can be considered to be primarily informative, explaining to readers how the project went, what it produced, and, sometimes, what actions they should take. Completion reports that give readers recommendations have a strong secondary persuasive aim. Because they constitute the final and most complete written record of the project, completion reports are often kept on file as reference documents.

For our last suggested assignment and genre focus, we will discuss a variation of a completion report that we call the **evaluation report.** Like other completion reports, this assignment discusses and reflects on how the project went and what it produced. This report differs from most completion reports in that it is primarily persuasive rather than informative, however; your main job is to make a credible and compelling assessment of your project, ending with a grade request. This evaluation report could be considered a hybrid of a completion report, **personnel report,** and **recommendation report.** Like a personnel report, it also evaluates the performance of workers (you) for a supervisor (the instructor), only in this case the writers are evaluating themselves. It also ends with a recommendation—that the reader accept the writers' evaluation and reward them with a specific grade. One common type of recommendation report, a **feasibility report,** documents a study that evaluates several alternative courses of action according to specific criteria (e.g., cost, feasibility, efficacy) and then makes a recommendation about which course the readers should take.

Your audience likely will use your report in several ways that reflect its hybrid nature. First, your instructor will use it to obtain a final update on your project and an overall picture of how it went from your perspective. Using it this way, your instructor may compare this report to your earlier proposal and progress report. More importantly, your instructor will use your report as an evaluation tool, considering your argument and recommendation along with her/his own evaluation and probably that of your agency contact person. Because your instructor will grade your evaluation report as its own assignment *and* use it to evaluate your larger service-learning project, the stakes of this final major assignment are doubly high.

The evaluation report builds an assessment around a set of audience-centered criteria, in this case about what makes a successful, effective project. We suggest writing this report for just one of your supervisors—your instructor—because your contact person probably won't care about your self-evaluation. Keep in mind your instructor's roles as a professional/technical communication instructor as well as your project supervisor. Unlike many managers, your instructor will likely be as concerned about your learning process and project management as she or he is with your final written product for the agency.

To develop audience-centered criteria for your evaluation, you should spend some time discussing how your instructor would define a successful project in terms of both the process and the product. For example, we define a successful process as one in which our students proactively respond to their obstacles and learn how to collaborate productively and manage their projects efficiently. We assess our students' products largely by how well they seem to achieve their intended purposes and how well suited they seem for their audiences and uses. You can identify other qualities of effective professional writing—such as accessibility, cohesion, and verbal–visual integration—by reviewing the key concepts discussed in the specific genre sections of this textbook's chapters. After identifying your instructor's likely criteria, you might rank these criteria to decide which ones to focus on. In addition to considering your instructor's hierarchy of values, consider too which criteria seem most relevant to your project and in which ones you excelled. Your instructor may make this discussion a whole-class activity.

In addition to audience-centered criteria, a persuasive evaluation report is built on good, well-supported reasons. Your reasons should address how well your process and product meet the criteria; your brochure is effective, for example, because the information it contains is accessible and follows a reader-centered arrangement. Beyond just asserting these reasons, you will need to provide ample evidence for them, pointing to specific aspects of your group process or the texts that you produced. To continue with our example, what specific elements of your brochure make it accessible? You can further create a credible ethos for yourselves by being honest about your assessment. You will seem more trustworthy and your evaluation will be more persuasive if you address your project's weaknesses as well as its strengths and make a realistic grade recommendation. After all, your instructor will probably already have some idea of where you could have performed better. This is not to say that you must give equal presence to your strengths and weaknesses, but that you should not ignore or gloss over the latter. You could even use a discussion of your shortcomings to your advantage, explaining how you proactively addressed them or at least learned from them. Consider discussing how you responded to the problems you mentioned in the progress report, in particular.

Evaluation Report Parts

In our earlier discussion of the user-test report, we mentioned some of the conventional parts of many reports, including the introduction, methods, results, discussion, and conclusions and recommendations sections. Although not all formal reports have these sections, most include some version of them. The following list provides a short explanation of what these parts generally do for readers.

1. Introduction—sets the context for the report and, in a managerial arrangement, previews the major findings, conclusions, and recommendations.
2. Methods—overviews how the writers arrived at the results; sometimes folded into another section such as the introduction or results.
3. Results—describes what the writers produced or found and its significance; sometimes combined with the discussion section.

STUDENT VOICES 9.A Getting Feedback on Your Project

Ashley: In one of my early college English classes I was required to create a figure out of Legos. After I finished making it, I had to write a manual describing how to remake my figure. The instructions in the manual had to include each step I wanted the person remaking the figure to perform. Although I had made a simple Lego figure, writing directions was anything but simple. When I finished and my partner tried to reconstruct my figure, the results were a completely different Lego man (actually it looked nothing like a man). Subsequent to that experience, I have a strong respect for people who write manuals. All people interpret things differently, which does not make writing instructions/manuals any easier. I am looking forward to testing our document. I know that we have put a great deal of work into our web page and, therefore, we can no longer look at it objectively. It will be interesting to get feedback from users, because I am sure that they will be able to tell us how we can improve it.

Jim: When I write songs, I only think of what pleases me and my ear. When I started out in college, I was solely writing for the teacher because I thought the teacher would be the only one who read what I wrote. Now, through taking technical and professional writing classes, I am forced to see a bigger picture. First of all, an audience will read my words. Secondly, they will be asked to follow a direction or be guided through a task by this writing. This is especially important in a service-learning course.

Fred: My first experience in peer analysis and user testing happened during my tech writing class last semester. My group was putting together a "how to master Napster" guide book. Not only did I find out that creating software documentation was hard, I found that writing instructions and telling someone instructions were two totally different things. Our tester had a bunch of questions, some of which I thought were a little ridiculous until I went back and reread what I'd written. I was writing to explain to myself how I would master Napster—I was ignoring my reader. Software documentation has to be written in basic terms; you have to keep in mind that whoever reads the writing may have absolutely no idea what you are talking about.

Kyle: My past experience with usability testing has been somewhat frustrating. I tend to become excited about projects I'm really involved in and I always expect a more enthusiastic response than what I receive. During usability testing, I've asked about the document's strengths, weaknesses, aesthetics, etc. Instead of a detailed response that would assist me in my project, I get a lot of yes's and no's and other one-word answers that don't really provide any feedback. For this project, we've designed a much better test that we believe will elicit meaningful feedback. I'm looking forward to that.

4. Discussion—helps reader interpret results and their implications.
5. Conclusions—tells reader what results mean, what they should take from them; sometimes combined with recommendations.
6. Recommendations—calls readers to particular actions based on the results and conclusions (not all reports lead to recommendations).

In addition to the main sections of the actual report, many formal reports include front and back matter. Front matter can include a title sheet and table of contents (some reports have separate tables of contents that list figures and tables), and an abstract or executive summary. Back matter is usually comprised of appendixes containing supplementary information referred to in the report but not integral to its discussion. You may not need appendixes. If your report is not too long (say, under ten pages), it may not require a table of contents. You should begin with a title page similar to that used with the proposal, however, and you might also consider using this assignment as an opportunity to practice writing an abstract. In what follows we take you through the parts of your evaluation, explaining how they relate to the parts listed above. We'll refer to the sample report in Figure 9.2 throughout. This report was written by three professional communication students who wrote a patient information packet and Relay for Life press release for their local chapter of the American Cancer Society. We chose this example because of its thorough assessment of both process and product and its informative subheadings that reflect the evaluation criteria.

Evaluation Report of
American Cancer Society Press Release
and Patient Information Packet

Submitted by
Team Cancer-Fighting Gators:
Jason Law
Katie Plumlee
Matt Lilley

Submitted to
Professor Blake Scott

April 19, 2001

(continued)

Figure 9.2 Team Cancer-Fighting Gators' evaluation report

Figure 9.2 *(continued)*

ABSTRACT

This evaluation report overviews and evaluates the process and products of our American Cancer Society service-learning project. Within the time frame of the project, Team Cancer-Fighting Gators successfully produced two informative, audience-centered texts: a Relay for Life press release and a patient information packet.

Our team faced several difficulties during the course of the project, including having our assignment changed, receiving little guidance from the agency, and starting late on document design. Not only did we respond productively to and compensate for these difficulties, we learned valuable project management skills from them. We are especially proud of the way we wrote an entirely new proposal in less than two days. We also improved our writing and design skills and learned how to accommodate team members' weaknesses and strengths at performing various tasks.

More importantly, our final products will benefit the American Cancer Society and the communities it serves. The press release will inform the public about a worthwhile fundraising event and inspire them to participate through its persuasive style. The patient information packet will most likely be used as an informative, accessible, and handy resource by physicians and cancer patients around Florida.

INTRODUCTION

The purpose of this report is to provide a self evaluation of our group's efforts in writing and designing the press release and patient information packet for the American Cancer Society (ACS). The packet, which overviews and provides contact information for many of the services, programs, and fundraisers made possible through the agency, will be used by health care providers as a resource for cancer patients.

We set out to make the packet informative, accessible, and engaging when developing it. We wanted to provide enough information so that patients could easily determine whether or not each program is appropriate for their needs. We also wanted patients to be able to move easily through the text to find information about a particular program; most patients will not want to read the details of programs not relevant to them. The writing needed to be to the point but also engaging to the reader.

The following report will first overview the steps and results of our project, then evaluate our process and main text according to a set of criteria, and finally recommend a grade for our performance. Although we will describe the press release briefly, we will base our evaluation on the longer, more complex information packet that comprised the bulk of our project.

PROJECT DESCRIPTION

We collected information from the ACS literature room, the ACS website, and agency personnel after determining our assignment with Diana Child and Lauren Dean. We then drafted a Relay for Life press release announcing the upcoming fundraising event for the

(continued)

Figure 9.2 *(continued)*

Ocala area. This text went through several revisions based on input from Diana and Lauren and from Professor Scott, mainly toward making it more audience centered. For our second, more complex assignment, we wrote and designed the patient information packet, first coming up with a general design, then writing and revising the written sections, then refining and implementing the packet's design, and finally editing and proofreading the final text using our revised style sheet. Although we split up the writing, we gave each other extensive feedback and collectively designed and edited the packet.

The Relay for Life press release was a relatively short and simple text that described the event, including logistical details, and encouraged people to come out. The information packet, which health care providers will give to cancer patients to introduce them to various support programs available, took the form of a folder containing sections with their page sizes staggered so that all of the major headings could be visible at once. The sections included educational groups, support services, outside supported services (affiliated with but not offered by ACS), and fundraisers.

Although the road to producing the information packet was rocky at times, both texts turned out to be successful, garnering much praise from Diana and Lauren, our contact people. Most importantly, we helped fulfill the ACS's goals of educating the public about cancer, attracting more people and financial support to the fight against cancer, and improving the lives of those who have cancer.

EVALUATION OF PROCESS

PLANNING

We were tested as a group from the outset of the project when we had to change our entire focus two days before our final proposal was due. Our initially agreed-upon project was to design a cancer prevention cookbook, a fundraising text that our contacts at the ACS had fully endorsed during our initial visit about a week and a half prior. However, when we returned to iron out the details of this project, full of ideas and ready to begin research, we were informed that the national ACS had recently released a similar cookbook, making our project moot. We were naturally discouraged upon receiving the information at such a late date, but our ability to adjust on the fly and not remain disgruntled was certainly a tremendous accomplishment in itself.

Fortunately, the organization did not lack other project suggestions for us. Along with a comprehensive information packet, they suggested we might revise a prostate cancer brochure or write press releases for upcoming fundraisers. The information packet would be given to health care providers, cancer patients and their families, and individual and corporate donors. Remarkably, we managed to revise our proposal fully within two days, putting us only a little behind on preliminary research.

RESEARCHING

We did most of our research for the project in the literature room at the American Cancer Society office. The national website and numerous brochures and other printed materials

(continued)

were also valuable resources. It was not hard to find up-to-date descriptions of ACS programs and achievements for the information packet. The only problem we had in the research process was deciding what information was most relevant and what should be omitted. So many programs and services are offered through the American Cancer Society that it could be overwhelming to present all of them without careful organization. It was at this stage that we decided, along with our contact people, to focus on programs for patients and to tailor the content to an audience of health care providers and cancer patients. One of our contacts, Lauren Dean, helped us decide what information would make the final cut and be placed among the packet's four main sections.

COMMUNICATING AND COLLABORATING

With so many persons involved in the project, including American Cancer Society contacts, Professor Blake Scott, and our three group members, our ability to collaborate effectively and efficiently was tested as well. Our different schedules as full-time students made it difficult to find common times to meet and work on the project.

The only major difficulty with the project was the lack of communication between the American Cancer Society and the Cancer-Fighting Gators. After pointing us to research material, our two contact people took a "hands-off" approach until the very end of the project, when they gave us some comments on the next-to-the-last draft. To some extent, though, our group took advantage of this by experimenting with the basic form and the design of the packet, which ultimately led to a more accessible, professional-looking text. In addition, we addressed the problem of little guidance by seeking extra feedback in the drafting and revising stages from our professor and classmates. Professor Scott, in particular, looked at numerous drafts of the writing and design. If we could do the project again, however, we would be more assertive and persistent about asking for more agency direction.

Overall, we worked well together as a group, especially toward the end of the project. At first we found it difficult to establish leadership roles and express our opinions and criticism openly because none of us are assertive people. As we became more familiar with one another and the project, however, we were better able to take initiative and facilitate each other's full participation. In dividing tasks, we learned to take advantage of each group member's strengths, such as Jason's organizational skills, Katie's editing and design skills, and Matt's writing skills. We also adjusted each others' workload at various times to accommodate scheduling conflicts and outside work pressures.

WRITING, DESIGNING, AND REVISING THE TEXT

Writing the text was perhaps one of the strengths of the collaboration. After compiling our research we divided the sections fairly evenly. Then, after completing initial drafts, we exchanged texts, gave each other comments, and together made the first round of revisions. The bulk of our writing was finished ahead of schedule, about midway through the project's time frame.

Creating the design, on the other hand, was our main weakness. In addition to the lack of guidance and supervision from the American Cancer Society, we had little experience with

(continued)

Figure 9.2 *(continued)*

designing a document using a computer. We felt unsure of ourselves and, as a result, started slowly on the design. We didn't take full advantage of the style sheet and document grid assignments and therefore had to cram these tasks into the last part of the project. In addition, being behind schedule on design prevented us from having a complete, polished draft for the in-class writing workshop. We still gave two other groups helpful feedback, however, and set up an additional workshop for a more complete version of our text a few days later. In the meantime, we worked overtime to establish design guidelines, implement them using the computer, and refine them. We learned what a time-consuming process designing a document and putting it together can be. Next time we will start the design process earlier and seek more direction throughout this process.

The Cancer-Fighting Gators were able to overcome several obstacles and even parlay some of them into positives. As a group and as individuals, we learned more about managing a large-scale, collaborative writing project, in the process refining our writing skills, learning new design skills, and improving our troubleshooting ability.

EVALUATION OF FINAL PROJECT

AGENCY EXPECTATIONS AND PURPOSES

The final product more than met the expectations of the agency. We produced an informative, readily accessible, engaging, and professional information packet that is well suited to its audience of physicians and cancer patients. We are especially proud of the handiness of our basic form, a folder that enables the reader to see all of the major categories of information at once and that also enables the ACS to include additional material about individual programs. Our contacts were not sure differing page sizes would be cost or time efficient in mass producing the packet, but we showed them how easy it was to print out all the pages on regular-size paper and then simply cut certain sections to size.

ACCESSIBILITY

One of the most important objectives for our group was to provide a great amount of information about services and programs in an easy-to-read, accessible manner. Dividing the information into four main sections, using big, bold headings, and staggering the page size of these sections make them easy to find upon opening the folder. Within each section, we created more accessibility by using subheadings and horizontal dividers, using color for all headings, keeping the paragraphs short, leaving extra space between subsections, and arranging the subsections alphabetically. This last organizational pattern should be easy for readers to identify.

CONSISTENCY AND COHERENCE

Our color design and text layout are consistent throughout the packet and its four main sections. Using only the organization's colors of dark blue and red enabled us to maintain the professional ethos of the ACS and keep the design focused on the primary purpose of relaying information. We also made sure spacing and fonts are consistent in the cues they give readers. In terms of style, we carefully revised and edited the written portion of the folder in accordance with the

(continued)

Figure 9.2 *(continued)*

ACS's house rules for grammar and style as well as our own style sheet guidelines. We spent a great deal of time editing each other's work and collectively implementing changes.

ENGAGEMENT OF READERS

While creating an informative, accessible text was our foremost concern, we also wanted to motivate patients to read it and keep them engaged. We accomplished this by several means, including using color, using photographs of program users, and making the writing interesting and well suited to the readers' needs. Instead of simply putting photographs at the ends of certain sections, we integrated them into the descriptions using the text-wrap technique. We also varied the sentence structure some without making any of the sentences too long or complex. Finally, we adjusted some of the program descriptions to make them more suitable for patients. For example, we explained the specific ways cancer patients can become involved in fundraisers such as Relay for Life.

CONCLUSION AND RECOMMENDATION

We have worked on this project for several months, and the final result is a packet that potentially will help hundreds of cancer patients become connected to life-enhancing services in our area. Without our efforts, this project may never have been realized, given that the ACS is so strained for time and resources. If we make even one person's ordeal with cancer a bit more bearable, we will consider the project a success. It is our hope and belief, though, that many lives will be helped by the time and energy that we have contributed on behalf of the organization and in honor of those who have struggled courageously in the fight against cancer.

We are proud of our effort as a whole, but especially of the packet's handy form and accessible design that we created with little guidance and supervision. We have learned a great deal about professional writing, design, editing, and collaboration from this project. We are also glad that our initiative, creativity, and problem-solving skills were tested and improved throughout. The usefulness of our products and our ability to learn from and make the most out of the obstacles we faced should be rewarded with a high grade. We believe our performance should earn us a grade of A-. Team Cancer-Fighting Gators has excelled in fulfilling the requirements of the American Cancer Society and the class, and we trust that you will reward us accordingly.

Executive Summary or Abstract

Longer professional reports sometimes begin with an **executive summary,** typically a one- or two-page thorough summary of all of the report's major parts. The parts of the summary should be roughly proportional to the parts of the larger report. As the name implies, an executive summary is written for an audience of managers who may want to read only an overview of the report before deciding whether to read certain sections for more detail. An executive summary should be complete enough to stand on its own for such an audience, though, and is often placed on a detachable page(s).

Abstracts are shorter than executive summaries, spanning only about 100–200 words. Rather than providing a comprehensive summary, an abstract highlights the most important information in the report, often not mentioning less important sections. An abstract of your evaluation report, for example, might stick mainly to the strengths of your project's process and product. Abstracts are often read by readers to decide whether the report is relevant or of interest to them.

There are two types of abstracts: descriptive and informative. A descriptive abstract tells the reader the kinds of things the report covers, acting almost like an extended forecasting element. Team B.A.R.K., a group of University of Florida students working with the Humane Society, included such an abstract as part of their evaluation report. An informative abstract actually provides (rather than forecasts) the most important information from the report. Most readers find informative abstracts more helpful than descriptive ones. The group of students writing for the American Cancer Society—Team Cancer-Fighting Gators—provides an example of such an abstract in Figure 9.2. We've reproduced this abstract with that of Team B.A.R.K. in Figure 9.3 to facilitate your comparison of the two types. Notice how the first three sentences of Team B.A.R.K.'s descriptive abstract are essentially forecasting elements. The next few sentences do mention some characteristics of their manual and production process but in a cursory way and still in forecasting mode. After an initial forecasting element and minimal background information, the abstract of the American Cancer Society group gets right to the details about the writers' main difficulties and accomplishments (paragraph two) and about the most impressive features of their texts (paragraph three). As a result, the abstract is both more informative and persuasive.

If you decide to write an executive summary or an abstract, the latter of which might be more suitable for a report of this length, we advise you to write it last so that you know precisely what you are overviewing or highlighting.

From the evaluation report of Team B.A.R.K.:

Abstract

The evaluation report is our final project and presents a detailed account of the design and creation of our volunteer training manual. An introduction contains background about our project and involvement with the Humane Society. The body of the report overviews the major components of our handbook and then evaluates our process and product of the volunteer manual. Our report focuses on the strengths of the manual, especially its accessibility and expandability. Our report also provides a frank account of the problems we encountered during the writing process. These include computer problems, communication breakdowns with the agency, and time constraints. However, we are confident that in spite of the problems we faced, our final project is our best work.

(continued)

Figure 9.3 Two abstracts from evaluation reports—the first descriptive and the second informative.

Figure 9.3 *(continued)*

From the evaluation report of Team Cancer-Fighting Gators:

ABSTRACT

This evaluation report overviews and evaluates the process and products of our American Cancer Society service-learning project. Within the timeframe of the project, Team Cancer-Fighting Gators successfully produced two informative, audience-centered texts: a Relay for Life press release and a Patient Information Packet.

Our team faced several difficulties over the course of the project, including having our assignment changed, receiving little guidance from the agency, and getting a late start on document design. Not only did we productively respond to and compensate for these difficulties, but we learned valuable project management skills from them. We are especially proud of the way we wrote an entirely new proposal in less than two days. We also improved our writing and design skills and learned how to accommodate team members' weaknesses and strengths at performing various tasks.

More importantly, our final products will benefit the American Cancer Society and the communities it serves. The press release will inform the public about a worthwhile fundraising event and inspire them to participate through its persuasive style. The Patient Information Packet will most likely be used as an informative, accessible, and handy resource by physicians and cancer patients around Florida.

Introduction

After the abstract, begin the report proper with an introduction. As with the introductions you wrote for the proposal and progress report, this one should perform several functions for the reader, namely these: State the report's purpose; reiterate your project assignment and its rhetorical situation; possibly outline your criteria and preview your overall assessment; forecast the remainder of the report.

The first two paragraphs of the sample introduction in Figure 9.2 overview the writers' assigned documents, their rhetorical situations, and their main goals for them. Although this introduction doesn't preview the writers' evaluation, it does suggest some of the criteria in its overview of their goals. It ends with a fairly informative forecasting element.

Project Description

You might think of this section of the evaluation report as a combined and abbreviated methods-and-results section. Although the results section of many reports is an extensive or even the longest section, in this case it will be relatively short to emphasize the next section—the evaluation. In this case you can consider the process your methods and the final product your results. Here you might just summarize the major steps of your project, including the ups and downs of your group process, and highlight the main elements of what you produced.

In Figure 9.2, the Cancer-Fighting Gators first take the reader through the main steps of their report and then overview the form and main parts of the documents they produced, especially the packet. This not only reminds the reader about the project, but also sets up the elements of the process and product that they evaluate in the following section.

Evaluation

Some reports integrate the discussion with the results, presenting one important finding and then discussing it, moving to the next finding and discussion, and so on. For this report, the discussion will be your evaluation, the most important section of the report where you apply your criteria to the different parts of your project. If you haven't overviewed your criteria in the introduction, you should probably do so in a short introduction to this section. Following the suggested pattern in the project description section, we recommend you evaluate your process and product one at a time. This makes sense given that the two sets of criteria are probably somewhat different. If you produced more than one major document, you could also address these in separate sections.

After breaking this section into process and product, create subsections that revolve around the parts of the project, the criteria you have discussed, or both. The sample report in Figure 9.2 first divides the section into *Evaluation of Process* and *Evaluation of Final Product,* and then further divides the former section chronologically according to the parts of the process and the latter section categorically according to the criteria it fulfills. It seems wise, in our view, to at least highlight your criteria for evaluating the texts in subheadings, as the writers in the sample do with such subheadings as *Accessibility* and *Engagement of Readers.*

You might use the following questions about your process and product as starting points in developing the evaluation. Basic criteria are embedded in the questions. Try to supplement these questions with your own, more specific ones.

Process

1. What did you learn about collaboration from this project? What were the strengths and weaknesses of your collaboration with each other, agency personnel, and your instructor?
2. How effectively did you adapt to assignment changes, deadline changes, lack of guidance, and/or other obstacles? How well did you manage ethical dilemmas?
3. What project management strategies did you develop over the course of the project? If you could start the project over again, what would you do similarly and differently?
4. What did you learn about the process of planning, designing, writing, and revising and editing a professional text? At which parts of this process did you excel, and through which parts did you seem to stumble? Which parts will you take more seriously next time?

Product

1. How well does the product meet the agency's expectations and achieve its purposes? To what extent does the project embody the agency's mission and ethos?
2. How would you forecast the effects of your text in the community? How will your texts help people? What else could you have done to make them more helpful and better suited to their audiences? What else could you have done to make them more ethical?
3. What are the texts' main weaknesses and strengths, and how do you account for these? Of which aspects of the texts are you most proud and why? Which parts would you change or refine if you had more time or different constraints?

In deciding how to organize the parts of your evaluation, consider how different arrangements might affect your reader. Although you want to address your weaknesses as well as your strengths, you may want to buffer explanations of the former with discussions of the latter. If you arrange subsections by criteria, consider which criteria you can make the strongest argument for achieving, as well as which criteria your instructor will view as the most important. Then arrange the subsections in a strategic way. You may want to begin with and spend the most time on your readers' most valued criterion, for example. In the section evaluating their process, the Cancer-Fighting Gators in Figure 9.2 usually follow descriptions of their problems with more positive discussions of how they effectively addressed and learned from them. In the section evaluating their packet, they begin by addressing the two most important criteria—achieving the agency's purposes and creating an accessible, reader- centered design.

Whatever criteria you address and however you arrange them, remember to provide specific evidence for each part of your evaluation. Strive to demonstrate rather than simply to assert. Because your instructor likely will have a copy of the final product already, you should feel free to refer to specific parts of this product. The Cancer-Fighting Gators support their reasons throughout, providing such specific examples as the way they quickly revamped their proposal in response to the assignment change and the way their packet's form makes its major sections immediately accessible. These writers probably could have developed and supported their discussion of their packet further, as this section seems less thorough than the one about their process.

Because this is the most important and lengthiest section of the report, we recommend that you at least brainstorm for it and edit it as a group. This brainstorming could include composing an outline and subheadings for the section. As you did with the proposal, orchestrate ways to make this section and the larger report accessible to your instructor. In addition to using informative subheads, this might include writing in short, manageable paragraphs and using lists. We've even had students end this section or begin the concluding section with a table that visually summarizes their evaluation of each part of the project in a few words. Your instructor will likely be reading several of these reports. In addition, she or he will likely refer back to different parts when factoring it into her or his overall evaluation of your performance. Therefore, it will be to your advantage to make the report easy to follow and the pieces easy to locate.

Conclusion and Recommendation

The final section, which should flow smoothly out of the preceding one, is where you make your overall evaluation and recommend a specific grade for your performance. Make sure you connect the evaluation to the grade. If you think it might work to your advantage, grade your process and your product separately before combining them into an overall grade.

In concluding, emphasize, as the Cancer-Fighting Gators do, how well your documents are suited to their rhetorical situations and what their beneficial effects likely will be. Also, summarize your assessment of the process, including how well you managed the project, how well you dealt with obstacles, and what you learned from the project. You might be tempted to use phrases such as *we feel* and *we believe* in this section to tone down your conclusion and recommendation; avoid this temptation. As the report in Figure 9.2 illustrates, such phrases can create a less-than-confident tone and unnecessarily wordy prose.

Style Focus: Emphasis

You have probably already been experimenting with ways to emphasize key ideas in your service-learning documents through visual cues such as color, font size, extra space, and boxes or rules. You may not have had as much experience adjusting verbal style for emphasis, though. For our last style focus, we will outline several ways to emphasize important ideas more effectively by adjusting words and sentence structures. Creating more strategic emphasis on the word and sentence levels can aid you in both your informative and your persuasive aims. If you are presenting readers with a substantial amount of new information, you might need to direct them to the most important ideas or threads. If you want to persuade readers with a particularly powerful idea, you'll want to make sure readers notice and remember it.

In what follows we take you through five easy-to-remember strategies, some adapted from Williams's *Style* book, for adjusting words and sentences to convey emphasis better. We supplement our brief explanation of each one with illustrations from the first page of the sample evaluation report we discussed earlier in this chapter. This excerpted first page, Figure 9.4, shows the before and after through the *tracking changes* function in Microsoft Word. Old text is crossed out and new text is underlined.

Use Punctuation and Signal Words to Indicate Important Ideas

Certain words and punctuation marks are intended to make the reader pause and take note of what follows. Think about the difference between a clause set off by commas and one set off by dashes; the dashes indicate that you should pay close attention to the clause rather than treat it like a peripheral idea. Colons can also have this effect, announcing that something noteworthy follows. The writers of Figure 9.4 use a colon in the second sentence to announce the two texts that will be the focus of their evaluation.

ABSTRACT

This evaluation report overviews and evaluates the process and products of our American Cancer Society service-learning project. Within the time frame of the project, Team Cancer-Fighting Gators successfully produced two informative, audience-centered texts: a Relay for Life press release and a patient information packet.

Our team faced several difficulties during the course of the project, including having our assignment changed, receiving little guidance from the agency, and starting late on document design. Not only did we respond productively to and compensate for these difficulties, we learned valuable project management skills from them. We are especially proud of the way we wrote an entirely new proposal in less than two days. We also improved our writing and design skills and learned how to accommodate team members' weaknesses and strengths ~~at performing various tasks.~~

More importantly, ~~O~~our final products will benefit the American Cancer Society and the communities it serves. The press release will inform the public about a worthwhile fundraising event and inspire them to participate through its persuasive style. The patient information packet will most likely be used ~~as an informative, accessible, and handy resource~~ by physicians and cancer patients around Florida, providing them with an informative, accessible, and handy resource.

INTRODUCTION

The purpose of this report is to provide a self evaluation of our group's efforts in writing and designing the press release and patient information packet for the American Cancer Society (ACS). The packet, which health care providers will use as a resource for patients, overviews and provides contact information for many of agency-sponsored ~~the~~ services, programs, and fundraisers ~~made possible through the agency, will be used by health care providers as a resource for cancer patients.~~

When developing the packet, we set out to make ~~the packet~~ it informative, accessible, and engaging ~~when developing it.~~ We wanted to provide enough information so that patients could easily determine whether or not each program is appropriate for their needs. We also wanted patients to be able to move easily through the text to find information about a particular program; most patients will not want to read the details of programs not relevant to them. The writing needed to be to the point but also engaging ~~to the reader.~~

(continued)

Figure 9.4 Sample from revised evaluation report

Figure 9.4 *(continued)*

The following report will first overview the steps and results of our project, then evaluate our process and main text according to a set of criteria, and finally recommend a grade for our performance. Although we will describe the press release briefly, we will base our evaluation on the longer, more complex information packet that comprised the bulk of our project.

PROJECT DESCRIPTION

After determining our final assignment with contacts Diana Child and Lauren Dean, ~~W~~we collected information from the ACS literature room, the ACS website, and agency personnel ~~after determining our final assignment with Diana Child and Lauren Dean.~~ We then drafted a Relay for Life press release announcing the upcoming fundraising event for the

Words can also signal readers to take note. The writers of our sample provide us an excellent example of this in first sentence of the third paragraph (abstract included), which they revised to begin with *More importantly,* a transitional phrase that quite explicitly alerts readers to their important assessment of their texts. Other words (often used as transitions) that can signal emphasis in many contexts include the following:

- Above all
- In conclusion
- Indeed
- First and foremost
- Thus
- In fact
- Take note
- Clearly
- In short

Though we encourage you to use these phrases in some strategic contexts, don't fall into the habit of overusing them. This waters down their impact and the power of your prose.

Vary Sentence Structure

Have you ever read an extended passage comprised of all simple or all complex sentences? No doubt this quickly became tedious. Varying your sentence structure can help you create a more interesting, engaging rhythm in your prose. You can also create emphasis through changes in sentence structure, especially when a change stands out. If you wanted to emphasize a key idea, for example, you could express it in a short, simple sentence occurring at the end of a string of longer, complex and compound sentences. This is what the writers of our sample passage do in the second paragraph of Figure 9.4. In addition to using the signal word *especially,* they emphasize the aspect of their process of which they are most proud by expressing it in the only short, simple sentence of the paragraph.

Repeat Key Words or Ideas Throughout a Paragraph or Passage

In Chapter Five we discussed how repeating key words or ideas can create coherence in a passage, pointing readers to its main topic thread(s). Such repetition can work similarly to make key ideas more memorable in readers' minds. To keep the repetition from becoming tedious, you can use synonyms or otherwise express the same idea in a slightly different form.

The writers of Figure 9.4 use repetition to emphasize several aspects of their work. Note first how they repeat variations of the words *informative* and *accessible* at the end of paragraph three, the beginning of paragraph five, and throughout the rest of paragraph five. These words highlight the most impressive qualities of their texts and forecast the criteria the writers will use later in their evaluation. In the second paragraph, the writers emphasize how much they learned from their service-learning project, obstacles and all.

Shift Important Ideas Toward the End of the Sentence and Peripheral Ideas to the Beginning or Middle

For the most part, readers tend to remember the end of a sentence more than the beginning, making the ending a more strategic place for the sentence's main idea. After reading your sentences and identifying the most important information, determine whether this information occurs in the latter parts and, if not, experiment with reordering the sentence's parts to get it there. A common mistake in emphasis is ending a sentence with modifying information that could be placed elsewhere. Dependent clauses, for example, usually can be moved easily from the end to the beginning or middle of the sentence.

Our sample in Figure 9.4 contains several sentences that the writers rearranged to take better advantage of the end slots. They revised the last sentence of the third paragraph, for instance, to end with the description of their packet as "an informative, accessible, and handy resource," an ending that corresponds better to the preceding sentence's ending of "persuasive style." The writers decided that describing their packet's most impressive features was more important to the sentence and the paragraph than explaining its users. Moving down the page to the second sentence of the fourth paragraph, we observe that the writers switch the order of the middle dependent clause and the ending. Once again, it's less crucial to clarify the packet's users and uses (which have already been mentioned) than its specific parts. Moving down to the beginning of the fifth paragraph, we see that the writers moved the last few words into an opening clause to end the sentence with the packet's main qualities. This change also has the benefit of creating better cohesion by connecting the revised sentence, now referring to the packet, with the end of the preceding one. Finally, at the beginning of the last paragraph on the page, the writers simply move the ending clause beginning with *after* to the beginning of the sentence. The revised sentence better emphasizes the various sources the writers consulted and presents the information in a more cohesive chronological order.

You may have a good reason for not shifting the most important idea to the end. If it connects to the end of the previous sentence, for example, you may want to leave the main idea at the beginning to create coherence through a given–new information chain. In addition, you may want to end occasionally with modifying clauses to make your prose more varied and interesting.

Trim the End of the Sentence

Writers waste the most memorable parts of their sentences when they end them with fluff words rather than important ideas. Trimming the ends of such sentences, a technique that overlaps with our advice about concision, simply requires reading the ends of your sentences and asking, "Are the last few words crucial to the meaning here?" You might be surprised at how often your answer will be "No."

When the writers of Figure 9.4 revised this page for emphasis, they found three unnecessary endings that detracted from their intended emphasis. First, the last sentence of the second paragraph ends with the unnecessary words *at performing their tasks.* In the second instance, the writers revise the clause of the introduction's second sentence to end with the list of *services, programs, and fundraising* instead of the words *made possible through the agency.* This last idea is shortened and placed earlier in the clause as the compound adjective *agency-sponsored.* The third and final instance of trimming occurs in the last sentence of the fifth paragraph where the writers eliminate the easily inferred phrase *to the reader* and instead end with the more persuasive and important description of their packet as *engaging.*

Deliberating about Your Project: Some Final Considerations

Because the main goal of your evaluation report is to persuade, writing it may not give you a full opportunity to deliberate critically about your service-learning experience and its effects, including its effects on you. For this reason we end the chapter with a heuristic for thinking critically about and discussing what you have experienced, learned, and produced. Your instructor may use these questions as prompts for informal discussions or may ask you to respond to them in a short memo or in a presentation of your project to the class (as a way to frame your presentation, for example).

As we've stated several times, critical reflection about one's service is one of the crucial components of service-learning. We prefer the term *deliberation* to *reflection* to denote activity that moves beyond individually and passively recalling or celebrating one's experience. As we're defining it, *deliberation* is a self-reflexive but also critical and social activity. We usually have our students deliberate about the questions below in their service-learning groups, on a class e-list or bulletin board, and/or in a class discussion. In our view, deliberation should also involve ethical considerations. You may recall from our discussion in Chapter Three that such considerations can focus on obligations such as the accommodation of users' needs, ideals such as rhetorical appropriateness and honesty, and effects such as the benefits one's writing makes possible. The questions below ask you to deliberate ethically in all of these ways.

Some of the questions ask you to deliberate about the process of your project, including your collaboration, your project management, and your adaptation to a community-based workplace and writing context. Chris Anson and Lee Forsberg discuss this last aspect in their study of professional writing students making the transition to nonacademic writing in semester-long internships. In "Moving Beyond the Academic Community: Transitional Stages in Professional Writing," Anson and Forsberg describe three progressive stages through which most of their students traveled in making such a transition. When their students began their experiences they were highly motivated and had high expectations for working in their new settings. After they became involved in their projects, this stage gave way to one of disorientation, frustration, and sense of alienation, largely because they were receiving less direction and nurturing from their colleagues and supervisors. Finally, however, the successful students took more initiative and consequently achieved a sense of resolution. Discussing the stages of your experience may help you clarify what did and didn't contribute to your project's success, as well as what you might expect and adjust to in your next nonacademic professional or technical writing experience.

Deliberation about Process

1. How would you compare this experience with your previous academic and other writing experiences? How would you compare, in particular, your roles as a writer? Your motivation? Your relationship to your audiences?
2. How were your stages similar to and/or different from the three stages of motivation, disorientation/frustration, and resolution that Anson and Forsberg's students experienced (see description of this study above)? What lessons about adapting to nonacademic writing contexts will you take from this project?
3. How did your relationship with the agency change during the course of the project, and how did you respond to these changes?
4. What did you learn about collaboration from this project? How well did you attend to your obligations to the agency, the course, and your community audience?
5. What did you learn about project management from this experience? What will you do differently in your next long-term professional writing project?

Deliberation about Product

1. How well do you think your group's documents fulfill their obligations to their audiences? How respectfully do they represent their readers/users and their needs?
2. How effectively do your documents convey the ethos of the organization?
3. What other documents could you and/or the agency have produced to address the same community-based problems or needs? What would you change about the agency's approach or specific responses?
4. How would you define the qualities of a successful professional or technical document?

Deliberation with Self

1. How, if at all, have you changed your views of yourself as a student, a writer, and a community member throughout this project?

2. What did you learn about the social issues the agency addresses, and how did this affect your civic interests and commitments?
3. To what extent did your personal and professional values and goals mesh with those of your agency and your groupmates? How could you have better negotiated any conflicts here?

Activities

1. Exchange user-testing questions with another group in the class, and help each other revise the questions so that they will elicit fuller, more critical responses. Watch for closed and leading questions.
2. After compiling feedback from your user testing and/or your workshops with classmates and supervisors, compose a list of consistent patterns of responses and a list of conflicting responses. Ask your instructor for advice about how to handle the latter. After finalizing your list, formulate a revision plan that includes a list of specific changes to make and where to make them.
3. To help your instructor evaluate your individual performances in the group, fill out the self/peer evaluation form in Chapter Eight or your group's own version of the form.

Works Cited

Anson, Chris M., and L. Lee Forsberg. "Moving Beyond the Academic Community: Transitional Stages in Professional Writing." *Written Communication* 7.2 (April 1990): 200–31.

Polanyi, Michael. *Knowing and Being*. Chicago: University of Chicago Press, 1969.

Rude, Carolyn D. *Technical Editing*. 3rd ed. Boston: Allyn & Bacon, 2002.

Williams, Joseph M. *Style: Ten Lessons in Clarity and Grace*. 6th ed. New York: Longman, 2000.

Chapter 10

Presenting Your Project

If there is one thing that many students dread or complain about even more than working on group projects, it is giving oral presentations. Students often recall times when they were embarrassed or terrified while delivering their own presentations or bored to the point of distraction while listening to fellow students deliver theirs. In this chapter, we will give you some suggestions for making this part of your service-learning project a positive and productive experience that will help you to round out your repertoire of technical and professional communication skills. Later in this chapter we will discuss other aspects of presenting your project, including the process of preparing your final version for transmittal to your agency. Finally, we will invite you to complete your service-learning experience by revising your résumé to include the skills and experience you've gained while working on this project.

Presenting Your Project to the Class or Community

Before you begin the process of putting together and delivering an oral presentation about your service-learning project, it's important to understand the purpose of this exercise. Sharing your challenges, your documents, and your experiences with a live audience is a way to bring your project and your intellectual development full circle. Practice and training in making oral presentations is also a critical part of your preparation for communicating in a technical or professional field. Most professionals find themselves making presentations to clients, coworkers, supervisors, and community members on a regular basis. Often, workers who are particularly adept at making these presentations are rewarded in the workplace through leadership roles and promotions. As we've stated repeatedly in this book, clear communication is essential in the workplace. Without it, resources are wasted, errors are made, relationships are damaged, and, sometimes, disasters take place.

On a less dramatic level, oral presentations give students a great opportunity to showcase the hard work they've done throughout a course. If your class has featured regular draft workshops, oral progress reports, and class discussions, most of the members probably have a general sense of projects on which other groups or individuals are working. But even if you have seen and heard bits and pieces of your classmates' work, you probably don't have a clear sense of their project goals, their processes for reaching them, and the results of their efforts. Putting together a presentation to tell classmates about those aspects of your project can actually help you to clarify them for yourself. Also, as the comments from Maggie Boreman in Other Voices 10.A indicate, participation in the presentation process often gives everyone in a class a profound sense of pride in what they've accomplished together. At the end of most of our classes, students indicate a sense of identification with their classmates' successes. When they hear about and see the impressive documents that the group has produced, students have a better sense of what the group has contributed to its campus or community. Consider the suggestions below as you move toward creating your own class presentation.

OTHER VOICES 10.A Maggie Boreman

Maggie Boreman is a technical and professional writing graduate student at the University of Central Florida. She is also a freelance journalist and an avid horse rider and trainer.

Class Presentations in a Service-Learning Course

As a nontraditional graduate student (older than many), I was unfamiliar with the idea of service-learning in the classroom. It's an amazing concept—letting students discover the relevance of their efforts in the "real world."

I had my first service-learning experience in a class with Dr. Melody Bowdon. The course was actually an introduction to service-learning for most of us, and we students approached our seemingly poorly defined semester projects tentatively at first, unsure of which agencies we might choose and what projects we might undertake.

The end-of-semester class presentations told an incredible story. We had moved from the most ephemeral of concepts to concrete assistance for an animal shelter, a community jazz/dance school, a home for pregnant unwed teens, and Big Brothers/Big Sisters. Although we began the semester with no clue (and considerable concern) about what we might accomplish, by semester's end we found ourselves speaking excitedly about a community's needs.

Classroom presentations were integral to our service-learning course because they connected us to each other, just as we had become connected to the people and agencies that our projects benefited. Each of us grew from presenting to the class: one discovered an innate talent for filmmaking, another used the service-learning experience as a springboard for some awesome poetry, and, as one who suffers from stage fright, I can attest to increased confidence before an audience. Through our presentations, we introduced our fellow students to our passions, expressed our convictions, and articulated our achievements, which let us realize the magnitude of our impact. We had shared frustrations all along the way and our presentations left us cheering for the significant difference we had made in communities in need.

Preparing Your Presentation

In the last stages of your project, you probably discovered how much time and effort go into preparing a persuasive text. The same applies to an oral presentation. We cannot stress enough how important it is to prepare thoroughly for this assignment, from reanalyzing your rhetorical situation to designing your visual aids to polishing your delivery. If you will be giving a team presentation, be sure that all team members are involved in the preparation and rehearsal. In assigning presentation roles, take advantage of each member's strengths, and give all members feedback about their parts ahead of time.

The tips below will start you thinking about this slightly different rhetorical situation involving face-to-face communication with an audience of listeners. Remember that your audience will be able to absorb only so much in an oral presentation. In addition, they will be assessing your ethos based on your physical delivery as well as the delivery of your text.

- *Focus on your audience.* By now you've read this phrase enough times to know that we consider it to be the center of effective communication. Whether you are presenting to your class, to the staff of your sponsoring organization, to community members, or to a combination of these groups, you must begin your preparation by thinking about your listeners. Ask yourself the following questions:
    - What do my listeners already know about the project and the subject matter it addresses?
    - How much background information will they need to understand the context, the process, and the results?
    - Why should my listeners care about this subject? How might it relate to their own experiences or concerns?
    - What information or ideas can I offer to my listeners in this presentation that might help them with their own work or enhance their lives in some way?
- *Clarify your purposes and main points.* Think of your purposes in terms of what you want to do for your audiences rather than what you have to do for the class assignment. Are you making this presentation to persuade your listeners, to inform them, to entertain them? If your purpose is to give your classmates a clear sense of what was involved in your entire process, you likely will have to explain things that you may take for granted, such as specialized terms or a short description of the agency. Isolate three or four major points you would like your audience to take from the presentation and practice discussing these in casual conversation. If you become nervous and flustered, or if something that you're not expecting happens, you can always continue if you have internalized a real message to share with your audience.
- *Choose tools to augment your presentation.* Consider using options from the list below to either augment your presentation as you go along or to give your audience members supplementary information. Because your message should be the central focus of your presentation, don't begin working with technology until you have a clear sense of where you're headed. Use technology to

enhance your message, not to substitute for it or to distract your audience from receiving it.

- *Print handouts.* Handouts can serve several functions. They can provide an outline or map of your presentation, motivational material such as an intriguing visual or persuasive quotes, or such supplementary information as a bibliography. As one student group presented their project to create an on-campus daycare center, they distributed copies of the blueprint of the future building and a list of the project's phases from the idea to the reality. Because it would be hard for most audience members to make sense of something like a blueprint on a projection screen, this was a good strategy. Use a handout when you want to give your readers something to consider after the presentation ends, such as a list of things they can do to improve the environment or to prevent campus crime.
- *PowerPoint and other presentation software and overhead projector transparencies.* In the last few years the use of presentation software has become commonplace; it seems that many people can't stand before an audience without it. Such software can indeed be used to make your presentation clearer, more persuasive, and more memorable, but it can also lead to a vacuous presentation. Plan your presentation slides *after* you have formed a more detailed script or outline, and use these slides only to augment rather than to replace your presentation. When preparing PowerPoint, web, overhead, or other slides, pay particular attention to clarity. Use readable sans serif font and a large font size (at least 20 point). Don't put more than four or five lines of type on the slide, and don't overuse cutesy effects such as moving transitions or distracting backgrounds. Be sure your slides *add* power to your points rather than take it away. See the sample project in Appendix B for an example of an effective slide presentation.
- *Audio/video clips.* As Fred Reynolds notes in the Other Voices box on the next page, sometimes nothing captures a point better than an audio or video clip. If you choose to incorporate such tools into your presentation, be sure that they are brief, relevant, and easy to see or hear.

- *Become aware of your physical presence.* Your facial expressions, posture, gestures, body language, and voice will all be part of your audience's experience of your presentation. Unless you regularly give speeches, you may not have given these things much thought. Consider making a video- or audiotape and watching it to identify tics or strengths. Don't focus on this kind of thing so much that you become overly self-conscious, but do keep in mind that you may not always be aware of how your audience sees you.
- *Plan strategies for combating nervousness.* Public speaking is a nerve-wracking experience for many of us. To prepare for this nervousness, go into your classroom or other setting beforehand to develop a feel for the surroundings, and imagine yourself speaking to your audience. Remember that a class audience is almost always supportive. Everyone knows that she or he will have a turn at being on stage and that most people tend to treat others as they'd want to

OTHER VOICES 10.B Fred Reynolds

John Frederick Reynolds, Ph.D., is professor of English at City University of New York's City College. He has authored and edited several books and articles on rhetoric and technical and professional communication.

Fifteen years ago, while teaching a course in technical and professional writing for English majors at another university, I asked each of five working groups in my class to focus their term projects on "campus problems" of their choice. One group chose to report on campus sidewalks, how they had almost nothing to do with actual student foot-traffic patterns on our campus. Near end-of-semester, the group's leader came to my office and asked if I would allow (!) them to include as part of their final report and class presentation a videotape one of them had shot during the course of their work. "I don't think our written report really makes our point very well without it," she said. The end of this story is, of course, that that group's twenty minutes of videotape proved far more communicative than their several dozen pages of text, charts, graphs, executive summaries, bulleted lists, and front and back matter. We forwarded the group's final report (with video) to our university president, who subsequently had many of our campus sidewalks re-routed. This story underscores the fact that sometimes the best community service we can promote is service to our own campus community and reminds us that delivery/presentation(s) matters.

be treated. Identify in advance people in class who are good audience members and who will offer you support.

- *Plan memorable introductions and conclusions.* In Chapter Three we introduced the classical concept of kairos, or the opportune moment. It's only logical that two kairotic moments in an oral presentation are its beginning and its end. Plan an opening that engages and motivates listeners and perhaps previews your main points. Consider beginning with an anecdote, but avoid trite attention-getters such as jokes or quotations. Plan an ending that drives home your main points and, if appropriate, moves your audience to some kind of action. Some students find it helpful to memorize these segments so that they go especially smoothly.
- *Don't overload your audience with information.* You don't need to give your listeners a complete narrative or description of what you produced in the project. Instead, think of the oral presentation as an extended abstract in which you highlight your main accomplishments. With your visual aids as well, be sure not to include so much information that your audience is overwhelmed or too distracted to pay attention to what you're saying.
- *Make your organization clear.* Creating an easy-to-follow arrangement might be even more important for an oral presentation than for a written text. As with the other texts you've produced, set the context for your readers at the beginning with a forecasting statement or list. Use your slides to not only emphasize key points, but to help readers identify the main sections of your presentation, giving them headings.
- *Plan ways for your audience to participate actively.* You may want to give your listeners jobs to do or roles to play. One student group making a presentation

about the prenatal health guide they produced for a clinic passed out balloons at the beginning of their session, asking their classmates to wear the balloons like bellies. They conducted the first session of their presentation as a mock prenatal health seminar geared toward pregnant women. They even incorporated contests and prizes into the presentation. This gave them an opportunity to demonstrate what they had learned about their subject and spiced up the presentation considerably.

- *Treat the presentation as an opportunity to receive input.* Identify specific questions you have about your final product or about a communication challenge you've faced, and assign certain audience members to consider them throughout the presentation. You might ask one group of people to respond from the perspective of possible donors; others might listen critically from the point of view of a client or a staff member. You can even prepare forms for your listeners to complete and return to you at the end of the session.
- *Invite guests to attend the presentation.* Many of our students have invited their project contact persons to attend or even participate in their class presentations. Sometimes an agency representative will want to use the forum as an opportunity to recruit volunteers or raise consciousness about an issue. As long as this is handled professionally and represents only a small part of the presentation, it can be a worthwhile addition to the process. You may also want to invite other professors or classmates who are interested in the subject matter you're discussing. Check with your instructor before making such invitations to avoid overcrowding or scheduling problems.
- *Practice your presentation.* This is crucial to ensuring a smooth, professional delivery within your allotted time frame. Before you rehearse with the group, practice your own segment of the presentation several times. Have at least two full-team dress rehearsals of your presentation, preferably with a volunteer practice audience. This will help you to anticipate possible problems, avoid unnecessarily repeating each other, improve your transitions and overall coherence, and improve your pace and delivery. Consider ways to help each other do things such as advance slides or transparencies, pass out handouts, and turn off the overhead when not in use.

Delivering Your Presentation

You may remember from Chapter Three that one way to create a persuasive ethos is to show your goodwill toward your audience. In an oral presentation, this can involve connecting with your audience's values, speaking directly to them, and speaking at an accessible volume and pace. You can also create goodwill by dressing appropriately and otherwise showing that you're taking them and your presentation seriously. You certainly wouldn't chew gum or crack your knuckles or wear a baseball cap in a presentation for a workplace audience. Here are some more specific tips.

- *Make a connection with your audience early in the presentation.* When you take the podium, try to draw in your audience with a friendly greeting, an engaging

STUDENT VOICES 10.A Public Speaking

Hannah: First of all, I hate public speaking! Second, I hate public speaking a lot! I got through all of high school without having to speak in front of class, but then on my last day as a senior I had to give my salutatorian speech in front of 500 plus people at my graduation. No one ever bothered to tell me that it would be terrifying. But I managed to speed read and stumble through it and I survived somehow. It worked out okay, because the valedictorian did even worse. ☺ I have been told that public speaking is the number one fear among people, and death is number two.

Gina: I love public speaking. I do sometimes get a little nervous before I go up in front of the class, but it all goes away as soon as I start speaking. I used to be really scared to face any group, but my public speaking class helped me in a tremendous way. Now I like to use visual aids and answer questions from others. Another thing that helps relieve the pressure is practicing in front of a friend or even a mirror. I know it looks funny if someone walks in on you when you're talking to a mirror, but it can definitely help.

Carolyn: For some strange reason I have the hardest time breathing while presenting in front of a group. Either I focus all of my attention on my breathing technique and forget what I am saying, or I don't pay any attention to it and I wind up out of breath. I have given so many presentations in school that I should have this down by now. The most comforting thing is knowing that I have survived each speech and done fairly well. One thing that has helped me is waiting tables. Sometimes I have to go through a specials list in front of 20 or so hungry customers, which can take a few minutes, and I have gotten over that . . . so I am sure that I will survive my last presentation as a college student, too.

opening, and an appeal to their values. For classmates, this might include referring to the sometimes stressful assignment in which you have all been engaged. For contact persons, this might involve emphasizing your shared community mission.

- *Make eye contact.* Speak directly to your audience. Many people tend to gravitate to one side of the room or one segment of the audience when they are speaking to a group. If possible, move your eyes around the room in a natural pattern so that you interact with everyone present.
- *Don't read to your audience.* Most listeners prefer an engaging, seemingly extemporaneous discussion to a passively read script. They want to see your mind working and feel that you are connecting with them. Though most of us give better presentations if we have a few cue cards to keep us on topic, don't rely on such aids too heavily. Also, don't make the mistake of speaking to your visual aids instead of to your audience members, no matter how much more comfortable this may be. We recommend that you memorize the text on your slides.
- *Project and speak slowly.* As you begin to speak, adjust to the room's acoustics and your audience's reactions. Speak up if your listeners seem to be straining to hear or if you are competing against the buzz of computers. Don't be embarrassed to ask if people in the back of the room can hear you or to test a microphone if you're using one. A common delivery problem even with experienced

speakers is speaking too fast. Much of what you're presenting will be new to your audience, and, therefore, it will take them time to digest it. If you have to speak quickly to stay within the allotted time, cut some information.

- *Speak with confidence.* Whether you are an expert in your area of study or not, you should feel confident that you know considerably more about your project than others in the room. Remind yourself, too, of your project's success—you have much to be proud about. Finally, remember that you and your audience(s) have been working toward the same general goals; they are already in your corner and will be listening more as supporters than critics.
- *Plant a "mole" to monitor your pace and tone.* If you are concerned that you may speed up or slow down your speech or that you may start to ramble or lose your train of thought, ask one of your classmates or other audience members to sit in an easily visible spot in the room and discretely cue you. If you are presenting with a group, offer to do this for one another.
- *Don't take yourself too seriously.* Try to be relaxed and calm, and don't become flustered if something unexpected happens. Your audience will look to you for cues about how to respond to things. If you seem stressed out and upset, they will likely feel something similar. If you are able to dismiss a minor (or even major) goof or glitch calmly, your audience will follow suit.
- If you include a question-and-answer segment, *give questions some thought before you answer.* Don't assume there is a right answer to every question, and feel free to ask the audience member to repeat or rephrase the question. Avoid becoming defensive when you are being questioned. Take the questions seriously; do not just quickly throw some response out there to escape from the hot seat.
- *Avoid using noninclusive language.* Think carefully about words you use that might be racist, sexist, classist, ableist, homophobic, or otherwise alienating for audience members.
- *Don't talk too long.* You know from your own experience as a student that even the best listeners have a finite capacity for processing new information or staying focused as listeners. Pace your presentation so that it includes some variety; move from a short narrative to a slide, a handout, a video clip, or a presentation from another group member. If your subject is so engaging as to demand more time than you're allotted, refer your listeners to additional resources in a bibliography, or invite them to read your entire project or to visit your web page.

Serving as a Good Audience Member

An odd phenomenon often takes place in classrooms and other presentation spaces: Audience members sometimes imagine that because they are not the center of attention, they are practically invisible. No one will see them reading a newspaper, cramming for a final, or whispering to classmates. Although you will likely feel most invested in the presentation process when you are in the spotlight, your work as an audience member is also important. Before the presentations begin, note audience behaviors that you find comforting, and then try to follow these when you are in the role of listener.

Above all, show the speaker(s) that you are engaged with the presentation. Pay attention to what the speaker is saying. Jot down notes or questions if you're so inclined, but spend most of your time connecting with your speaker through eye contact, nodding your head, and providing other nonverbal cues. Remember that classmates are looking to you for support and feedback during their presentations and that you are likely to reap what you sow in this area. When a presentation ends, ask questions that show your interest and that are designed to elicit meaningful responses. Trying to stump the presenters won't be productive for anyone. If the presenters ask you to provide constructive criticism in assessment forms or debriefing sessions, give them your honest response, however.

The suggestions for preparing, delivering, and serving as an audience for oral presentations in the sections above are meant to be an introduction to this process. We invite you to check out more detailed texts on this subject, especially Laura Gurak's *Oral Presentations for Technical Communication.*

Presenting Your Project with a Memo/Letter of Transmittal

Whether you present your final product to your sponsoring organization in a formal oral presentation or instead deliver it or mail it, you'll want to include an accompanying cover memo or letter called a **memo/letter of transmittal.** Many longer, more formal professional or technical documents (e.g., proposals, research reports, feasibility studies) are accompanied by such a memo or letter. You can think of a transmittal memo/letter as the written *announcement* to readers (often supervisors or clients) about what you're presenting to them. As such, it should probably remind your readers about the larger project and its final product, especially if they are involved in multiple projects.

If you send a text to your instructor or contact person as an email attachment, the email transmitting the attachment could be considered an electronic transmittal memo and should include a description of what you're sending. When you deliver or send the print version of your final products to your organization, use a transmittal letter if you want to be more formal or if you and your teammates functioned more like consultants than provisional members of the organization. If, however, you feel more like part of the organization, as most of our students have, you can use the memo form. Either way, include a specific subject line that signifies your delivery of the final product.

Your transmittal letter/memo can have several functions besides announcing and explaining the accompanying document(s). First, it can help shape how your audience will approach what you're transmitting. You can help predispose them to read favorably by emphasizing how your texts and project were successful and by highlighting the most effective, impressive features of those texts. Direct your readers to the elements to which you want them to pay the most attention.

Second, your transmittal letter/memo can reinforce your relationship with your readers, especially in terms of the values you share. Because this may be the last correspondence you write to them, it may be one of the last chances you have to end the project on a positive note. One way you can do this is by reaffirming your shared goals for your final texts, the beneficial effects that both you and your sponsors want the texts to have in the community. Another way you can end on a positive note is to thank your contact person and others at the agency for facilitating your learning and for giving you the opportunity to contribute to their valuable work and to make a difference in your local community. Some transmittal memos/letters acknowledge all of the people who helped the writers; in addition to your contact person, you may want to acknowledge the head of the agency (if this person is not your contact) and all of the other agency personnel who helped you. Some of our students have also written their sponsor a separate thank-you letter or email when the sponsor was especially helpful. In turn, your contact person might write you a thank-you letter for your writing portfolio.

Reinforcing your relationship with your audience might be especially helpful if you want to continue volunteering at the agency and/or if you would like to use your contact person as a future job application reference who can address your writing and other relevant qualifications. You may even want to ask your contact person in the transmittal memo/letter if she or he would be willing to possibly serve as such a reference. We advise asking this only if you know your readers are pleased with your performance.

Figure 10.1 shows the cover memo one group used to transmit their service-learning project to their two contacts at the American Cancer Society. This sample demonstrates that a transmittal memo/letter need not be long to fulfill its multiple functions. The opening paragraph both announces the final delivery of the two texts and reaffirms their desired uses and effects. The writers' description of their goals for the texts efficiently reminds readers about the texts' rhetorical situations.

The next paragraph highlights the most effective features of the major text they produced, pointing to the packet's user-centered, accessible design. The memo ends with a thank-you paragraph that recognizes the different ways in which the agency helped the writers. Notice how the third sentence smoothly leads into their request for a reference by describing how the readers already helped them professionally.

The letter of transmittal may also be an opportunity for you to follow up on a concern you raised in your progress report. Perhaps you faced an ethical dilemma or logistical glitch and found a way to work through it during the final phase of your project. It would be appropriate to refer to this fact briefly to preserve continuity and to give the reader a full sense of what you've accomplished.

Updating Your Résumé

Updating your résumé is something you should do regularly, especially after completing a major academic or work-related project. Doing this while the project is fresh in your mind will save you from racking your brain for specific achievements later, although you should keep track of how your service-learning texts are used and what they accomplish so that you can add this information.

Although some of you might have had a hard time keeping the résumé you created at the project's beginning to one page, others might have struggled to

MEMORANDUM

TO:	Diana Child, Area Administrative Assistant of ACS
	Lauren Dean, Area Cancer Control Director of ACS
FROM:	Team Cancer Fighting Gators: Jason Law, Katie Plumlee, Matt Lilley
DATE:	April 12, 2001
SUBJECT:	Delivery of Patient Information Packet and Relay for Life Press Release

We are pleased to deliver the final versions of our Relay for Life press release and patient information packet to the American Cancer Society. We hope that press release not only informed the community of your worthwhile fundraising event, but also led to a number of additional participants and donors. We are confident that the information packet will prove to be a valuable resource for area health care providers to pass on to cancer patients in their time of sickness and beyond.

We are especially pleased with the "staggered" design of the information packet, which enables the patient to find basic and more detailed information about several types of resources upon first opening the folder. The individual sheets are clear and uniform but still engaging, and the folder design enables you to also include brochures and contact cards.

Thank you for allowing us the opportunity to work with your agency on such redeeming educational and client service efforts. You have helped us grow as community citizens; we know that we will continue to give back to whichever communities we live in. You have also helped us apply and enhance our knowledge about professional writing and design, and we would like to ask if we could possibly include you as a reference in the future. If you have any questions or comments, please do not hesitate to contact us.

Figure 10.1 Transmittal memo for final documents

come up with specific descriptions of relevant qualifications and accomplishments. Your experience with the service-learning project should be a valuable source, as it likely involved a number of tasks and skills relevant to most jobs, internships, or other positions. You might want to replace a dated, scant, or vague subsection of your résumé with one about your project.

You can incorporate your service-learning project into your résumé in various ways. We recommend developing an entire subsection about the project, complete with your title (e.g., Technical Writing Assistant), the agency's name, and a bulleted list of achievement statements. You could place such a subsection in one of several sections, including *Education, Related Work Experience,* or *Community Service.* After all, you were assigned the project as part of a course, you worked as a member of a professional writing team in a workplace setting, and your writing helped the agency serve the community. You should probably place your project description in whichever section seems the weakest and/or whichever section your résumé readers will view as the most crucial. Remember how you tailored the previous version of your résumé to highlight your writing and community service accomplishments.

You can also incorporate your description into other sections. In addition to including a mini-section on the project, you might want to mention specific skills you applied in a separate *Relevant Skills* section, especially if you're designing a primarily skills-based résumé. The revised résumés at the end of some appendixes show various ways our students have added the qualifications gained through their service-learning projects.

The typically transferable skills you can show in a specific description of your project include the following.

- document design
- professional/technical writing, including collaborative writing, to various internal and external audiences
- revising and editing
- project management, including reporting to multiple supervisors
- teamwork, including collaboration with other group members, classmates, agency personnel, and potential audience members
- leadership.

In describing these skills and the actual products of your project, remember to be specific and make them sound impressive as qualifications and achievements; for example, mention the different genres of texts you produced (e.g., proposal, progress report) to help you manage the project, the specific leadership roles you served, and the software, style guides, and document grids you used to design the texts. As we advised in Chapter Four, providing specific evidence of your qualifications will make you more credible and memorable. Remember to begin your achievement statements with specific action verbs.

The most important achievements to describe are the final products themselves. Your description should clarify their audiences, purposes, and uses; note their complexity, and emphasize the positive effects the texts will have on the community. Quantify their uses and/or effects wherever possible. Maybe, for instance, the instructions you produced will be used by more than 20 offices in the region, the grant proposal you wrote will likely procure $5,000 for the agency, or the brochure you produced will be given to 1,500 new students at your university. Document these facts in your résumé.

To illustrate what we've been discussing, let's turn to the example embodied in Figures 10.2 and 10.3, the *before* and *after* résumés of one of the students in the American Cancer Society group mentioned earlier. Figure 10.2 shows the résumé Jason submitted along with his letter of inquiry to the agency. This résumé is well suited to its rhetorical situation, persuasively emphasizing Jason's volunteer activities, including one that involved writing and editing. The subsection about the Cuddler Program doesn't describe as many transferable skills and qualifications as the others, however, and the *Special Skills* section seems to have some fluff (knowledge of Windows 95 and 98 is not really worth mentioning, for example, and the second bulleted item is too vague to be helpful).

JASON RYAN LAW

900 SW 99th Place Apt. 9 (352) 999-9999
Gainesville, FL 32608 jrl99@hotmail.com

OBJECTIVE

To donate my writing, planning, and editing skills to a local health-related service organization with communication needs.

EDUCATION

B.S. in Finance, University of Florida
Expected graduation May 2001
Grade point average, major: 4.0 Grade point average, overall: 3.61

EMPLOYMENT AND VOLUNTEER HISTORY

Revision Reader and Proofer, Gleim Publications, Inc. Gainesville, FL, 2000

- Revised and proofread various complex manuscripts and text-based computer documents for accounting test preparation/review books.
- Assisted in writing questions and explanations for accounting practice exams.

Volunteer in Cuddler Program, Shands Hospital Gainesville, FL, 1996–97

- Visited sick and lonely children in various areas of the hospital.
- Rocked, bottle-fed, and changed diapers of sick and premature infants.

Counselor and Group Leader, Florida Sheriff's Youth Camp Barberville, FL, 1996

- Supervised and took full responsibility for the activities and general well-being of 12–15 socially and/or emotionally challenged youth in a rustic camp setting.
- Interacted weekly with volunteer deputies.

SPECIAL SKILLS

- Fluent with Microsoft Office Suite, including Excel, Word, and PowerPoint
- Knowledge of both PC and Mac applications
- Knowledge of Windows 95 and 98

LEADERSHIP ACTIVITIES

- Omicron Delta Kappa National Honorary
- Golden Key National Honor Society
- University of Florida Honors Program
- Fellowship of Christian Athletes
- University of Florida Intramural Athletics
    - Participant, head coach, and team captain of more than 12 softball, basketball, and flag football teams.
    - Recruited team members, organized practices and games, assigned positions.

Figure 10.2 Résumé before service-learning project

JASON RYAN LAW

900 SW 99th Place Apt. 9
Gainesville, FL 32608

(352) 999-9999
jrl99@hotmail.com

OBJECTIVE

To donate my writing, planning, and editing skills to a local health-related service organization with communication needs.

EDUCATION

B.S. in Finance, University of Florida — Expected graduation May 2001
Grade point average, major: 4.0 — Grade point average, overall: 3.61

EMPLOYMENT AND VOLUNTEER HISTORY

Writer and Document Designer, American Cancer Society — Gainesville, FL, Spring 2001

- Leader of three-person team of University of Florida student volunteers who produced needed texts for the agency over a semester.
- Designed and wrote a patient information packet that will likely be used by hundreds of physicians as a cancer patient resource.
- Wrote a press release for Relay for Life that was published in local newspapers to encourage participation and donations.
- Wrote a series of professional documents—including a letter of inquiry, proposal, progress report, and final report—to manage our work and update agency and university supervisors.

Revision Reader and Proofer, Gleim Publications, Inc. — Gainesville, FL, 2000

- Revised and proofread various complex manuscripts and text-based computer documents for accounting test preparation/review books.
- Assisted in writing questions and explanations for accounting practice exams.

Counselor and Group Leader, Florida Sheriff's Youth Camp — Barberville, FL, 1996

- Supervised and took full responsibility for the activities and general well-being of 12–15 socially and/or emotionally challenged youth in a rustic camp setting.
- Interacted weekly with volunteer deputies.

SPECIAL SKILLS

- Fluent with Microsoft Office Suite, including Excel, Word, and PowerPoint
- Editing and proofreading print and online documents; developing editing style sheet
- Designing documents from document grids to final products

LEADERSHIP ACTIVITIES

- Omicron Delta Kappa National Honorary
- Golden Key National Honor Society
- University of Florida Honors Program
- University of Florida Intramural Athletics
    - Head coach and team captain of softball, basketball, and flag football teams.
    - Recruited team members, organized practices and games, assigned positions.

Figure 10.3 Résumé after service-learning project

Figure 10.3, Jason's revised résumé, shows how he removed the subsection about his volunteer work with the Cuddler Program, making room for a larger, more substantial, and more relevant subsection about his service-learning project, headed by his title of *Writer and Document Designer.* This subsection fits neatly into the larger *Employment and Volunteer History* section. After briefly overviewing the project in the first bulleted item, Jason presents a set of achievement statements that describe the texts he helped produce and that convey their significance. Notice how he gives their specific names and quantifies their readership. Jason follows this up with a list of the professional genres he produced throughout the project, which further demonstrates his experience as a professional writer and his ability to manage a large project from start to finish. Just as he refers to his project management skills in the last bulleted item, Jason refers to his leadership role in the first statement (although this could have been more specific).

Jason also incorporates his experience from the service-learning project into the *Special Skills* section, which needed some development. Replacing the two unimpressive skills in the previous résumé are editing and document-design skills. Instead of simply stating these skills, as anyone could do, Jason gives supporting details, such as developing a style sheet and document grids.

Updating your résumé is one way to make your unique service-learning experience work for you in the job application process. Of course, each member of your group should each keep both an electronic and a print copy of your project, including the proposal and other project management documents, to add to your writing portfolio.

Final Reflection

As you reach the end of your course, we hope that you will take some time to reflect on your service-learning experience. Many of you will be under stress-inducing deadlines as you rush to revise and submit your final projects to your instructor and contact person. You may be anxious about final presentations and concerned about accomplishing all of the other things that are required of students at the end of an academic term. We hope that when you have had a break from the work and some time to let the experience sink in, you will find it to be uniquely satisfying. Whether you produced a document that will secure funds for an important social program, that alerted members of the campus community to a danger, or that helped a business fulfill its philanthropic goals, your decision to use and develop your rhetorical skills in the interests of your civic commitments sets you apart from most students and citizens.

We recognize that many of you participated in this process because it was a course requirement, and from our point of view that does not in any way diminish the contribution that you have made to your community. As we, too, reach the end of the long process of writing and editing this book, we are thrilled that students across the country will be engaging in rhetorical activism as they read and employ the ideas in this book. Many of you may continue the relationship you have started (or continued) with your sponsoring agencies well into the future;

others may choose to work with another such group at some point in the future. We hope that all of you will find what you've learned here useful as you pursue your future careers. We wish you the best in all you do as effective and ethically engaged rhetors.

Activities

1. Design a listening guide for your fellow classmates to use while you are giving your presentation. Consider giving them questions to answer, points to consider, and areas in which you'd like feedback from them.
2. Participate in an online or in-class discussion about your worst feared or lived experiences in giving oral presentations. Also discuss positive experiences and strategies for recreating them.
3. Brainstorm as a group about ways to incorporate your accomplishments into your updated résumé. Exchange drafts with a group of classmates to provide feedback and share ideas.
4. Create an online version of your class presentation to which you, your agency, and your instructor can refer in the future.

Works Cited

Gurak, Laura J. *Oral Presentations for Technical Communication.* Boston: Allyn & Bacon, 1999.

APPENDIXES
SAMPLE STUDENT PROJECTS

Appendix A: Team B.A.R.K.

Team B.A.R.K. (Broadcasting Animal Rights Knowledge) consisted of four professional writing students—Sharon Sandman, Ryan Pedraza, Katie Roland, and Loretta Belfiore—from the University of Florida. For their service-learning project, they worked with the Alachua County Humane Society to produce a comprehensive Volunteer Training Manual that will be used as a training tool and reference handbook for new volunteers. The manual, which the students designed to be easily expandable, was readily adopted by the agency and will likely be used as a template for other Humane Society chapters as well.

The following texts might be thought of as part of the "document trail" through Team B.A.R.K.'s service-learning project. The group's progress report appears in Chapter Eight. Since two students wrote the agency in their letters of inquiry, we've included both here.

1) Sharon's Letter of Inquiry
2) Sharon's Résumé
3) Ryan's Letter of Inquiry
4) Ryan's Résumé
5) Trip Report
6) Proposal
7) Discourse Analysis
8) Sharon's Revised Résumé
9) Ryan's Revised Résumé
10) Transmittal Memo
11) Final Product: Volunteer Training Manual

In answering one of our final reflection questions in Chapter Nine about the stages their group went through, Katie said that the entire project was "invigorating" but that the group also experienced frustration, though not until the end of the project. "As we were working down to the wire, it was really hectic," she explained, due to time limits and "strained communication." She also wrote that this frustration was followed by "relief and a sense of accomplishment when Audrey [their contact person] said she was extremely satisfied with our final results." Sharon, another group member, reported a final stage of "exhaustion and relief."

In reflecting on how this project was different from others he had engaged in, Ryan explained how his motivation changed after the group began meeting with Audrey weekly and especially toward the end of the project. "During the last few stages of production," he wrote, "our team looked at our project in terms of, 'Is this going to help the agency and its volunteers?' more than, 'Are we going to get an A?' Initially I felt we [the group and agency] were both using each other," Ryan continued. "Now, when I look back, I see how we became part of the agency and they became part of our group."

Sharon Sandman
444 NW 4th Terrace
Gainesville, FL 32601

352.444.4444
ssand@ufl.edu

January 23, 2001

Audrey Holt, Volunteer Coordinator
Alachua County Humane Society
2029 NW 6th Street
Gainesville, FL 32609

Dear Ms. Holt,

I would love for the Alachua County Humane Society to be involved in a project we are working on at the University of Florida. This semester I am taking a Professional Communication course focusing on writing in the community. The class will get to assist a few select organizations with writing projects. I would like to help with the communication needs of the Alachua County Humane Society because I share the desire to prevent cruelty to all animals through education, sterilization, and legislation. Also, I noticed on your website that you are seeking writing volunteers.

The Humane Society produces many of the same texts that we are qualified to produce. We can help, for example, with the quarterly newsletter, news releases and public service announcements, information bulletins, mailings, or even media presentations to be used in the educational program. We can also assist in writing to the legislature. The scope of this project is between 10-20 pages of written text depending on complexity and time constraints. The time frame we have to work with is February 20th through April 10th.

The class will divide into teams of three to four students, and each team will select one organization. The individual expertise of each team member will enable us to encompass a very broad writing skill set. (The enclosed resume will illustrate my writing background and accomplishments.) The projects will also involve a contact person from the agency who will guide us. We will also be working closely with our professor to refine our skills and better attune them to your needs. We can make weekly visits to coordinate the types of assignments that will most benefit the Humane Society. Again, our focus will need to stay in the realm of communication needs.

I hope you are excited about this project my classmates and I are presenting to you. With our assistance, you will be able to distribute more information, which will help the Humane Society grow and accomplish goals. I expect the demand from contacted organizations to be high. Since we can only help a handful this semester, I urge you to take part in this opportunity. If you are interested, please contact me as soon as possible at the phone number or email address above. Our professional communication team along with the Humane Society can generate an immeasurable asset for the community. I look forward to discussing possible communication projects with you soon.

Sincerely,

Sharon Sandman,
UF Accounting Senior

Enclosure

SHARON L. SANDMAN

ssand@ufl.edu

Permanent
888 NW 8th Street
Longwood, FL 32779
(407) 888-8888

Local
444 NW 4th Terrace
Gainesville, FL 32601
(352) 444-4444

OBJECTIVE

To apply and adapt my writing, design, and leadership skills to the communication needs of a local organization.

EDUCATION

B.S./M.A. Joint Degree in Accounting, May 2002.
University of Florida, Gainesville, FL
GPA: 3.5/4.0

Related Courses: Technical Writing; Introduction to Computer Software; Public Speaking; Professional Communication; Expository and Argumentative Writing; Advanced Expository Writing; Leadership

Computer Skills: Microsoft Office Suite (Word, Excel, PowerPoint); Word Perfect; Adobe Photoshop; HTML

RELATED WORK EXPERIENCE

Creative Tax Services, Total State Tax Solutions Summer Associate, Deloitte & Touche LLP, Atlanta, GA, May 2000-August 2000

- Prepared multiple PowerPoint presentations used for client engagements
- Organized and coordinated research projects to be completed by multiple associates in various offices nationwide
- Communicated through email, memos, and telephone daily
- Prepared detailed matrices

Intramural Chair, Alpha Chi Omega Sorority, Gainesville, FL, January 1999-January 2000

- Prepared a weekly newsletter, calendar of events, and team rosters
- Designed flyers promoting upcoming events
- Coached intramural tennis, basketball, and flag football
- Compiled a database of over 150 records to monitor participation
- Delegated tasks among a committee

VOLUNTEER EXPERIENCE

Cuddler Program Volunteer, Shands Hospital, Gainesville, FL, January 2000-May 2000
S.P.A.R.C Fundraiser Assistant, Alpha Chi Omega Sorority, Gainesville, FL, August 1998-August 2000

ACTIVITIES/ HONORS

Golden Key National Honor Society
Florida Academic Scholar
Fisher School of Accounting Dean's List
Fisher School of Accounting Council Member

Ryan Pedraza
3333 SW 33 St. #33
Gainesville, FL 32608
(352) 333-3333
rp33@ufl.edu

January 30, 2001

Melissa Gilkes, Director
Alachua County Humane Society
2029 NW 6th St.
Gainesville, FL 32609

Dear Director Gilkes:

My name is Ryan Pedraza and I am an English major at the University of Florida. I am writing to inform you of a terrific opportunity. My Professional Writing class is hoping to collaborate with a non-profit organization on a major writing project. The class's goal is to prepare documents for an organization while gaining real-world experience. Your organization interests me because I also believe that the public needs education about animal rights. In the past, I have helped collect pet food for the Highlands County Humane Society and now want to be of service to the Alachua County branch.

According to the Alachua County Humane Society's website, your organization needs volunteers in its Public Relations department. My class can fill this need by producing newsletters, fundraising materials, news releases, or educational pamphlets. I would work with two or three fellow students to create these documents. The course requirements stipulate that we produce ten to twenty pages of text. Our group can also produce documents with a heavy visual element. We can begin in late February and must complete the project by early April. Our services are completely free of charge. We only require someone from your agency to act as our supervisor and liaison to our professor. Our group would visit the Humane Society on a weekly basis to share our progress and obtain feedback from the supervisor.

Each student in our class is writing to an organization they are interested in working with. Only three or four organizations interested in this collaboration will be selected. We expect a large response, so please contact me soon if you feel the Alachua County Humane Society could benefit from our services. I believe our class would be a great asset in preparing documents for projects such as Body Revolution and the Spring Garden Festival. I look forward to talking with you and hope to assist your organization by raising money and educating Alachua County about animal rights.

Sincerely,

Ryan Pedraza

Enclosure (1)

RYAN PEDRAZA

3333 SW 33 St. #33
Gainesville, FL 32608

352-333-3333
rp33@ufl.edu

Objective

Apply my technical writing, editing, computer, and design skills to a collaborative writing project with a local non-profit organization.

Education

Bachelor of Arts in English and Political Science
University of Florida, Gainesville, FL
Anticipated Fall 2001
Grade Point Average: 3.72/4.0

Related Coursework
- Advanced Argumentative Writing
- Professional Writing
- Scholarly Writing

Employment

2/00-present

Leasing Consultant
Lexington Crossing Apartments, Gainesville, FL
- Explain apartment amenities to customers and sign leases
- Handle rent monies for property with over 1000 residents
- Prepare monthly newsletter for residents detailing coming events using Microsoft Office 2000 and Print Shop Premiere
- Write business letters, memos, and emails to potential residents
- Provide excellent customer service

8/99-2/00

Event and Technical Services
- Stephen C. O'Connell Center, Gainesville, FL
- Organized Security crews for large events such as football games
- Set up lights and audio equipment for concerts
- Provided assistance to patrons attending events

Awards

- Florida Academic Scholar Scholarship – 100% of tuition
- Alfred Dupont Scholarship 2000-2001
- Ronald E. McNair Scholar 2000-2001 (only 14 selected from pool of nearly 100 applicants)
- Golden Key Honor Society
- Alpha Lambda Delta (English Honor Society)

Team B.A.R.K. Memorandum

To: Professor Blake Scott
From: Ryan Pedraza, Sharon Sandman, Loretta Belfiore, Katie Roland
Date: February 19, 2001
Subject: **Humane Society Trip Report**

Team B.A.R.K (Broadcasting Animal Rights Knowledge) met with Audrey Holt, the volunteer coordinator for the Alachua County Humane Society, on February 15th at 3:30 p.m. Our purpose of the trip was to meet Audrey and to discuss possible projects. We also wanted to find out about the organization's resources, expectations, and time frame. While we were there, we toured the facility and met other personnel including the agency's director, Melissa Gilkes. We also gathered multiple sample texts and previously used brochures. Audrey suggested that we view related websites such as ASPCA, HSUS, and AHS to gain background information on animal rights.

The Humane Society's main need is for a comprehensive volunteer brochure. The brochure will highlight the organizational philosophies, volunteer procedures, and sample jobs, and will conclude by making volunteers feel valuable and encouraging their continued support. It would include heavy visual design elements. In addition to creating the brochure, we could perform a cost analysis of its production. Team B.A.R.K. could produce several versions of the brochure for the agency to use as needed.

The organization also needs parts of a grant proposal produced. The grant is financed by the Maddie's Fund. The Humane Society would form a coalition with Gainesville Pet Rescue, the Alachua County Animal Shelter, and the University of Florida School of Veterinary Medicine. If the coalition is awarded the grant, it will provide nearly $1 million to establish an extensive spay and neuter program.

Team B.A.R.K. plans to meet with Audrey on a weekly basis (Thursday afternoons) to share our progress and obtain feedback. The agency has no specified time frame, so it is willing to work around our class's due dates. Audrey is excited about working with Team B.A.R.K., and we are encouraged by the agency's nurturing environment.

Producing Volunteer and Educational Brochures for Alachua County Humane Society: A Project Proposal

Team B.A.R.K.:
Ryan Pedraza
Sharon Sandman
Loretta Belfiore
Katie Roland

Submitted to:
Professor Blake Scott
University of Florida

February 26, 2001

Introductory Summary

The Alachua County Humane Society is the local chapter of a national, non-profit organization that relies solely on the help and support from volunteers throughout the community. Volunteers are extremely valuable to the Humane Society not only because they do not require monetary compensation, but also because they perform a wide range of necessary functions within the organization. Without them the Humane Society would be unable to carry out its mission of educating the public about animal rights and population control measures.

The Gainesville branch of the Humane Society is in desperate need of a comprehensive brochure that would not only outline the philosophy and procedure for all volunteers, but also serve as a means for recruiting new volunteers and graciously thanking current ones. They have also inquired about help with drafting Maddie's Grant, a proposal comprising four local animal organizations and over $1 million for the adoption of strictly regulated animal reproductive measures.

Team B.A.R.K. (Broadcasting Animal Rights Knowledge) recognizes the need for a strong volunteer base in our local community, and for support of the philosophy and education of reproductive control for our domestic pets. We want to help the Humane Society produce a text better defining the role of the volunteer, and to assist in the construction of Maddie's Grant.

Our project, which consists primarily of producing a comprehensive volunteer brochure, will involve assessing the cost needs for production, locating funding and printing sources, and designing, writing, and revising multiple drafts, all in coordination with our project director Audrey Holt. Assisting with Maddie's Grant will require our writing and drafting skills and will allow us to gain invaluable collaborative skills. The proposed project will be completed between the dates of February 26 and April 10, 2001, under the supervision of Audrey Holt, volunteer director for the Alachua County Humane Society.

The remainder of this proposal will outline and designate the problems and solutions the project addresses, and will explain in greater detail how Team B.A.R.K. will successfully complete this undertaking. With this, our contact information, assigned deadlines, and designated tasks will be further discussed.

Problem: Raising Funds for Operational/Capital Needs

The Alachua County Humane Society is for the most part a no-kill shelter (i.e., shelter that does not practice animal euthanasia) in Gainesville. The shelter will only consider euthanasia when all efforts to work with an overly aggressive animal have failed. The Humane Society has the following mission: to educate the community about animal rights and to provide spay/neuter services and information. The shelter also provides housing and veterinary care to discarded dogs and cats until they can be adopted by new, responsible pet owners who will spay or neuter them.

The expanding number of stray or abandoned animals in Alachua County has caused area shelters to become overcrowded, as more and more animals are being euthanized by county animal control facilities. As a result, the Alachua County Humane Society has been forced to operate beyond its capacity and its current funding sources.

Funding Problem

The Alachua County Humane Society is a non-profit organization that operates through private donations by individuals (raised mostly through special events, letters, and the organization's website and newsletter). These funding sources are not meeting the shelter's current operational and capital needs, listed below:

Operational Needs

- Dog and cat food
- Cleaning supplies (e.g., litter, racks, pails)
- Medical and veterinary expenses
- Office supplies (e.g., file cabinets, copier)
- Other supplies (e.g., cages, collars)

Capital Needs

- Cooling system and fans
- Rent, overhead costs
- Plumbing and other repairs
- Vans for transporting
- Traps

Writing Problem

Brochures

The Humane Society needs more volunteers to help it fulfill its mission of educating the larger community about animal rights and protection and about the different spay/neuter programs available in the community. Although the agency has a website and newsletter that can be used to inform and persuade potential volunteers, it needs a print text dedicated to this function, specifically a comprehensive volunteer brochure that can explain the agency's volunteer opportunities, recruit new volunteers, and thank past and present volunteers. The brochure must be of professional quality and include visuals. None of the current staff has the time or desktop publishing skills to revise or produce new brochures, however. The staff is too busy with the demanding task of running the shelter.

Grant Proposal

Because of the expanding operational and capital expenses, the Humane Society needs a way to raise larger sums of money. This funding can be raised through grant proposals, but none of the current staff has the time to write one. The shelter relies on volunteers and has only a small staff of paid workers who are already overworked and lack grant-writing expertise. The shelter is aware of a large grant called Maddie's Fund and hopes to be able to find able volunteers who can assist in the research and writing of this project. For this grant, the Humane Society would join other local animal organizations to form a large coalition. If successful, this grant would provide the coalition with approximately $1 million toward a large spay/neuter/adoption facility in Alachua County. This much-needed grant would help solve many of the community's animal shelter and hospital needs, including responsible animal care and protection.

Solution: Creating Brochures and Grant Writing

After consulting with the Humane Society, Team B.A.R.K. determined that the best way to satisfy the agency's needs is to produce two separate brochures. Although the agency proposed a single brochure, they required it to include a lot of information. Including too much information in a single brochure may compromise its usefulness. Therefore, one brochure will be targeted at potential volunteers, while the second will focus on thanking current and past volunteers and encouraging their future support. Team B.A.R.K. will perform the following tasks to ensure an adequate brochure design.

Task 1: Compile a list of information to be included in each brochure,
Task 2: Assist Audrey in performing a cost analysis to determine the types of brochures that can be created,
Task 3: Research other literature produced by animal rights groups and sample texts from the Humane Society to determine layouts and visuals,
Task 4: Create a first draft of the brochures,
Task 5: Revise and edit brochures.

Task 1: Compile a list of information to be included in each brochure

Team B.A.R.K. has met with the Humane Society to determine the information to include in each brochure. Our meeting provided us with a preliminary list that will be supplemented by future meetings and brainstorming sessions. The first brochure will be targeted at potential volunteers. Our preliminary list of information to include in the brochure includes:

- Contact information for potential volunteers
- Skills the Humane Society needs (writing, office work, etc.)
- A list of typical jobs performed by volunteers
- Testimonials from past volunteers who had positive experiences
- A heavy visual element including pictures of volunteers in action and animals awaiting adoption

The second brochure will thank current and past volunteers and encourage their continued support. Our preliminary list includes:

- Text thanking volunteers
- Statistics showing how volunteer support helps the organization grow and the number of animals that have been adopted

Task 2: Assist Audrey in performing a cost analysis to determine the types of brochures that can be created

We will consult the agency to determine their budget for the two brochures. Team B.A.R.K. will also contact Renaissance Printing (who have donated free services to the Humane Society in the past) to determine if any portion of the production costs can be donated. Our team is also contacting other printing services in the area. After determining the budget, we will be able to choose the best layout for our text.

Task 3: Research other literature produced by animal rights groups and sample texts from the Humane Society to determine layouts and visuals

Audrey gave us websites to several prominent animal rights agencies and many samples of brochures. After gathering the necessary information and making basic decisions about document design and visuals, we will experiment with different document grids. We will get preliminary feedback on the most effective grids from Audrey and Professor Scott. Audrey's input regarding cost will be especially helpful at this stage.

Task 4: Create a first draft of the brochures

Next we will create a first draft of each brochure. At first we will split up the writing of the brochure according to section. Loretta will provide relevant research information to all members before

according to section. Loretta will provide relevant research information to all members before writing, and Katie will make the first draft of all visuals. Then we will exchange drafts via email, make suggestions, and go over those suggestions during our meetings. To ensure that we all follow the same design, we will make our document grid into a template and distribute this among group members.

Task 5: Revise and edit brochures

Finally, Team B.A.R.K. will workshop, revise, and edit the brochures. One workshop will be an in-class one with our peers as part of the course. Another one will be with Audrey, before which we will give her a list of questions to address in her feedback. Audrey's help will be especially valuable because she is more familiar with the intended audiences. If possible, we will also ask a couple of actual volunteers for comments. After getting feedback from these various sources, we will as a group come up with a revision plan and then split up revision tasks. Once revisions are made, we will edit the design and writing of the brochures all together, looking for minor mistakes and inconsistencies. Finally, we will give copies to the Humane Society for distribution and to Professor Scott for evaluation.

As a smaller project, Team B.A.R.K. will also assist in writing a grant for the Humane Society. Known as Maddie's Grant, it will provide nearly $1 million for capital improvements. The money will be divided between the Humane Society, Gainesville Pet Rescue, the Alachua County Animal Shelter, and the University of Florida College of Veterinary Medicine. This coalition will use the funds to develop a more comprehensive spay/neuter program. Our roles will be limited in scope; we will help with research, editing, and a small portion of the writing.

Management: Collaborating & Completing Tasks

Now that we have explained our projects' components, we will outline how we will conquer our tasks within the given time frame. The first part of this section will introduce everyone who is involved with the project and highlight relevant qualifications and experiences of Team B.A.R.K.'s members. We will also explain the roles of each team member and how we will collaborate with each other, the agency, and our professor. The second part of this section will present our timeline for completing the project and its parts as well as highlight all critical dates.

Project Participants and Related Work Experience

A list of contact information for all members involved with this engagement is shown in Table 1 below. Team B.A.R.K.'s contact person at the Alachua County Humane Society is Audrey Holt, the volunteer coordinator. She has agreed to meet with us on a weekly basis—exact times and locations are presented in the "Meetings and Timeline" subsection of this report—to provide feedback and to advise Team B.A.R.K. along the way. The Team and Audrey will stay updated on recent accomplishments through email as well. Also, Professor Scott will supervise our work and refine our skills throughout the course of this project.

Table 1. Contact Information for Project Participants

Name	Phone	Email
Ryan Pedraza	271-1569	rp33@ufl.edu
Sharon Sandman	336-9968	ssand@ufl.edu
Loretta Belfiore	846-5518	loree1@ufl.edu
Katie Roland	378-4274	yamarkt@ufl.edu
AudreyHolt	373-5855	zot@fdt.net
Blake Scott	392-6650 ext. 280	jbs148@english.ufl.edu

The members of Team B.A.R.K. possess a broad range of skills. The following is a list of our relevant qualifications:

Ryan Pedraza, Project Coordinator and Liaison to Agency

- English major
- Great communication skills
- Coached a UF intramural football team and demonstrated excellent leadership skills

Sharon Sandman, Writing Leader and Liaison to Course

- Organized and coordinated research projects to be completed by multiple associates in various offices nationwide
- Designed multiple PowerPoint presentations used for client engagements
- Prepared a weekly newsletter, calendar of events, and team rosters while intramural chair for Alpha Chi Omega sorority

Loretta Belfiore, Research Leader

- English major
- Very familiar with numerous library resources, including databases on the network
- Managed three companies and headed one company through chapter 11

Katie Roland, Design Leader

- English major
- Layout editor for class yearbook
- Amateur interior design specialist

Organization

We strategically divided up the leadership roles based on the above qualifications of each member. Ryan has experience leading a team of people along with great communication skills. Since he also has the closest relationship with the agency, he will be the project coordinator and main liaison between the team and the agency. He will be the one to contact Audrey through email and set up meeting times for the group. Each member will communicate with Ryan to ensure that he gathers all necessary information from Audrey so the rest of the group can be productive.

Sharon also has taken on leadership positions in the past and has done well in several relevant English courses. She has strong communication skills as well, so she will be the writing coordinator and liaison to the course. Sharon will be in charge of compiling the parts of the project and putting them together. She will make sure that there is consistency among the style and make any necessary adjustments. Sharon will also be the main liaison to Professor Blake Scott. All projects turned in through email will be her responsibility.

Loretta has a lot of real-world experience that has required a high level of knowledge. She has utilized many research tools to obtain such knowledge. She is also very efficient in using the library databases at the University of Florida. This is why Loretta is the Team's research coordinator. She will gather all of the pertinent information that will go into our brochures and conduct research about the grant.

Katie has been interested in design elements ever since she was the layout editor for her school's yearbook. Over the years she has improved her skills by giving interior design advice to friends and throughout her major English projects. Katie is our design leader. She will make sure the layout of our projects is sufficient and add most of the graphics to our project. Figure 1 below is an organizational chart of project participants that shows the communication links among each member.

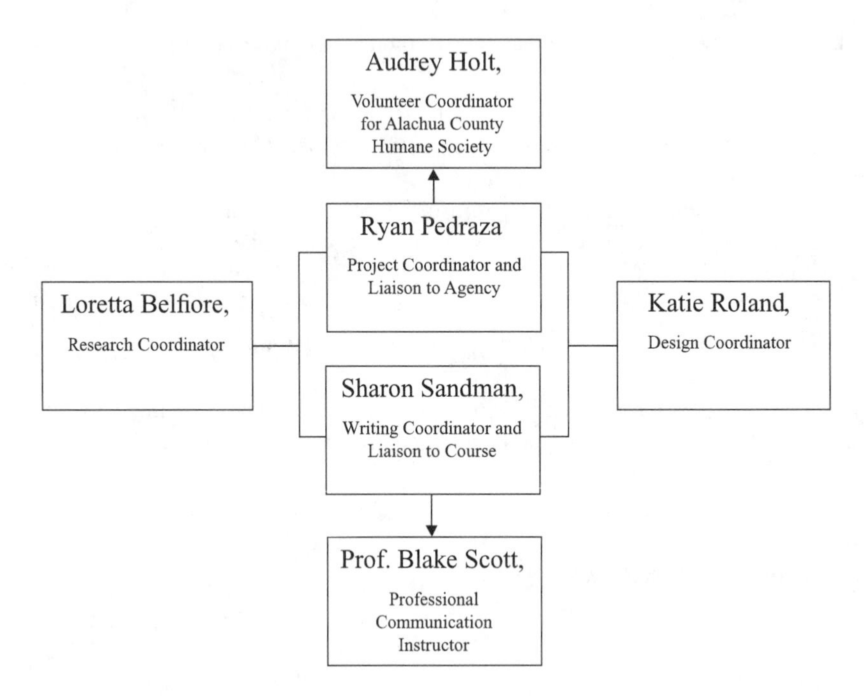

Figure 1: Organizational Chart of the Team and Our Supervisors

Our collaboration will be a hybrid of the three different schemes. Most of the initial writing will be divide and conquer while much of the research and design elements will be sequential. We are going to implement a sequential collaboration technique for our revising and editing. This will allow each member to have input on each part of the project. During our class meeting times, the team will set goals for the next group session.

Team B.A.R.K. has discussed available meeting times and gone over the requirements for this project thoroughly. The following section will detail how we will use our time to implement our solution.

Meeting and Timeline

In addition to class time designated for working on our projects, Team B.A.R.K. plans to meet on Friday afternoons from 3:30-5:30 p.m. at Loretta's residence as needed. Loretta has a computer accessible for our use, so this will enable us to design, draft, and revise online, thereby saving time and effort. When we have less to work on, we may decide to conduct meetings through email or by phone. All in-person contact with Audrey at the Humane Society will take place on Thursday afternoons from 3:30-5:30 p.m. Our schedules are flexible in case conflicts arise and we need to reschedule a meeting time. We also understand that we will need to hold additional meetings at certain points throughout the project, especially when critical dates are upon us.

Figure 2 is a milestone chart showing the major tasks and their time frames, critical dates, and the group coordinator of each task. Critical dates are bolded in the chart and listed below.

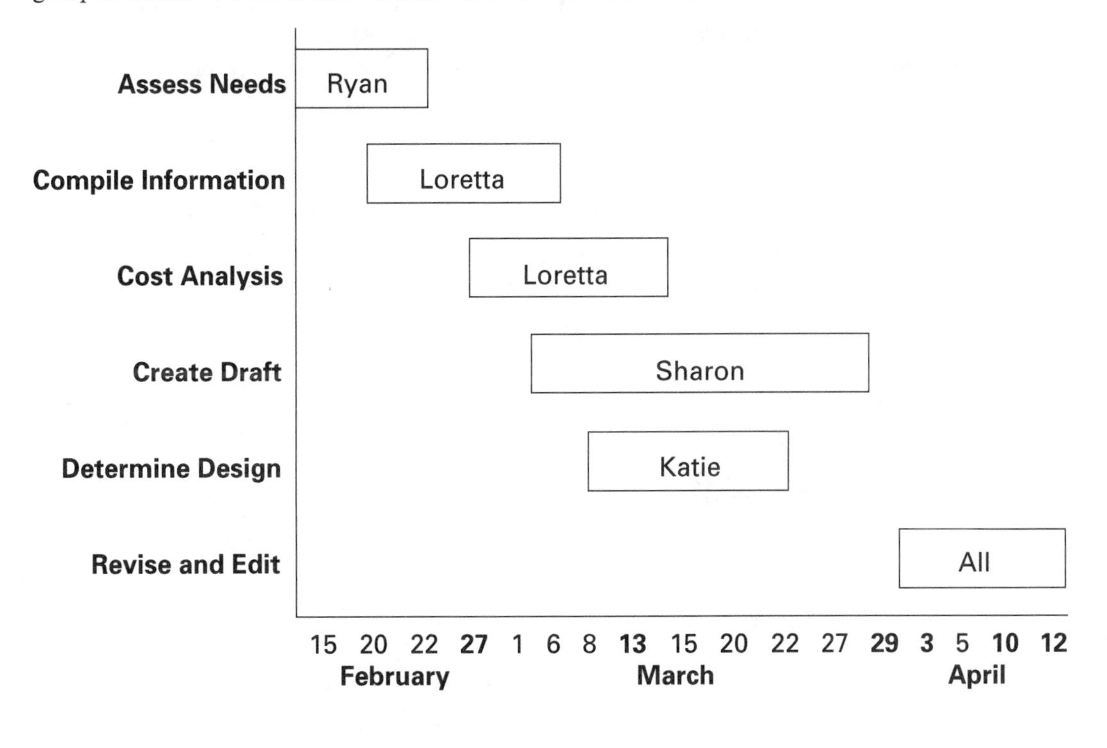

Critical Date	**Assignment**
Feb. 27	Proposal Due
March 13	Progress Report
March 29	Writing Workshop with Humane Society
April 3	Writing Workshop with Class
April 10	Evaluation Report
April 12	Presentation of Final Project in Class and Delivery to Agency

Figure 2. Milestone Chart Showing Major Tasks and Timeline

Team B.A.R.K. is confident that we have budgeted ample time to make all necessary revisions for a polished piece of work. We have already completed the first task of assessing the agency's needs. Although we will not be completely finished gathering information for our brochures when we begin writing, we should have enough to initiate the writing process. We have centered our tasks on the six critical dates and assignments highlighted in the chart. The agency expects to receive our finished project on April 12th.

Conclusion

Team B.A.R.K. is very excited to begin the bulk of this project. Our commitment to the agency and our parallel views in preventing cruelty to animals through education and sterilization will enable us to prepare very feasible brochures and influence the grant in a positive fashion. We have the necessary leadership, design, and writing skills it takes to produce these outcomes. We also expect to meet the urgent needs of this community by making this project a successful endeavor.

Team B.A.R.K. Memorandum

To: Professor Blake Scott
From: Ryan Pedraza, Sharon Sandman, Loretta Belfiore, Katie Roland
Date: March 2, 2001
Subject: Discourse Analysis of Humane Society Texts

The purpose of the forum analysis is to help our group analyze our agency's needs and constraints, the characteristics of our audience, and the needs of the discourse community as a whole.

Background

The discourse community consists of current volunteers, people interested in animal rights, and pet owners seeking information. Our texts will be available throughout Alachua County, thus making the discourse community quite large. Our discourse community's ethos is centered around animal rights issues.

Audrey Holt has been responsible for creating the Humane Society's text such as brochures. Previously, the volunteer brochure consisted of one half-page note thanking volunteers for their time and brief instructions on what volunteering entails. The text is difficult to read and contains only one visual. No contact information or request for continued involvement is included.

The Humane Society's current text does not include any information about its mission or values. However, it is made to educate the community. Education is an important aspect of the Humane Society.

Cost and constraints will depend on the agency's budget. Team B.A.R.K. will present several options for the Humane Society to choose from. Type of paper, colors, and other aesthetic options will be determined by the agency. The agency's time constraints will coincide with our course's requirements. All texts will be completed by April 10, 2001.

Discourse Conventions

Who Supervises the Discourse?

Audrey Holt, Volunteer Coordinator, is supervising our project. She is a former doctoral candidate from the University of Florida. Her writing experience is quite extensive. She taught a college composition course for several years and currently teaches English at a local school.

Audrey expects a professional and error-free brochure. She also expects several versions of the text in order to give the agency options. She will watch us very closely and take an active role in the editing process. Her desired means of communication is through our weekly meetings (Thursday afternoons) and by telephone.

Who is the Audience?

The primary audience is potential volunteers and the secondary audience consists of all those who are interested in animal rights and want to learn more about the agency's mission and activities. The knowledge level of the audiences ranges greatly. Both audiences will want the texts to be informative. Potential volunteers will be motivated to read our text in full because it will provide pertinent information about their duties. Motivation of other users of this text will vary.

What do they write about?

The Humane Society writes about:

- Preventing cruelty to animals
- Spay and neuter programs to prevent pet overpopulation
- Education
- Adoptions
- Pet-friendly legislation and funding
- Programs such as Pet-Sharing and Pet Meals on Wheels
- Reporting dog bites and acts of cruelty towards animals

Our text will include:

- Instructions for volunteers
- Agency's goals and objectives (education, animal rights, etc)
- Testimonials from past volunteers to encourage others to become involved
- Thank yous to past volunteers

Sources for information:

- Past brochures from the Humane Society (both Alachua county and national chapters)
- University of Florida College of Veterinary Medicine
- ASPCA website
- Alachua County Animal Shelter
- Gainesville Pet Rescue

How do they write it?

The Alachua County Humane Society produces simple, easy-to-read documents with a sense of humor. The small brochures and flyers are relatively simple, lacking visuals and color. The text is written mostly in present tense and second person. It consists mostly of short, active sentences and only a few technical terms, namely "no-kill," "spay," and "neuter." The documents are designed so that laypeople can easily comprehend.

The Humane Society's brochures follow a tri-fold form. Every brochure contains the agency logo. Headings are used and important information is presented in lists to promote easy access. However, the contact information is difficult to find in some of their publications.

SHARON L. SANDMAN

ssand@ufl.edu

Permanent
888 NW 8th Street
Longwood, FL 32779
(407) 888-8888

Local
444 NW 4th Terrace
Gainesville, FL 32601
(352) 444-4444

OBJECTIVE

To apply and adapt my writing, design, and leadership skills to the communication needs of a local organization.

EDUCATION

B.S./M.A. Joint Degree in Accounting, May 2002.
University of Florida, Gainesville, FL
GPA: 3.5/4.0

Related Courses: Technical Writing; Introduction to Computer Software; Public Speaking; Professional Communication; Expository and Argumentative Writing; Advanced Expository Writing; Leadership

Computer Skills: Microsoft Office Suite (Word, Excel, PowerPoint); Word Perfect; Adobe Photoshop; HTML

RELATED WORK EXPERIENCE

Creative Tax Services, Total State Tax Solutions Summer Associate,
Deloitte & Touche LLP, Atlanta, GA, May 2000-August 2000

- Prepared multiple PowerPoint presentations used for client engagements
- Organized and coordinated research projects to be completed by multiple associates in various offices nationwide
- Communicated through email, memos, and telephone daily
- Prepared detailed matrices

VOLUNTEER EXPERIENCE

Public Relations Volunteer, Alachua County Humane Society,
Gainesville, FL, February 2001-April 2001

- Prepared a weekly newsletter, calendar of events, and team rosters
- Produced a volunteer manual to be used by all future Alachua Humane Society volunteers and possibly adapted by numerous other chapters around the nation
- Conducted extensive field and web research and document analysis
- Coordinated collaboration among a four member team and two sets of supervisors
- Assisted in creating the style and design guidelines for the manual, including typography, spacing, visuals, and overall design scheme

ACTIVITIES/ HONORS

Golden Key National Honor Society
Florida Academic Scholar
Fisher School of Accounting Dean's List

RYAN PEDRAZA

3333 SW 33 St. #33
Gainesville, FL 32608

352-333-3333
rp33@ufl.edu

Objective

Apply my technical writing, editing, computer, and design skills to a collaborative writing project with a local non-profit organization.

Education

Bachelor of Arts in English and Political Science
University of Florida, Gainesville, FL
Anticipated Fall 2001
Grade Point Average: 3.72/4.0

Writing Experience

1/01-4/01

Volunteer Writer
Alachua County Humane Society, Gainesville, FL

- Assisted with the creation of a 25-page Volunteer Manual
- Manual will be used to train all new volunteers and will be given to other Humane Society chapters as a model
- Manual's production required research skills, writing skills, visual design and layout skills, and desktop publishing skills.
- Utilized teamwork and leadership skills while facilitating a four-person group and communicating with several competing supervisors

Employment

2/00-present

Leasing Consultant
Lexington Crossing Apartments, Gainesville, FL

- Explain apartment amenities to customers and sign leases
- Handle rent monies for property with over 1000 residents
- Prepare monthly newsletter for residents detailing coming events using Microsoft Office 2000 and Print Shop Premiere
- Write business letters, memos, and emails to potential residents
- Provide excellent customer service

Awards

- Florida Academic Scholar Scholarship – 100% of tuition
- Alfred Dupont Scholarship 2000-2001
- Ronald E. McNair Scholar 2000-2001 (only 14 selected from pool of nearly 100 applicants)
- Golden Key Honor Society
- Alpha Lambda Delta (English Honor Society)

Team B.A.R.K. Memorandum

To: Alachua County Humane Society
From: Loretta Belfiore, Ryan Pedraza, Katie Roland, Sharon Sandman
Date: April 10, 2001
Subject: Transmittal of ACHS Volunteer Manual

Attached is the completed volunteer manual. The purpose of our project was to combine a wide array of volunteer information into a user-friendly manual.

First we researched and selected the most pertinent information to include in the manual. We sorted and organized the information into a user-friendly, manageable, and multi-dimensional text. Many documents and designs had to be updated and rearranged to better suit the text's primary audience. Several major design prototypes were created and combined to produce our final text. Interaction with agency staff members and volunteers was essential to the manual's production.

Our completed project is an appealing, well-organized manual that we hope will greatly assist you in training volunteers as well as serve as a valuable handbook for the volunteers themselves.

Our wish is for this text to provide a strong foundation for the future of the volunteer programs at ACHS, and to ultimately promote the humane treatment of all animals. The members of Team B.A.R.K. have gained invaluable experience in working with the Alachua County Humane Society. We appreciate the opportunity to have contributed to such an important mission for our community.

Volunteer Handbook

Welcome to The Alachua County Humane Society!
We genuinely appreciate your interest in our volunteer program and are excited about helping you find your special niche within our community. With your help we can profoundly impact the lives of the animals here at the shelter and around the region. Since we are a non-profit organization, we rely upon your commitment, time, and talents to provide quality care and resolutions for our animal population.

As we expand and grow, so does our need for dedicated volunteers who will assist in providing a safe and comfortable environment for our residents, promoting humane education within our community, and contributing to our advocacy goals.

This handbook was designed with you in mind! It will familiarize you with the principle structure of our program and will enable you to make an educated decision about how to contribute as a volunteer. The following pages will be your tools in creating rewarding and beneficial experiences for yourself and the animals that you assist.

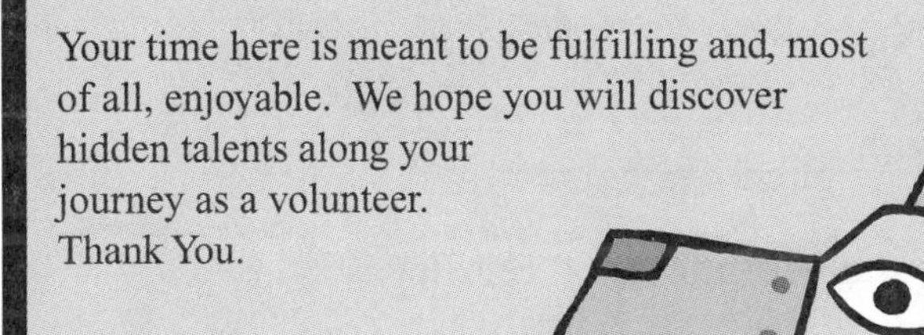

Your time here is meant to be fulfilling and, most of all, enjoyable. We hope you will discover hidden talents along your journey as a volunteer.
Thank You.

Sincerely,

The Staff of the
Alachua County Humane Society

Table of Contents

Introduction to Volunteering

Goals for the Volunteer Program

- To maximize the potential of our volunteers through continuous training and education.
- To establish a personal schedule with the volunteer that will combine the needs of the Humane Society with the talents of the individual.
- To empower the volunteer through goal-specific programs, providing them with valuable skills that can be applied outside of the Humane Society.
- To engage the volunteer in government and legislative action.

Staff

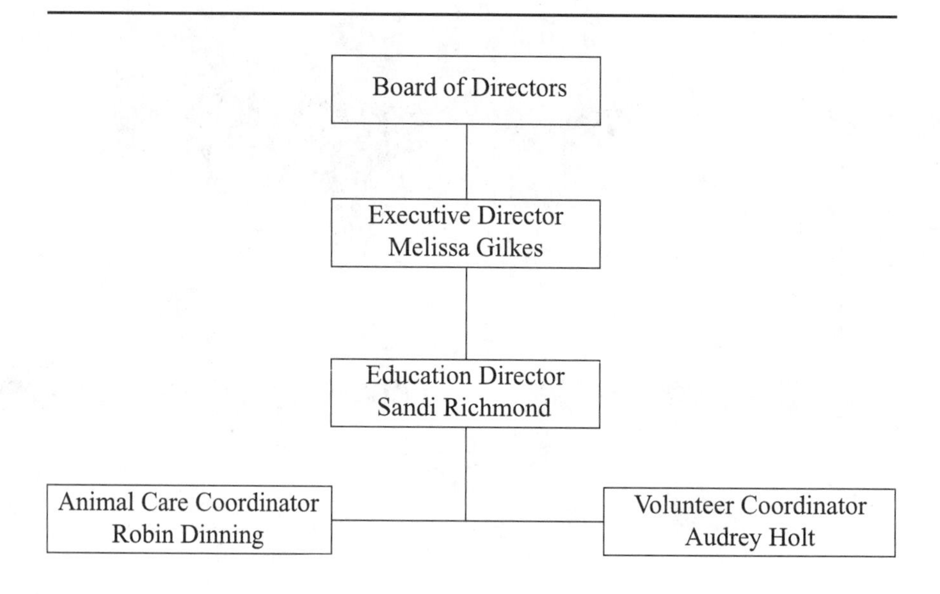

Figure 1. Organizational Chart of the Alachua County Humane Society

Policies for the Volunteer

During your time here at the Humane Society, we ask that you always abide by the guidelines we impart to you for your and the animals' safety. The care, feeding, and medical treatment of the animals is regulated by *trained* volunteers. If you notice an animal in distress or with signs of illness, please notify a staff person immediately. Your socialization with the animals is key to our adoption process. Interaction will allow you to become familiar with their personalities, therefore enabling us to make a better adoption match.

If you are *injured* in any way while volunteering, please notify a staff member *immediately* and *apply* the appropriate *first aid*. The Humane Society needs to have a report on file for all confirmed injuries.

In the event that you are unable to meet a scheduled shift, always notify your supervisor ahead of time. Your time commitment to us is valuable, and we expect you to consider volunteering as a job, regarding your work schedule as you would anywhere else. We ask support for our philosophies inside and outside of the Humane Society. Be realistic about your needs and availability. Your well being is fundamental to the success of our volunteer program. Remember to always **have fun!**

Future Plans for the Humane Society

Currently our primary focus is the relocation of our headquarters, which will be located one block north of our present location. Since 1981, Alachua County Humane Society has been operating out of our present site, an 1800 sq. ft. former gas station deficient in design and ventilation.

We need a floor plan of 6000 sq. ft. to accommodate our expanding program and animal population. We must raise over $1 million to meet our goal of a modern shelter. This will provide us with various utility spaces for the animals and our staff including a Pet Adoption/Public Area, Retail Area, Office/Administrative Area, and real-life rooms for the animals to replace their kennels.

Your presence will have a major impact on our ability to accomplish this objective. We need your efforts within the areas of fundraising, reinforcing our philosophy, and solidifying our relationship with the community.

Facts and Statistics

- In Alachua County, 10,000 animals were euthanized last year.
- A dog can have two litters a year (two separate heat periods).
- A cat can have up to three litters a year (a continuous heat period).
- Dogs and cats can become pregnant while nursing.
- The gestation period for dogs and cats is approximately 63 days.

2 Cats x 2 Litters/Year x 10 years = 80 MILLION CATS !

- We sell approximately 3000 vouchers a year through Gullencamp spay and neuter program.

Adoption Policies

The Alachua County Humane Society is committed to adoptions that are in the best interest of both the prospective adopting family and the pet. To achieve this goal, we consider each proposed adoption carefully and on an individual basis. We have also established several guidelines. First, a signed application must be approved by two or more ACHS adoption team counselors. Second, prospective adopters such as those listed below will not be approved.

Common Reasons For Not Being Approved

- Renters not providing the landlord's name and phone number
- Renters without the landlord's explicit consent given to the ACHS counselors
- Persons who plan to move soon
- Minors without their parents
- Persons without a steady source of income
- Persons who want to give the animal as a gift
- Families with children under the age of six years for cats and under the age of six months for dogs (HSUS studies show a child under six cannot determine the difference between a live animal and a toy)
- All persons living in household must see and approve of the pet being adopted
- Persons whose present animals are not sterilized or not current on vaccinations
- Persons who do not want the animal inside the house
- Persons who have previously given away animals without good reason
- Persons who have had animals killed by a car or unknown cause, or who have had animals disappear (There are exceptions. For example, if you can determine that it was an accident, not just from roaming, or if you can determine that they tried to find their pet or feel remorse.)
- Persons who intend to declaw
- Persons who intend to keep dog on a chain

Reasons Why We Have These Guidelines

Compiled from our years of experience, the following list states reasons for giving up a pet:

- The landlord won't allow pets that they tried to sneak in
- Persons moving to an apartment that does not allow pets
- Person has more animals than they can afford or not enough time to deal with behavior issues necessary for proper pet stewardship
- Person can't find homes for the pet's offspring
- A small child is mistreating the animal
- Someone in the household is allergic or doesn't want to put up with fleas, shedding, litterbox, furniture scratching, etc.
- The pet was given to them and they can't keep it

TO GO HOME, CATS MUST BE IN A CARRIER AND DOGS MUST BE ON A LEAD. NO DOGS ON CHAINS ALLOWED!

Self Assessment and Job Descriptions

Self-Assessment Checklist

The following checklist will assist the Alachua County Humane Society in placing its diverse population of volunteers. Please check the different positions and skills that you believe will be an asset to your volunteer work at the agency.

Job Descriptions

- ❑ Answering telephone
- ❑ Working in pet supply store
- ❑ Cashier
- ❑ Handling dogs/cats (circle if preference)
- ❑ Walking dogs
- ❑ Maintenance of facility
- ❑ Cleaning and grooming animals
- ❑ Fundraising and promotions
- ❑ Writing/editing
- ❑ Counseling for adoption
- ❑ Pet sharing
- ❑ Public relations
- ❑ Meals on wheels for pets

Abilities

- ❑ Past experience working with animals
- ❑ Enjoy working directly with and caring for animals
- ❑ Willing to work in areas other than direct care of animals
- ❑ Experience with fundraisers
- ❑ Writing experience (newsletters, grants, etc.)

Animal Caregiver

This involves socializing, feeding, cleaning, and caring for the pets with one-to-one attention. This is one of our most popular programs, but keep in mind that this may also involve such tasks as sweeping the floors, taking out garbage, etc. Time spent with our animals is precious.

Objective: Provide the best care possible to the animals while using the utmost safety precautions.

Specific Duties:

- Walk dogs
- Feed animals
- Play with the animals
- Sanitize living quarters by changing blankets, sheets, and litter boxes

Qualifications:

- Knowledge of animal behavior and breed characteristics
- Good communication skills
- Love for animals

Available Times:

- M, T, R, F: 9:30 am – 5:00 pm
- W, Sat: 10:00 am – 2:00 pm

Thrift Shop Helper

Our bustling little Thrift Shop is overflowing with an ever-changing array of donated goods ranging from the practical, such as small appliances, linens, tools, luggage, and lamps, to collectible treasures, including old and new ceramics, prints and paintings, and fine and vintage jewelry. Books, patterns, records, toys, and magazines are also among our wares. Our volunteers greet and wait on customers and answer the phone. The major source of income for the animals comes from our Thrift Shop.

Objective: Provide exemplary service to customers, and maintain accurate records for all transactions.

Specific Duties:

- Answer phone inquiries
- Greet customers
- Cashier
- Make sure all sale items are in their respective places
- Assist customers with purchases

Qualifications:

- Cash handling experience
- Good organizational skills
- Good communication and interpersonal skills

Available Times:

- R, F: 10 am – 4 pm
- Sat: 10 am – 2 pm

Pet Sharing

One of our most exciting programs involves exploring the therapeutic relationship between pets and people. Volunteers take pets to visit residents at local nursing homes. Often those who have not moved or talked have petted and whispered to the animals. This wonderful program needs caring, committed volunteers with a special interest in working with seniors.

Objective: To brighten the spirits of people, especially the elderly and handicapped, with a chance to interact with companion animals on a regular visitation basis.

Specific Duties:

- Choose a place to visit from our site list and schedule a visit with the Activities Director
- Provide one-to-one attention to pet and resident
- Clean up after animals

Qualifications:

- You must provide your own pet!
- Pet must be up to date on vaccines, be spayed or neutered, and must pass a temperament test and evaluation conducted by an ACHS staff member
- Pets must be predictable, controllable, and friendly

Available Times:

- Visiting times are usually for 30-60 minutes, 1 or 2 times a month

Fundraising

Interested in volunteering on community projects, such as pet walks, dinners, auctions, sidewalk sales, car/dog washes? We are ALWAYS in need of great ideas and people to work in various capacities on fundraising projects throughout the year. We also need volunteers with great people skills to approach businesses and people for memberships.

Objective: Raise money to promote growth for ACHS, and ensure that all fundraising events run smoothly.

Specific Duties:

- Post flyers to inform community of upcoming events
- Seek out donations from the community
- Wash/Walk pets

Qualifications:

- Great communication skills
- Enthusiasm when speaking to people
- Not afraid to ask for money or speak out about ACHS and education issues
- Energetic and creative

Available Times:

- Varies depending on scheduled events

Public Relations

How would you like to have an active role in solving some of our most important issues? From time to time, issues arise about which the Humane Society in Alachua County must speak out. Volunteers help by writing to the legislature, assembling our newsletter and information bulletins, and preparing materials for mailing. As a non-profit organization, we are dependent on volunteers with skills in writing stories, news releases, and public service announcements for print, radio, and television.

Objective: Keep the community up to date on ACHS happenings through mailings and media use. Also, address key issues to the legislature and fight for our philosophies.

Specific Duties:

- Design monthly newsletter, flyers, and bulletins
- Assist in writing grants, proposals, and other texts
- Research hot topics concerning animal rights
- Any experience in writing for telecommunications or advertising is a plus

Qualifications:

- Great written and oral communication skills
- Strong commitment and acceptance of ACHS guidelines and philosophies

Available Times:

- M, T, R, F: 9:30 am - 5:00 pm
- W, Sat: 10:00 am – 2:00 pm

Meals on Wheels for Pets

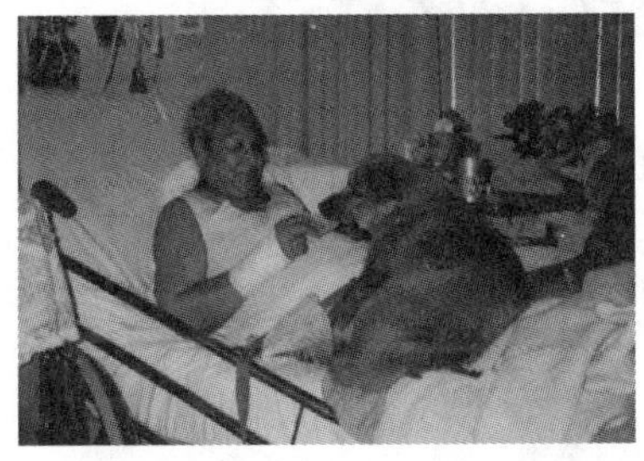

We recognized the need for Meals on Wheels for Pets when Elder Care caseworkers reported that their clients were giving half of their meals to their pets! This program provides food, supplies, and basic veterinary care to the companion animals of the elderly and disabled. These people can then continue to benefit from their pets' company when health and finances might otherwise force them to give up the animals.

Objective: To preserve the bond between people and animals, and provide food for pets whose owners could not otherwise provide.

Specific Duties:

- Monitor the needs of the individuals, and schedule monthly or bi-monthly visits to deliver food and check on the well-being of the animals
- Ensure effective interactions and safety for all involved
- Research hot topics concerning animal rights

Qualifications:

- Enjoy working with both people and animals
- Able-bodied volunteers are also needed for weekly food pick-ups

Available Times:

- M, T, R, F: 9:30 am – 5:00 pm
- W, Sat: 10:00 am – 2:00 pm

Volunteer Rights, Responsibilities, and Privileges

Schedule

You should work between two and three hours per week. Your hours will be scheduled at the same time each week to create regularity. Please keep in mind that you must commit to volunteering for at least six months. So if you are just trying to pick up a few volunteer hours for a college class, this is probably not for you. The ACHS does not accept court-ordered community service.

Sign-In Procedure

When you arrive for a shift, please sign the time sheet at the front desk. You should also sign the calendar on the wall. It's located on the "Volunteer Information" bulletin board. The time sheet helps us keep track of how many hours you are working a week.

Appropriate Attire

The ACHS does not have a strict dress code. The key is to look professional. If you ever have doubts about the appropriateness of something, keep in mind this simple rule: You are a representative of the ACHS and should dress as such. You shouldn't wear tank tops, T-shirts with profane language or inappropriate graphics, or shorts or skirts that are too short. For extra safety, you may want to secure long hair or remove any piercing (not a requirement, just a suggestion). A cat may be tempted to paw at anything dangling.

Customer Service

The amount of contact you have with the public will greatly depend on your specific duties. For now, you should let the full-time staff or more experienced volunteers handle the public. As your contact with the public increases, so will your training and instruction.

Eating and Drinking Policy

The ACHS does not have too many rules about eating and drinking. Feel free to help yourself to soda, water, or anything else in our refrigerator. Just remember one key concept: Always clean up after yourself. If you are a smoker, please do not smoke inside the building. We also prohibit throwing cigarette butts on or near our property. Please help to keep our facilities beautiful.

Grievances

If you find yourself having a problem with another volunteer or staff member, please notify Audrey immediately. We want to create a happy environment and make sure everyone feels comfortable. Disciplinary action will depend on the situation. Most problems can be worked out if the parties involved simply talk about the situation.

If someone is having problems with you, you will be given a verbal warning. Audrey will let you know what's going on and how you can improve the situation. A second written warning will be given if needed. Finally, if both warnings do not resolve the situation, you'll be asked to leave the ACHS.

You may also have a problem with a difficult customer. These situations can be stressful. Please remember that you are a representative of the ACHS. No matter how difficult someone is being, do your best to treat him or her with courtesy and respect. Our main goal is education and you can't educate someone by being rude. Please let a full-time staff member know if you are having continued difficulties with a customer. Customer service is key!

Injuries

No matter how slight you believe an injury to be, please alert a staff member immediately. If an animal scratches or bites you, tell us! Don't be embarrassed or think a scratch is too small to be concerned about. We must fill out a bite report if a bite occurs while you are volunteering. Our staff members will make sure any injury is treated correctly. Remember, report any injury to a staff member immediately.

Rewards

Aside from working in a well-respected environment and meeting others with similar interests, volunteers walk away from their experience with unique memories. You have to create your own experience, but we do provide incentives:

- Annual volunteer recognition parties
- The opportunity to travel to conferences and government agencies
- Flexible scheduling
- Reduced prices for pet supplies

You will also gain immense satisfaction in contributing to the lives of our orphan animals.

Pet Store Information

Sale at Register

1. Figure sales tax at 6%.
2. Please put your name or initials on invoice.
3. Mark invoice paid "**cash**" or "**check**."
   Customer receives yellow copy—**not white copy**.

Cash Donations at Register

When customer pays for items and wants us to keep the change, please put the overage in the donation box.

Keep Taxable Items Separate

When customer wants to adopt a cat or dog and wants to purchase pet supplies, make sure they write two separate checks…one is taxable, one is not.

Animal Rehabilitation Items

Robin must authorize all items taken out of pet shop for animal rehabilitation.

Refunds

Refunds **must be authorized by a staff member.** Please separate invoice and mark "**refund**" on top of (large) invoice. Remember to refund sales tax on the returned item.

Keep Spay/Neuter Money Separate

Please do not make change for the pet store out of the spay/neuter money.

Voided Invoices

When voiding invoices, write "**void**" across the sales slip and leave in the sales book. **Please do not throw any invoices away.**

Pet Store Sales

Writing A Sales Slip

1. Specify *quantity, simple item description,* and *net price.* (Net price *does not* include sales tax.)
2. Figure the sales tax (6%) and write "tax" to the left of the figure.
3. Add the two amounts (net and tax) together and put a double line underneath the new total. Put your initials and the date in the designated "salesperson" area.
4. This total should reflect the exact amount received from the customer.

Paying With Cash

If paying with cash, please count the change back to the customer. Specify in the body of the sales slip "paid cash." Give the customer the yellow copy.

Paying With Check

If paying with check, please designate "pet store" at the bottom left corner of the check in the "description" area. Circle the words "pet store." Specify in the body of the sales slip "Pd ck # (write check number)." Give the customer the yellow copy.

Donation on Same Check as Sale

The customer may write the check for over the amount (e.g., $10 over) and ask for the remainder to be donated. Take the donated amount in cash $10 out of the cash box and put into the donation jar. If the donation is a substantial amount, notify Mel or Rita.

Paying With Credit Card

If paying with a credit card, ACHS accepts Visa, Master Card, and check cards. We do not accept Diner's Club, American Express (AMEX), or Discover.
Swipe the card through the credit card machine, magnetic strip facing out, and follow the prompts. Figure the sales as stated above and give the customer the yellow sales copies of both the sales receipt and the credit card receipt.

Voids

If a mistake is made or the customer changes his/her mind, write in large, capital letters "**VOID**" across the sales slip. Do not tear the white sheet out of the book. The yellow copy can be thrown away. To avoid a credit card sale, press "void" on the credit card machine and then follow the prompts. A void transaction is performed the same day as the sale only!

Cash or Check Returns and Credits

A return is considered a separate transaction from a sale. In other words, if a person returns a 15-pound bag of food priced at $15 and wants a 20-pound bag priced at $20, write the returned item up as if a sale were being made except write the word "return" at the top of the sales slip and highlight in yellow. The item *amount, tax,* and *total* are to be put in parentheses. (Parentheses always designate a credit when working with numbers.) The customer is refunded $15.90. Turn to the next page in the sales book and write up the 20-pound bag of food as a regular sale.

Credit Card Return

A credit card return is transacted by pressing the "credit" button and then following the prompts.

Other Information

Animal Rights Laws

Here are some important pet laws that every responsible citizen should know. You might find these laws useful as you progress in your training and deal more and more with the public. According to Alachua County Animal Services County Ordinance 99-20:

Section 72.10 An owner shall treat a companion animal in a humane manner and shall provide humane care for an animal. Humane care includes but is not limited to adequate food, adequate water, adequate shelter, adequate space, and veterinary care to maintain health and prevent or cure diseases.

Section 72.11 Confinement of domestic animals in heat (estrus). Animals in heat must be confined indoors or in a locked structure that prevents entry of a male domestic animal and prevents the female from escaping. Confinement solely by a leash or in an open fence is not in compliance.

Section 72.12 An owner shall have physical control of a dog at all times when off the owner's property, unless the property owner designee consents to the removal of the physical control. Off the owners property means any public, or private property, including streets, sidewalks, schools, parks, or privately owned property.

Section 72.17 It is illegal to fail to quarantine a biting/scratching animal.

Section 72.20 All dogs and cats must be vaccinated by a veterinarian against rabies at three months of age. Dogs and cats are to be re-vaccinated 12 months after initial vaccination and bi-annually thereafter. Ferrets are to be re-vaccinated annually.

Section 72.21 All dogs and cats 3 months of age or older, must wear a current Alachua County License Tag at all times. New residents have 30 days to comply with this section.

There are also two state statutes that deal directly with animals:

828.12 Cruelty to Animals

(1) Unnecessarily overloads, overdrives, torments, deprives of necessary sustenance or shelter, or unnecessarily mutilates, or kills any animal, or causes the same to be done, or carries in or upon any vehicle, or otherwise, any animal in a cruel or inhumane manner, is guilty of a misdemeanor of the first degree.

(2) Any person who intentionally commits an act to any animal that results in the cruel death, or excessive or repeated infliction of unnecessary pain or suffering, or causes the same to be done, is guilty of a felony of the third degree.

828.13 Abandonment of Animals

(3) Whoever:

(a) Impounds or confines any animal in any place and fails to supply the animal during such confinement with sufficient quantity of good and wholesome food and water.

(b) Keeps any animal in any enclosure without wholesome exercise and change of air, or

(c) Abandons to die any animal that is maimed, sick, infirm, or diseased, is guilty of a misdemeanor of the first degree.

If you have any questions about these laws, check with a staff member or contact Alachua County Animal Services (352-955-2333). Even though spaying and neutering are not mentioned in any of these laws, you should always encourage pet owners to do so.

ALACHUA COUNTY HUMANE SOCIETY
VOLUNTEER AGREEMENT

This agreement is intended to indicate the seriousness with which we treat our volunteers. The intent of the agreement is to assure you of both our deep appreciation for your services and to indicate our commitment to do the very best we can to make your volunteer experience here a productive and rewarding one.

AGENCY

The Alachua County Humane Society agrees to accept the services of:

________________________________ beginning ________________.

And grants each volunteer the following rights:

1. To respect the volunteer as an equal partner with agency staff, jointly responsible for completion of the agency mission.
2. To be receptive to any comments from the volunteer regarding ways in which we might mutually better accomplish our respective tasks.
3. To offer the volunteer careful placement with consideration for personal preference, temperament, life experience, education and employment background.
4. To provide adequate information, training, and assistance for the volunteer to be able to meet the responsibilities of their position.
5. To ensure diligent supervisory aid to the volunteer and to provide feedback on performance.

VOLUNTEER

1. To fulfill my commitment of six months to the Alachua County Humane Society.
2. To perform my volunteer duties in good spirit and to the best of my ability, and to seek guidance when in doubt.
3. To be prompt and reliable in attendance, to contact my supervisor if unable to work as scheduled.
4. To take advantage of educational opportunities that will allow me to maintain an ongoing competence in the performance of my job.
5. To respect the paid staff and strive to maintain a smooth working relationship with staff and other volunteers.
6. To accept Alachua County Humane Society's right to dismiss a volunteer for poor performance, including poor attendance.
7. To work safely, adhering to the Alachua County Humane Society's list of safe practices for all parties including volunteers, staff, and animals.
8. To represent ACHS as ambassadors to the public in a manner that's consistent with ACHS guidelines and philosophies and to adhere to all ACHS employment policies and procedures.

I acknowledge that I have read and fully understand the terms and conditions of the Volunteer Agreement and that I will comply with the same.

______________________________	______________________________
Signature of Volunteer	**Signature of Volunteer Coordinator**
______________________________	______________________________
Volunteer Social Security Number	**Date**

Appendix B: HomeEase for Humanity

This group of texts takes you through the service-learning project of HomeEase for Humanity, three professional writing students who worked with Alachua County Habitat for Humanity (thus the dual connotation of their name). Marisa Lopez, Sarah Hummer, and Alexandra Simpson produced two complex documents for Alachua Habitat: 1) a grant proposal to the City of Gainesville for funding new houses as well as stipends for a full-time construction supervisor and office manager; 2) a PowerPoint presentation to be used in fundraising and volunteer recruitment presentations to different regional groups. These two texts have the potential to raise thousands of dollars for the agency and, ultimately, help dozens of families in need move into new homes.

Since Marisa already had a connection with the agency before the project began, she wrote the letter of inquiry. Here are some of the documents HomeEase produced over the course of their project:

1) Marisa's Letter of Inquiry
2) Trip Report
3) Proposal
4) Agency and Document Profile
5) Style Sheet
6) Evaluation Report
7) Final Product 1: Grant Proposal to City of Gainesville
8) Transmittal Memo for Final Product 2
9) Final Product 2: PowerPoint Slides for Fundraising Presentation

In comparing her group's progress with the stages outlined by Anson and Forsberg—excitement, frustration, and resolution—Marisa wrote that her group went through these stages, though in a more recursive way. In the beginning they were excited and ambitious about the project. As the demands of both the project and the class increased at different pressure points in the semester, they began to get frustrated, but this motivated them to take more initiative. The final editing stages were tiresome and sometimes frustrating, she explained, but as they approached their finished texts, this gave way to excitement and pride in their work.

Although at the end of the project the group was "tired of working," in Marisa's words, both Marisa and Sarah plan to continue working for Alachua Habitat and to take additional professional writing courses. "I'm glad I can still be a part [of the agency]," wrote Sarah. "In the two and a half years I have been at this university, I haven't found anything that drives me this way."

Marisa M. Lopez
987 College Hall 352-123-4567
Gainesville, FL 32612 marisa123@ufl.edu

January 22, 2001

William Wagner, Chief Executive Officer
Habitat for Humanity of Gainesville, Inc.
2317 SW 13th Street
Gainesville, FL 32608

Dear Mr. Wagner,

I am writing to propose an idea to you. I am currently enrolled in an exciting Professional Communication course at UF that enables students to write for community agencies like yours. As you know, I am already volunteering at Habitat for Humanity, helping to write City of Gainesville grants. The professional communication project would expand my work with your organization to incorporate two other service-minded students from my class.

Our assignment can involve designing and writing any of a number of texts, including monthly grant reports, grant proposals, the organization's website, the newsletter, and basic correspondence. In terms of scope, the project should consist of about 15-20 written pages and include a large text and perhaps a couple of smaller ones. I would like to suggest a grant proposal for our main project. Because this project would be so extensive, only one small additional assignment would be necessary. For the second project, perhaps we could create a fundraising brochure or PowerPoint presentation. All of our work would be supported by our class and professor. The dates for the group's service would be approximately February 20, 2001, to April 10, 2001.

Each student in the course is writing a letter of inquiry much like this one to a community organization. Because of the large number of inquiries being made, I encourage you to reply as soon as possible, as we will not be able to serve every organization that we contact. As a group we would bring strong communication skills to your organization (see my attached resume). We are all motivated and care about helping your organization and the problem of substandard housing in the Gainesville community. I will call you in a few days to discuss potential service projects for the group. We hope to be working with you soon. Thank you for your consideration.

Sincerely,

Marisa Lopez
Enclosure (1)

HomeEase for Humanity

TO: Professor Blake Scott
FROM: Xan Simpson, Marisa Lopez, Sarah Hummer
DATE: 19 February 2001
SUBJECT: Trip Report—Habitat for Humanity of Gainesville

This memo is a summary of our group's initial impression of our first trip to Habitat for Humanity of Gainesville. The purpose of the visit was to define the role we will play in the organization and meet the people we will be collaborating with for the next two months.

Summary

The three of us met William Wagner, CEO of Habitat for Humanity and a local judge, during our first visit on February 12. We also met Carrie, Office Manager of Habitat for Humanity. Although we were not able to completely establish the specifics of our project, we agreed that our large text would be a grant proposal. Possible ideas for smaller texts include a PowerPoint presentation or a brochure, which we will begin to work on immediately upon finishing the grant proposal.

Discussion Topics

Contact and Collaboration Information:

- Mr. Wagner will oversee, supervise, and also write an evaluation of the project.
- Carrie will be our main contact person.
- We will meet with Carrie at least every Monday until the conclusion of this project.

Projects and Deadlines:

- The grant proposal is an application to the City of Gainesville, which must be turned in to the City Commission no later than March 5.
- Due to personal time constraints, our group set a goal for completing the proposal by March 2.

Information Gathering:

- Due to the short amount of time we have to work on the grant proposal, we asked Carrie to forward all of the crucial information to us immediately.
- We received copies of and reviewed various relevant documents such as applications from previous years, the city requirements for this year, and the Habitat for Humanity bylaws.
- We discussed the possible concerns of the Grant Committee. Our proposal should address the fact that Habitat for Humanity is the only organization that provides housing for *low-income* people and is, therefore, more deserving of funds than other organizations. We may also need to address the concern of the City that Habitat for Humanity has not yet used any funds from their current grant.

Conclusion

This initial meeting was extremely productive because we were able to establish a tentative schedule of what the next few weeks will entail. We discussed what is expected of both our group and of the organization. Mr. Wagner and Carrie were extremely courteous and welcoming, and appear more than willing to answer questions and help us as much as possible. Though the time constraints of the grant proposal are very tight, our group is more than capable of achieving our completion date of March 2—especially with the help of others involved. Once we receive the necessary information to begin work on the grant, we will divide up the tasks and start on the project.

Securing Necessary Funding for Alachua Habitat for Humanity: Project Proposal

By HomeEase for Humanity:
Marisa Lopez
Alexandra Simpson
Sarah Hummer

Submitted to:
Professor Blake Scott

February 26, 2001

Introductory Summary

Since 1976, Habitat for Humanity, a nonprofit organization, has helped build over 30,000 homes for low-income families throughout the United States, as well as 10,000 more in over 60 different countries. Virtually the only organization in the nation that assists extremely low-income families in their search to find affordable housing, Habitat for Humanity would not be able to continue to assist these families without the aid of community donors.

This is the case for the Alachua County Habitat for Humanity, which needs the approval of a grant proposal to build eight to ten more homes this coming year for families deeply in need of assistance with their living conditions. This year Alachua Habitat is in need of $120,000, which it will request from the City of Gainesville through HOME/CDBG City Grants. HomeEase for Humanity proposes to write this grant proposal.

One requirement for the City of Gainesville HOME/CDBG grant is that the funds be matched from sources other than Alachua Habitat, who hopes to receive donations totaling $511,426 from outside sources (e.g., Haile Plantation Association, Lions Club) for the upcoming fiscal year. Included in these donations are various items such as parts, tools, and money. Often the money donated supports the building of entire houses.

It is essential that Alachua Habitat maintain its strong ties within the community, continuing to inspire donors to ease the negative aspects that arise in low-income housing. To this end, HomeEase for Humanity proposes to create a PowerPoint presentation that will be used by Dave Feathers, Chief Executive Officer of Alachua Habitat, in speeches throughout the community. The purpose of the PowerPoint presentation will be to inspire donors to continue aiding Alachua Habitat with its mission of building affordable housing.

This proposal will discuss the writing problem of Alachua Habitat, as well as the steps we intend to take to solve this problem. In doing so, the proposal will cover the specifics of our timeline, our work distribution, and our contact information.

Problem: Securing City Grants and Community Donations for Alachua Housing Projects

In this section we will discuss in turn the overall community problem, the funding problem, and the resulting writing problem that we will attempt to solve in our project.

Nearly 30 million U.S. households are paying more than 30 percent of their monthly income for rent and utilities. For their money, these 30 million Americans are still suffering from overcrowding and severe physical deficiencies. They lack such "luxuries" that most Americans take for granted, such as hot water, electricity, and bathtubs or showers.

Despite this urgent problem, government aid is incapable of helping many of these people. Government housing programs involve subsidies that do not actually benefit very low-income families because of high interest and mortgage rates. This is where Habitat for Humanity steps in, providing families throughout the nation with affordable homes that they might not otherwise be able to have. However, Alachua Habitat will have to address, and then correct, two fundamental problems that currently hinder the organization from successfully helping these families: funding and writing.

Funding Problem

Because Habitat for Humanity is a non-profit organization and sells all of its homes at cost, it relies on outside donations and grants for all of its funding. The grants are obtained through proposals to the city, while other donations are obtained through the goodwill of various community members.

In order to meet the goal of building eight to ten new houses next year, Alachua Habitat needs to raise $511,426 from outside sources and $120,000 from the City of Gainesville HOME/CDBG grant. A combination of both community donors and city grants is a necessary component in building these homes.

HOME funds would be used for:

- Purchase of trailers and equipment,
- Real property acquisition,
- Building costs, including permits, utilities, insurance, and closing costs.

CDBG funds would be used to pay stipends of:

- Construction Supervisor,
- Office Manager/Volunteer Supervisor.

Writing Problem

Alachua Habitat currently needs volunteers to assist with two writing projects that will support the work of the organization. The writing needs of Alachua Habitat are:

1) A five-page grant proposal in order to secure the $120,000 from HOME/CDBG,
2) A fifteen-minute PowerPoint presentation in order to reach out to the community.

Though Alachua Habitat is very aware of its writing needs, the organization is somewhat limited in its ability to develop these much-needed projects for a number of reasons. First, Alachua Habitat has little expertise in the writing of grant proposals. Past proposals could have been clearer and more cohesive. Second, the staff members' busy schedules prevent them from paying attention to design details, which can affect the proposals' accessibility and professional appearance.

Lastly, the past few months have proven to be quite chaotic for all staff members due to some drastic changes in both management and office staff positions. Last week saw the transition of Chief Executive Officer from William Wagner to David Feathers. These changes are keeping the current staff busy trying to organize the office in order to ensure that Alachua Habitat is able to aid families.

Solution: Writing a Grant Proposal to the City of Gainesville and Creating a Fundraising PowerPoint Presentation

To solve the funding and writing problems of Alachua Habitat, HomeEase for Humanity will write a grant proposal to the City of Gainesville and create a PowerPoint presentation for use at speeches throughout Alachua County. The grant proposal will request HOME/CDBG funds to support the building of eight to ten new houses in Gainesville. The PowerPoint presentation will be used to inform the community of Alachua Habitat's mission and to inspire the community to support Alachua Habitat. Our main tasks for each of these projects include:

Task 1: Gathering necessary information from the organization,
Task 2: Identifying and analyzing the concerns of our audience,
Task 3: Writing and revising the text.

City of Gainesville Grant Proposal

Task 1: Gathering necessary information from the organization

HomeEase for Humanity is proposing to write a grant for Alachua Habitat because it is an immediate and basic need of the organization. Houses cannot be built without money from this grant. The HOME/CDBG grant proposal application is due by March 5, 2001. Because of the urgency of this assignment and its priority to Alachua Habitat, HomeEase for Humanity will begin this project at once. Our group will accomplish this through the genres learned in our class assignments and textbook. Our professor, Alachua Habitat staff, and past grant formats from the organization will also aid us.

We have received the budget for the HOME/CDBG grant from Carrie Reppert, Alachua Habitat Office Manager. The budget outlines the funds that Alachua Habitat will be requesting for this coming year. We have also received a copy of the bylaws, last year's successful proposal application, and proof of insurance. We are still awaiting a copy of the current organizational budget, a list of board members, and notes toward the current HOME/CDBG grant.

Task 2: Identifying and analyzing the concerns of our audience

In writing the grant proposal, our audience will be the Gainesville City Commission Grant Review Board. Our first step was to talk to Alachua Habitat staff about the possible concerns of the City Commission. They outlined the following points:

- The Commission may be concerned with the inconsistency of Alachua Habitat in dispersing funds from previous HOME/CDBG grants.
- The Commission may be concerned with offering funds to other low-income housing organizations.
- The Commission requires a short and simple application (any additional information should not exceed five pages).

Currently, we are reading and analyzing the information given in the grant's application packet. From this information we are hoping to find additional concerns as well as the preferred organization.

Task 3: Writing and revising the proposal

After receiving all necessary information from Alachua Habitat staff, HomeEase for Humanity can begin writing the documents. We are currently working on the application for the grant proposal as a group, of which each of us will write about a half page.

We will collectively edit the application and then collaborate on the remaining two sections of the proposal. Marisa will write the first draft of one section and Sarah the other. Alexandra will be in charge of design and editing, although we will all exchange drafts and make editing suggestions. After workshopping the proposal ourselves and getting feedback from our professor and from Dave Feather and Carrie Reppert, we will collectively make revisions and then individually proofread the entire packet.

PowerPoint Presentation

Task 1: Gathering necessary information from the organization

Since this is an original project, we will need to gather old brochures and promotional information that will give us a basis from which to start. We will need to determine the necessary length, the main message that needs to be conveyed, and any other specifics that need to be included. We will rely on our previous experience using PowerPoint and on information already gathered for the proposal to help ensure that matching funds are attained through this presentation.

Task 2: Identifying and analyzing the concerns of our audience

When creating the PowerPoint presentation, we will obtain a list of possible groups who will view the presentation. From this information we may be able to tailor different presentations toward different groups of viewers (religious, service, educational, government, community, and previous donors) by simply altering a few slides. By catering to various audiences, we will have greater overall appeal.

Task 3: Writing and revising the PowerPoint presentation

HomeEase for Humanity will collaborate as a group and with the organization to form an outline of the presentation based on past speeches. This outline will then be divided among group members. After discussing the overall purpose and background elements of the project, we will begin individual work on the project. From there the group will revise the presentation, focusing on both writing and design problems. HomeEase for Humanity will then present the project to Dave Feather and Carrie Reppert for final revision and approval.

Management: Collaborating and Completing Tasks

Now that we have explained what our main and smaller projects will entail, we will explain how we will successfully complete them. The Team Roles and Collaboration section will describe the group members' strengths, roles in the projects, and coordination efforts. The second section, Meetings and Timelines, will present our group schedule and explain how it meets the requirements of the class and organization.

Team Roles and Collaboration

The three members of HomeEase for Humanity are Marisa Lopez, Alexandra Simpson, and Sarah Hummer. Each member has a specific role in the group, tailored toward her personal strengths and experience.

Marisa Lopez

As a volunteer with Alachua Habitat from the beginning of the semester, Marisa has assumed the role of Agency Contact Person. Her established position will ensure a close bond between the group and the organization and will eliminate contact problems. Marisa is also working towards an English degree and has taken the title of the group's Technical Writing Professional. She has an extensive background in writing classes and has written for A+ Notes. Although each group member will contribute to the research and writing of the projects, Marisa will oversee these portions.

Alexandra Simpson

Alexandra, an enthusiastic Journalism major, has taken on the role of Senior Editor. Along with her experience working with the *Alligator,* she has a knack for making documents look professional, which will greatly benefit our group. Alexandra also has extensive layout and design experience from working with the school newspaper in high school, and has naturally become the group's Visual Design Artist. Although each member will take part in the process, her editing and design skills will help put the final touches on all projects created by HomeEase for Humanity.

Sarah Hummer

Sarah has taken on the role of Group Coordinator for HomeEase for Humanity. She is currently the Dance Marathon Delegate for her sorority, and has held numerous other positions that require her to keep records of all work and make sure that it is completed on time. She will form the link between the classroom, group, and organization. An Economics major with computer knowledge, she will also be in charge of all charts, graphs, and financial information for the projects.

It will require the skills of all group members, cooperation from Alachua Habitat, and expertise from our professor to make these projects successful. Table 1 below details the contact information of all persons involved with our project.

Table 1: Contact Information for Project Participants

Name	Phone Number	Email Address
Marisa Lopez	846.9831	Marisa341@aol.com
Alexandra Simpson	224.6627	Xannadu12@aol.com
Sarah Hummer	378.2740	Hawaii190@hotmail.com
Dr. Blake Scott	392.6650 x 280	Jbs148@english.ufl.edu
Carrie Reppert	378.4663	Clreppe33@hotmail.com

In addition to class time designated for working on our projects, HomeEase for Humanity will meet every Monday at 12:30 p.m. at Alachua Habitat's office on SW 13th Street. At this time Carrie Reppert will meet with us to provide any information or documentation that we may need. HomeEase for Humanity will maintain group contact primarily through email, and will arrange additional meeting times as necessary throughout the remainder of the project. Figure 1 below is a milestone chart detailing the project's major tasks and timeframe.

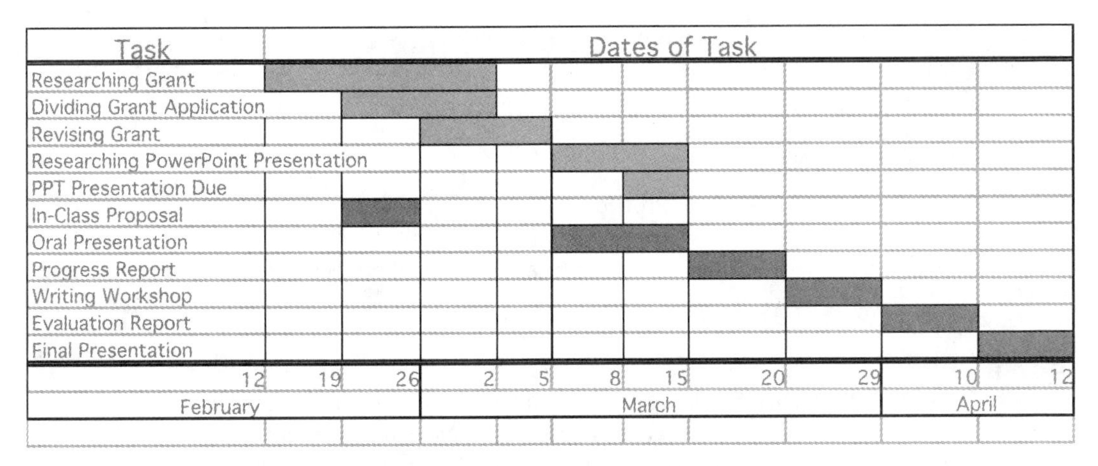

Figure 1: Milestone Chart Showing Project Timeline

As you can see, we have arranged our work to coordinate with course dates for the writing workshop, oral presentation, and progress report. This schedule will also enable us to complete the City of Gainesville grant before its due date of March 5, 2001, and will give us ample time to create a persuasive PowerPoint presentation.

Conclusion

In conclusion, HomeEase for Humanity is confident that our eagerness to work with Alachua Habitat, our combined writing skills, and our growing knowledge of the organization's needs will lead us to produce a professional, successful grant proposal to the City of Gainesville. We will also produce a PowerPoint presentation that will elicit emotional response from various donors, causing them to contribute much-needed funds to Alachua Habitat. Finally, our polished final products can be used as templates for future grant applications and different versions of fundraising presentations.

HomeEase for Humanity

TO: Professor Blake Scott
FROM: HomeEase for Humanity
DATE: 20 March 2001
SUBJECT: Agency and Document Profile

This memo will briefly profile HomeEase for Humanity's current project, a PowerPoint presentation to be used for soliciting donations.

Background

Alachua Habitat for Humanity consists of a close knit group of four main individuals: CEO Dave Feather, his secretary Pat, Office Manager Carrie Reppert, and Construction Manager Bill Smith. On a given day, additional volunteers may assist in answering phones and other office duties. Our group member, Marisa, currently volunteers to write texts for the organization.

Our project will be to create a ten-minute PowerPoint presentation to assist Alachua Habitat members while delivering speeches to solicit donations. This is a brand new idea that will hopefully inspire many donations to the organization. Our project will relieve much stress in delivering these speeches by providing a template and visual aids to the speaker.

Rhetorical Situation

Supervision

As this is a start-up project, we have full creative control of design elements. We will use past speeches delivered by Alachua Habitat staff to create an outline for the presentation. While the project is under construction, we will rely on our main contact, Carrie Reppert, to provide direction. Upon completion, we will present our project to Dave Feather, CEO and visionary for the PowerPoint presentation, for final changes and approval.

Presentation Objectives

The main objective of our project is to visually enhance current speeches delivered by Alachua Habitat. It will be both informative and persuasive. Our presentation will do the following:

- Relay general speech information in a concise fashion,
- Provide an outline for future speeches,
- Rely heavily on visual elements to elicit emotional appeal to the audience,
- Persuade community members to donate either their time or money to help Alachua Habitat.

Audience

Our basic template will be directed towards the general Alachua community. It will provide this audience with the main mission, goals, and activities of Alachua Habitat, as well as media clips from past home projects. This format may be used when delivering speeches to:

1) Schools,
2) Possible donors, and
3) Grant committees.

The presentation can also be tailored to specific audience groups who will receive the speeches. For example, we may add a few slides emphasizing the religious history and mission of Habitat for inclusion in a slide to a church or other religious organization.

Alachua Habitat for Humanity
PowerPoint Presentation
Style Sheet

Marisa Lopez | Alexandra Simpson | Sarah Hummer

Design	
Overall	Microsoft PowerPoint presentation; 10-20 slides to accompany a informative/fundraising speech that can be tailored to different audiences
Layout	Four slide variations on same color background: 1. *Text slide:* heading aligned with paragraph indent. 2. *Image slide:* heading centered on top with picture enlarged and centered. 3. *Text with image slide:* heading centered with text in left column, image in right column. 4. *Contact info. slide:* no heading, "Alachua Habitat for Humanity" centered with contact info. below.
Spacing	• Minimum one-inch margins on all sides. • Text and lists single-spaced. • One space between heading and text. • Text justified left, ragged right.
Type	• Headings: Arial 44 pt. • Main text: Times New Roman 32 pt. • Text in lists: Times New Roman 38 pt. • Text in contact info.: Arial 40/32 pt. • Diamond bullets • Numbers with periods in numbered lists.
Grammar and Mechanics	
Abbreviations	Write out "Alachua Habitat for Humanity" the first time; after that write "Alachua Habitat."
Commas	• Do not use a comma before the conjunction when there are only two items in the series. • Use a comma before a conjunction joining independent clauses. • Use a comma after long introductory prepositional phrases. • Use a comma after introductory adverbial clauses. • Use a terminal comma before 'and,' 'or,' or 'but' in a series of more than two. • Use a comma when necessary for clarity.
Hyphenated words	If possible, combine hyphenated words into one (for example, Microsoft Word accepts "healthcare" as one word).

Numbers	• Write numbers one through nine in words; write numbers 10 and higher as figures. • Use figures with percentages and dollar amounts.
Percentages	Use the percentage symbol (%) instead of writing out the word "percentage."
Quotes	Put commas and periods inside final quotation mark.
Semicolon	• Use a semicolon if it is necessary to further punctuate a sentence that is also punctuated by commas in a series. • Use a semicolon to connect independent clauses that are not joined by a conjunction.
Sexist language	Try to avoid sexist language, but also avoid using awkward phrases such as, "his or her." Try to reword phrases so that it is not necessary to express gender.
Spacing	Use one space after a comma. Use two spaces after a period. Use one space after a colon and semicolon. Use one space before and after a dash. Don't add spaces before and after a slash.
Style	Use an informal register. Write in an active style, avoiding passive voice. Vary sentence length and types, but be as concise as possible.
Visual Elements	
Charts and Graphs	• Allocation of costs. • Pie chart/graph with homeless info. • Contact info. on final slide.
Color Scheme	• Background: blue deepening into dark blue from left to right. • Headings: rustic orange. • Text: white. • Photographs: one black and white to accentuate substandard housing; all others in color.
Photographs	Three different designs, none with captions (speaker can elaborate if necessary): 1. Heading above large centered photo. 2. Collage with three photos of building in left diagonal pattern. 3. Text on left, photo on right.

Securing Necessary Funding for Alachua Habitat for Humanity: Final Evaluation Report

By HomeEase for Humanity:
Marisa Lopez
Alexandra Simpson
Sarah Hummer

Submitted to:
Professor Blake Scott

April 10, 2001

Introduction

The purpose of this evaluation report is to inform Professor Blake Scott of both the successes and disappointments that HomeEase for Humanity endured in the construction of a grant proposal and PowerPoint presentation. This report will evaluate the overall effectiveness of the completed projects and will rate how well the projects have accommodated their multiple audiences and uses.

HomeEase for Humanity received approval from Alachua Habitat in January to begin writing a grant proposal that would be used to obtain much-needed funds from the Gainesville City Commission. HomeEase for Humanity also received approval to create a PowerPoint presentation that would accompany speeches delivered to various audiences throughout Alachua County that were considering donating to Habitat's cause.

In this report we evaluate both our efforts in the process and the features of our final products. We evaluate our process according to how well we followed our plan and timeline, collaborated with the agency, analyzed our audience, and wrote and workshopped the texts. We also consider how we overcame problems in the process. Our criteria for evaluating our texts include how effectively we met the agency's expectations, influenced our desired audience, designed the texts, and wrote the texts.

HomeEase for Humanity is confident that we have exceeded the expectations of not only the organization, but of the course and group as well. This report will detail each step that was taken to complete each project, demonstrating how our group achieved its goals and overcame obstacles along the way.

Project Description

HomeEase for Humanity focused on two main projects over the course of the semester. When we first corresponded with our agency, Alachua Habitat for Humanity, it was in the process of applying for a grant from the City of Gainesville. We took on the main text portion of the grant and made Alachua Habitat aware of the portions of the application we were not able to complete, such as the anticipated annual budget. We then divided the text equally among group members, and set out to complete our portion of the application in proposal format.

As you can see from sections 13, 15a, and 15b of the proposal, headings are typed in larger san-serif font, while main text is in smaller serif font. In addition to creating accessibility and clarity within the grant, this proposal will provide the busy Alachua Habitat staff with a template for future proposals. Our portion of the application was delivered to the organization five days before it was due, and we eagerly await a response from the City of Gainesville grant committee.

Our next project involved a PowerPoint presentation. With our growing knowledge of the organization, we were able to work together and design an effective layout. The PowerPoint presentation will be presented as a slide show before fundraising and recruitment speeches.
Our group created emotional appeal in the presentation by scanning photographs of substandard housing environments as well as pictures of Habitat volunteers building houses together. The PowerPoint presentation came together easily, as we collectively edited the text and design elements

of all slides. Although we each had to compromise on the specifics of the presentation, we were able to create a polished product that satisfied the needs of our class, organization, and audience.

Process

We established the following processes to ensure that we would complete the projects in a timely and effective manner. Although we encountered some difficulties along the way, we managed to overcome them and meet our objectives.

Following Our Plan

Different sections of the project were divided up to ensure a more timely completion of the proposal. Aided by a group schedule that listed who was doing what and when it was due, our group was able to manage the project with unwavering assuredness.

Grant Proposal

As a group we completed the form application given to Alachua Habitat by the City Commission. We used previous years' proposals only to an extent to aid us in the creation of a new and more professional looking proposal that we hoped could be used as a template for future ones. From there, each group member took responsibility to finish a detailed section of the proposal. Each group member proofread the other members' sections, offering ideas and tips to make them stronger and more accessible. No obstacles were encountered in this division of the project. We actually finished the proposal five days early, which gave us extra time for fine-tuning.

PowerPoint Presentation

Though we did not follow our style sheet exactly when it came down to the design of this presentation, we finally agreed upon a consistent background, font, font color, and font size for the slides. At first each of us had slightly varying visions of the slides and presentation, but we were able to combine our ideas and come up with something we could all agree upon and be happy with.

Collaborating with the Agency

At our very first meeting with Alachua Habitat staff, HomeEase for Humanity established continuous dialogue with then-CEO William Wagner and Alachua Habitat Office Manager Carrie Reppert in order to keep our lines of communication continuously open. We met with Carrie each Monday at 12:00 p.m., which enabled us to get her feedback throughout the project. We also kept in touch through telephone calls and email.

Grant Proposal

Throughout the construction of the grant proposal, the major obstacle we faced was making sure Carrie was at the office at the designated times. Often she was out of the office or could not be reached, and this posed difficulty in meeting our tight deadlines. In our weekly meetings, we sometimes asked Carrie to supply us with various pieces of information for the following week. Though she did sometimes forget, we made sure deadlines were met by calling her or stopping by the office to remind her. The majority of the time we did the research ourselves to keep from worrying about falling behind.

PowerPoint Presentation

Alachua Habitat staff made it clear that they wanted a presentation that would elicit an emotional response from donors. In trying to meet their expectations, we continued our regular weekly meetings with Carrie, who supervised our work to her specifications and gave us ideas and tips to make the presentation a success. At the Habitat office, we had access to the international Habitat website and various pamphlets, brochures, and news articles that aided us in gathering the necessary information that would make the presentation effective.

Analyzing Our Audience

Working together as a team, we carefully identified and analyzed the concerns of multiple audiences, tailoring each project to its respective viewers.

Grant Proposal

Through careful research, we pinpointed any specific concerns the Gainesville City Commission might have. For example, the Commission was concerned with the inconsistency of Alachua Habitat in dispersing funds from previous HOME/CDBG grants, so we demonstrated that Alachua Habitat had reorganized its office and, as of February 20, had reported the use of funds for five homes. Because the proposal had specific directions listed in it, we were able to cater directly to the City Commission with no problem whatsoever.

PowerPoint Presentation

Realizing that the presentation would be viewed by a wide variety of audiences—including churches, schools, and community organizations—we made sure that any person, regardless of race, faith, or gender, would feel comfortable viewing it. Initially, each group member had a certain task to attend to, whether it was finding appropriate photographs or researching Habitat history. Fortunately, we were able to work together in a collaborative effort to come up with ideas we felt would effectively convey Alachua Habitat's desired emotional effect.

Writing and Workshopping

One of the most important aspects of revising our work had to do with the workshops. Probably the most effective way to make our projects understandable to the average reader, these workshops made all the difference in the final outcome of our work.

Grant Proposal

Although we wrote and revised several drafts before the workshop, we realized how important the workshops were as we kept finding additional mistakes that we had overlooked. From the workshops, we were able to make our proposal more cohesive, coherent, and accessible. The revision stage is also a time when our group collaboration was crucial. We tried many different ways to workshop as a group. For example, we used the "tracking changes" function on Microsoft Word to edit each other's work when we could not meet. This allowed us to make many revisions and ensure that each assignment was a reflection of the entire group. At times the tracking function was not sufficient for the work we needed to do, however. We often workshopped each other's work while at Habitat. We also had a final workshopping session at the library on Martin Luther King Day. One of the most important steps was workshopping with our agency, which allowed us to ensure that we were meeting their expectations at each step of the way.

PowerPoint Presentation

Though we had much information that we wanted to convey to readers, it was important that we not make each slide too lengthy. Otherwise, we risked the chance of having the slides be unappealing to the eye and thus ignored by readers. The in-class workshop assisted us in making the presentation more concise, as well as providing us with ideas to make it both consistent and clear throughout. The workshops enabled us to turn this initially cluttered and confusing presentation into something accessible and pleasing to the eye. We also continuously revised each other's work in class and at the agency.

Product

Our process throughout this project met the criteria set forward by our group, our professor, and Habitat for Humanity. Because of our successful process, our final product was also a success. Our grant proposal and our PowerPoint presentation can be evaluated as final products by the following criteria:

Effectiveness in Meeting the Agency's Expectations

Throughout our process, we made every effort to determine and exceed the agency's expectations for our final products.

Grant Proposal

For the grant proposal, Habitat for Humanity wanted us to create a professional document that communicated the necessity for city funding. We created a document that reflected the ethos and mission of Habitat for Humanity. We effectively communicated that Habitat is the only organization of its kind that builds housing for very low-income people. We also made a persuasive argument for $120,000 of HOME and CDBG money that is essential for the continuation of Habitat projects. Perhaps most importantly, we met the agency's expectations by finishing the grant proposal five days before the critical deadline. We foresee that our proposal will convince the City of Gainesville Commission to grant $120,000 to Habitat for Humanity and, in that way, meet the agency's most crucial expectation.

PowerPoint Presentation

We created a slide show that could be shown before a speech was delivered to people who might be interested in donating money or time to Alachua Habitat. To elicit an emotional appeal from these individuals, we strategically inserted stories and pictures showing the need for adequate housing. We foresee that the presentation will be a memorable aspect of Habitat's speeches, and will encourage people to donate money and time to Habitat's cause.

Persuasiveness for Desired Audience

During our process we carefully analyzed the audience that would be receiving the grant proposal and the PowerPoint presentation. This effort carried over into our final product in a number of ways.

Grant Proposal

For the proposal our audience was the Gainesville City Commission. By including key terms such as "very low-income" and "Community Housing Development Organization," we established our credibility and our knowledge of the subject. Knowing that our audience does not have time to read a lengthy document, we created a document that was accessible and concise. We kept all additional information within the five-page limit.

We also set up headings that allowed for quicker reading. We identified each section with the number of the question that it answered, maintaining the professional format of a proposal. Accessibility will greatly help our cause with the Commission because it will allow them to read more of our proposal and therefore be further inclined to give Habitat the grant.

PowerPoint Presentation

For the PowerPoint presentation our audience consisted of various community members and community organizations throughout Gainesville. For this project, our main goal was to move our audience to action, which we accomplished in a few effective ways. First, we began and ended the presentation with a personal story about Glenda Hamilton. We began by telling about her difficulties in life and ended on a note of triumph, letting our audience know that she is now living in her Habitat home.

One of the key pictures we included in the presentation was a picture of a family standing in front of their shack. This, along with the opening about Glenda Hamilton, set up the urgent problem of inadequate housing in the beginning of the presentation. We took the presentation a step further and showed the audience how they could get involved in helping people like Glenda Hamilton. We persuasively suggested contributions and volunteer work.

We also established our credibility by telling the audience Alachua Habitat's mission and processes and by listing the accomplishments of the organization. For example, we discussed how many houses Alachua Habitat has built and their plans for the future.

Text Design

In both the grant proposal and the PowerPoint presentation, we used design elements such as lists and headings to make our documents more accessible. In the PowerPoint presentation we also used pictures and graphs to drive home our message.

Grant Proposal

There was not much opportunity for design in the grant proposal. However, we did utilize lists, special fonts, headings, and effective white space to give our document a user-friendly look. Our lists include a task list, a list of current housing projects, and a list of goals and objectives. These lists allow the reader to easily see the information without searching for it in lengthy paragraphs. We also used sans serif font for all headings in order to attract attention to important information.

PowerPoint Presentation

Here we utilized our design skills more. As with the grant proposal, we used lists, special fonts, headings, and effective blank space. We also chose colors, pictures, backgrounds, and graphs that improved our presentation. Our background was a calming and professional-looking blue color. We removed the distracting curving line that was included in the template of the background in order to ensure that nothing took away from our message.

We chose pictures that had the greatest emotional effect. For example, our slide on volunteering with Alachua Habitat showed two young girls nailing a board together to help build a house. We also used a donut chart to demonstrate the circular nature of Habitat's work. The chart shows the movement from building a home to "sweat equity" labor and mortgage payments that finance future home projects. This visual will leave a lasting image in the minds of our audience.

Text Writing and Style

During our process we spent time improving our writing through group revisions and workshops and by applying what we learned to our work. The result was a product consistency, coherence, cohesion, and concision.

Grant Proposal

One of strongest elements of our proposal is its cohesion. It was difficult to find a way to organize the information in a way that answered all application questions in a logical sequence while fitting the template of a proposal. We accomplished this by splitting the body of the document into two parts: CDBG grant and HOME grant. Both parts had their own problem and solution section. Each major part had a section that discussed how the goals and objectives would be accomplished with the requested funds. Each goal or objective referred back to a previous section that introduced the overall goals and objectives of Habitat for Humanity. In this way we tied the whole document together while still answering the questions of the commission in the requested order.

PowerPoint Presentation

The most impressive element of writing within the PowerPoint presentation was its concision. Because this presentation will be used as a tool to get an audience's attention, the length was crucial. We had to provide information to build credibility and interest in the organization; we were very careful not to over-inform, however. We kept information to a minimum and only used it in persuasive ways. For example, when we discussed the facts about Alachua Habitat, we used a bulleted list to emphasize their successes.

Conclusion and Recommendation

HomeEase for Humanity has completed every task set forth in our proposal. We have done so in a timely manner and have created quality work. We worked well as a group and with our agency to help those living in underdeveloped housing. We also met all class requirements and deadlines. Although there were a few setbacks along the way, we did not let them get in the way of our success. Based on our diligent efforts and our outstanding final products, we recommend a grade of "A" for our work on this project.

Beyond any grade, we feel that the lasting results of our work for Alachua Habitat will be a 30-home community of Habitat homes that will provide people like Glenda Hamilton with a safe and comfortable place to live. Our work will also be an impetus for increased community support in the City of Gainesville that will grow exponentially as time goes on. Our final success is a personal one. We feel a sense of accomplishment in our ability to so profoundly help the community. From here we look forward to continuing to help Habitat and other organizations and to using our experience this semester in our future careers.

Securing Necessary Funding for Alachua Habitat for Humanity: HOME/CDBG Grant Proposal

Submitted by:
Alachua Habitat for Humanity, Inc.

Submitted to:
Gainesville City Commission
Grant Review Board

February 5, 2001

Introduction

Since 1976, Habitat has built around 10,000 houses in 60 different countries, 30,000 of which were built in the United States. Habitat coordinates the construction process, volunteer participation, and both the selection and involvement of the chosen homeowner in order to successfully build a home. Churches, academic institutions, and other organizations frequently agree to "sponsor" a home by providing all or most of the money and labor needed to build a single house.

Habitat for Humanity is part of a chain of international nonprofit, nondenominational, Christian housing organizations that work in conjunction with the community to build simple, decent, and reasonably priced homes in partnership with those in need of adequate shelter. The Alachua Habitat program is clearly distinguished from others of its kind because of the basis on which deserving persons are chosen. According to the official Habitat website, houses are purchased strictly by families chosen according to their need; their ability to repay the no-profit, no-interest mortgage; and their willingness to work in partnership with Habitat, also known as "sweat equity," to build these homes.

Alachua Habitat has currently constructed 53 homes for very low- to low-income families in the Gainesville/Alachua County area. There are six additional homes currently under construction. Alachua Habitat has also entered into a unique partnership with the county to build one of the first sustainable communities.

13. Why is this project needed?

Though all people should be able to live in safe and affordable housing, not everyone is fortunate enough to acquire the means to finance such places. Twenty percent of adults and 26% of children living in Alachua County are forced to live below the poverty level. According to projected housing needs conducted by Alachua County, people earning between $10,000 and $35,000 are in need of 543 new homes. People earning between $10,000 and $22,500 cannot afford a house that costs more than $44,950—including homes subsidized by city or county organizations.

If approved, the funds from this grant will assist Gainesville's very low- to low-income families in virtually the only housing initiative of its kind. Alachua Habitat will leverage approximately $43 of private dollars for each $1 received with grant from the City of Gainesville in order to further support potential homeowners. This private support includes aid from churches, contributions from individual tax-deductible charities, and professional and corporate sponsors who donate labor, land, and financing.

Because very low- to low-income families require support to a greater degree than that of other income households, ongoing assistance is needed to meet the inevitable financial emergencies that make home ownership difficult for low-income buyers. Much of Habitat's success is a result of its ongoing nurturing of homeowners, flexibility with mortgage loan repayments, and assistance with other financial obligations such as property taxes and home maintenance. Habitat also provides counseling on budget management, household repairs, how to be a good neighbor, and various other issues. Habitat holds mortgages for 20 to 30 years, making housing affordable for low-income people by substantially controlling housing production costs. As a result, most Habitat families surveyed spend less per month on housing than do regular owners, helping to eliminate the "severe cost burden" facing many extremely low-income families.

Alachua Habitat's emphasis is on creating a partnership with homeowner families. In order to assist families in need, Habitat also empowers people and communities who shape all aspects of the housing delivery process from construction and homeowner selection to loan servicing and default counseling. Habitat directs considerable efforts toward the broader objectives of strengthening Habitat families and helping to break the poverty cycle. Habitat homebuyers have benefited from home ownership both in monetary terms and in terms of qualitative lifestyle changes.

This project is consistent with section 3.1 of the City's Consolidated Plan. It will provide 10 of the 30 additional houses desired by the city for low-income households through new construction. The project continues to assist these families by relieving the severe cost burden of home ownership. The Habitat project will also meet the goal of reducing the energy costs of low-income families by creating Energy-Star homes. In addition, the Habitat project addresses the concern of the City regarding the overwhelming number of minorities in substandard housing. Habitat serves all people of the community without regard to ethnicity or family structure; for example, Habitat has several female homeowners who are heads of households.

14. Goals and Objectives of Project

1. Help to eliminate families from living in substandard housing.
2. To provide a full time Construction Supervisor and Office Manager/Volunteer Supervisor.
3. Construct five to eight new Energy-Star homes with greater energy efficiency.
4. Increase the scope of our projects by building 10 more houses this year.
5. Coordinate and train volunteers in a timely manner.
6. Provide homeowners with continued education on home ownership, including counseling on budget management issues and household repairs.

15A. Description of Proposed CDBG Project

Problem: Staff Funding

Habitat uses over 3,000 volunteers to raise the funds needed to build Habitat houses, to work in the Habitat office and thrift store, and most importantly to work the construction sites. Without our volunteers, Habitat would not be able to keep the cost of construction down. Our contracted services are usually limited to areas that require licensing, and even then Habitat generally relies on volunteer hours from these licensed professionals.

Habitat is committed to the City of Gainesville and meeting the community's need for affordable housing. However, as we continue to build more homes for very low- to low-income members of our community, Alachua Habitat can no longer rely on or expect part-time volunteers to hold the full time positions of Construction Supervisor and Office Manager/Volunteer Supervisor. The problem becomes how to fund these crucial positions.

Solution: CDBG Grant

Since Habitat is spending most community-donated funds on construction materials, we rely on city grants for other needs such as the funding for our full-time positions. Community Development Block Grant/HOME funds for Habitat staff would free up enough money to build two new houses. The CDBG would fund the direct costs to pay 75% of the salary and benefits of a full-time Construction Supervisor and a full-time Office Manager/Volunteer Supervisor.

Twelve of the past fourteen years Habitat has had a Volunteer Construction Supervisor. Two years ago, when Alachua Habitat began partially funding his salary through the CDBG program, Habitat was able to construct three more houses than in previous years. The Construction Supervisor has been able to train two Volunteer Supervisors and has gone from working on Habitat houses one day a week to three and four days a week. This in turn has given Habitat the ability to build homes in less time than was previously needed—from six to seven months, to four to five months.

How Goals and Objectives Will be Met

With the support of our full time staff, Alachua Habitat will meet its previously stated goals in the following ways:

- The Construction Supervisor will assist in demonstrating the importance of reducing utility costs through energy efficiency and solar technologies in all of our projects.
- The Office Manager/Volunteer Supervisor will help coordinate over 3,000 volunteers—including within the office, store, and on site, that will also help keep construction costs down.
- Full-time staff will work in cooperation with other housing programs to ensure that there is no unnecessary duplication of efforts.

Ultimately, the work of full time staff will help to eliminate families living in substandard housing by enabling Habitat to build around eight more houses this year.

Implementation

A Construction Supervisor and Office Manager/Volunteer Supervisor are already in place on the staff. They will be paid on a regular basis. For specific information on the implementation of the housing project, see section 15B, "The Description of the HOME Project."

15B. Description of Proposed HOME Project

Problem: Building Construction

An average of 900 volunteer days and some 5400 volunteer hours are spent constructing each home. Each home takes approximately six months to construct, and there are usually two to four houses under construction at any given time.

Alachua Habitat is currently facing a large shortage of building equipment and supplies. Not only does this hinder multiple site construction and the timely completion of homes, it forces Alachua Habitat to turn away crucial volunteers. Without these volunteers, Habitat would not be able to facilitate the building of homes for very low-income Gainesville residents and meet the community's need for affordable housing.

Solution: HOME Grant

The money received from the proposed HOME grant will enable Alachua Habitat to develop a Habitat Community by:

- Providing trailers and equipment for trailers
- Supporting survey, engineering reports, and appraisal costs
- Acquiring real property
- Purchasing building permits and Builder's Risk Insurance

- Satisfying hookup and lot clearing fees
- Providing utilities at construction sites
- Covering documentary stamps, title insurance, and other closing costs
- Building sheds, signs, driveways, sidewalks, and landscape
- Providing site control through cell phones with coordinators
- Covering salary cost of a part-time Warehouse Manager

How Goals and Objectives Will be Met

With the support of additional building equipment and volunteers, Alachua Habitat will meet its previously stated goals by:

- Providing ample equipment for volunteers at all sites, allowing for faster rates of completion
- Providing for support costs for planners, architects, and volunteers
- Purchasing tracts of land for additional home sites
- Utilizing funds for infrastructure of eight to 10 new Energy-Star homes
- Providing effective communication on site for all Construction Supervisors
- Allowing for smooth material flow to sites from the staffed warehouse

Implementation

The process of creating a Habitat Community involves the following tasks:

Task 1: The family is selected by a volunteer Family Selection Committee.

Task 2: The family performs over 200 hours of sweat equity work on other Habitat homes before they are eligible to begin work on their own home.

Task 3: A building site is designated for the particular family. For this project, the site selected will be located in Alachua County outside the incorporated limits of the City of Gainesville.

Task 4: The family meets with the Construction Supervisor to select a house design that meets Habitat's building criteria.

Task 5: Habitat secures the services of sub-contractors and obtains necessary approvals and permits.

Task 6: Under the oversight of the Building Supervisor and Construction Supervisor, the family works with other volunteer laborers (usually 25+ workers per house every Saturday) to construct the house. The Construction Supervisor coordinates sub-contracting, inspections, and acquisition of materials.

Task 7: Accounts are paid by bookkeeping staff in the office under the daily direction of the Executive Director and are then subject to a monthly review by the Treasurer of the Board of Directors.

In accordance with section 3.1 of the City's Consolidated Plan, Alachua Habitat will demonstrate the importance of reducing utility costs through energy efficiency and solar technologies. To accomplish this, the Construction Supervisor will seek the guidance of several leading scientific and research organizations and experts in the field of sustainable development, such as Energy-Star. We will also work in cooperation with other housing programs within the county, such as Neighborhood Housing Development Corporation, Community Action Agency, and Alachua Housing Authority, for referrals and information sharing. Alachua Habitat will also work with the Youth Build program to provide construction training.

16. Status of Previously Funded CDBG or HOME Project

As of February 20, 2001, Alachua Habitat has reported $5,417 for CDBG. This has been used at a rate of $1,666 per month to pay for the salary of full-time Habitat staff.

As of February 20, 2001, Alachua Habitat has reported $4,755 for Home grant. This has been used for the infrastructure of four new start homes (less than two months old) and one existing project.

The following is a list of Homes that have been reported under the current CDBG and HOME grants:

1. **Haywood Home**
   The Haywood home is an already existing house. The electrician has finished electrical rough in and Youth Build will finish the roof and siding.
2. **Kimmons Home**
   The Kimmons home is a new start. The permit has been issued and the stem wall has been built.
3. **Murray Home**
   The Murray home is a new start. The permit has been issued. Shingles, window installation, and trim installation are complete.
4. **Warrick Home**
   The Warrick home is a new start. The permit has been issued.
5. **Lockley Home**
   The Lockley home is a new start. The permit has been issued. We are preparing to clear the land and hook up water and sewer.

HomeEase for Humanity

TO: Carrie Reppert, Dave Feather, and William Wagner
FROM: Marisa Lopez, Sarah Hummer, and Alexandra Simpson
DATE: 10 April 2001
SUBJECT: Transmittal of Completion Report for Alachua Habitat PowerPoint Presentation

Attached is a print copy and disk of the PowerPoint presentation for Alachua Habitat. The presentation highlights three main points: 1) the need for housing in Alachua County, 2) the history of Habitat for Humanity, and 3) the ways in which the community can contribute to Alachua Habitat's efforts.

We familiarized ourselves with the agency's goals and expectations through talking to the office staff as well as researching and compiling information from brochures, fact sheets, and the international Habitat for Humanity website.

We hope the presentation will create an emotional response in its audiences, inspiring individuals, businesses, and community organizations to contribute their time, talent, and treasure to Alachua Habitat.

We sincerely enjoyed working with you and the other members of the Alachua Habitat staff and look forward to hearing about the outcome of our work. Thank you for giving us this great opportunity to learn new skills and serve the community.

Building Houses...Building Hope

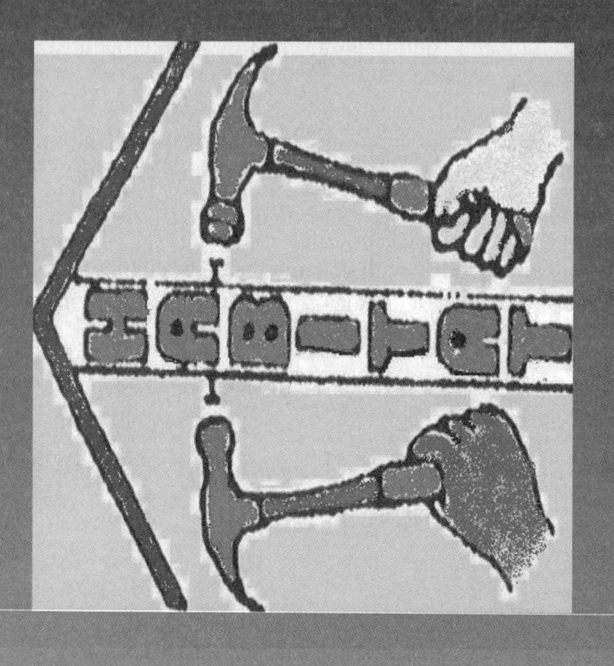

A New Lease on Life

It has been a long road for Glenda Hamilton of Jacksonville, Florida. She has been homeless for four years, and feels as though everything will cave in under the weight of illness and stress. Glenda is not alone. There are many others in our community suffering from poverty and feeling as if there is no way out.

Mission

Habitat for Humanity responds to the dire situations in our community by building simple, decent, and affordable homes in partnership with those in need.

Habitat is a grassroots Christian organization that welcomes people of all faiths and backgrounds to join in partnership with Habitat families.

The Problem of Substandard
Housing

Alachua County Facts

The problem of poverty and substandard housing is not something that we can ignore. It is a problem that exists in our own community.

- 14.4 % of people in Florida are living below the poverty level.

- Gainesville has an estimated 900 homeless people, and only two homeless shelters.

Habitat's Beginnings

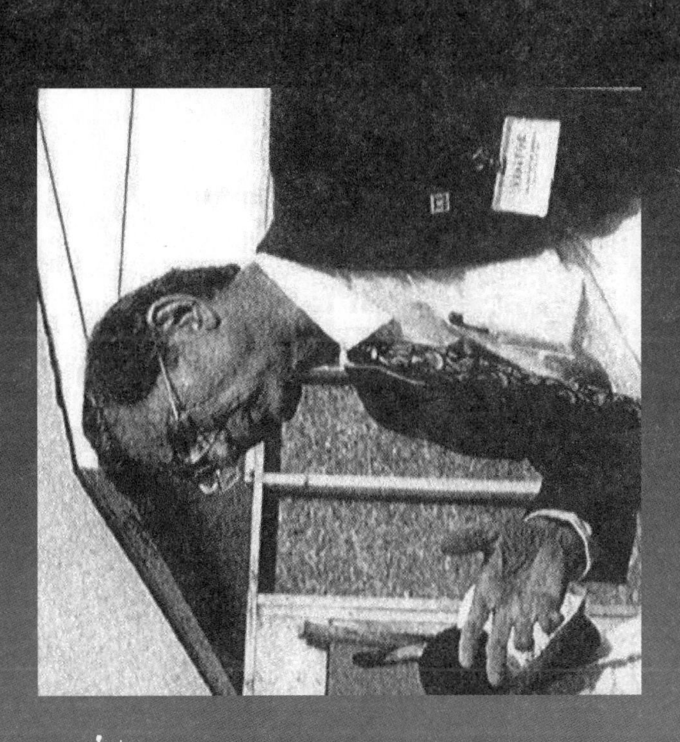

Founder Millard Fuller responded to this deficiency in 1976 when he gave up his fortune to make shelter a matter of conscience in today's world.

Continuing the Vision

Millard Fuller created Habitat for Humanity, an organization with now more than 1,400 independent affiliates in the United States that have already provided over 100,000 people with safe, decent, affordable shelter.

Alachua County Chapter

- The Alachua County chapter of Habitat for Humanity was founded in 1986.
- Since then Alachua Habitat has created 53 homes in addition to four current projects.
- Alachua Habitat plans to build a 30+ family community over the next several years.

Building Lives, Not Houses

Steps to Building a Home

1. Mutual selection process between homeowners and Habitat for Humanity.
2. Homeowners contribute 200 initial hours of "sweat equity" labor.
3. Homeowners select a Habitat site.
4. Homeowners contribute 300 final "sweat equity" hours.
5. Homeowners close on their new Habitat Home.

"Sweat Equity"
Homeowners and volunteers building communities, one house at a time.

Bringing Communities Together

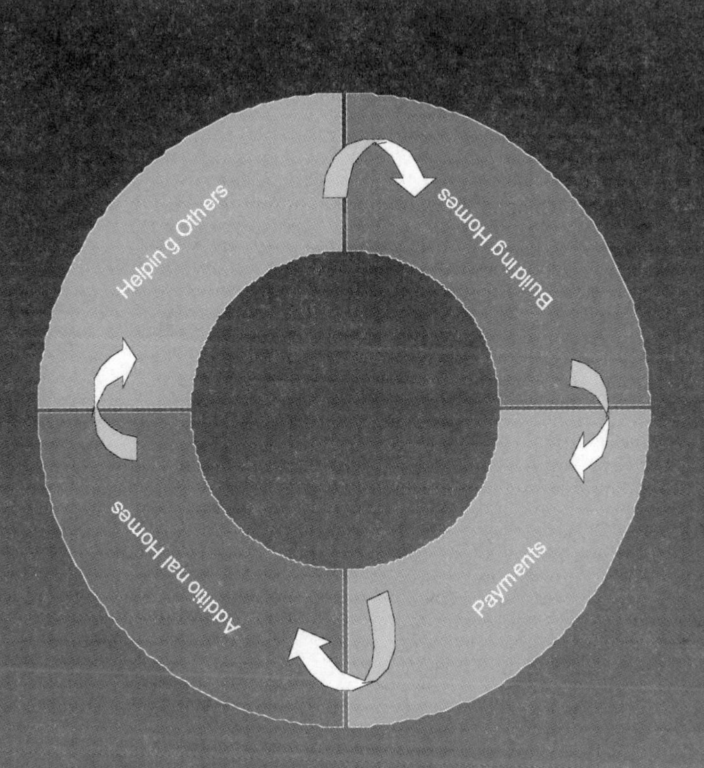

Habitat connects families in need with others in their community by using volunteers and resources to build housing at a reasonable price.

Volunteering

The driving force behind the organization is its volunteers – those who contribute a day, a week, a few months – who give Habitat's vision a tangible form. This can involve building a home, pulling weeds, answering phones, working in our thrift store, or applying any skill you have to contribute.

Getting Involved

Contributions

Individuals, religious organizations, community groups, and corporations that support the mission of Habitat for Humanity can make tax-deductible donations of money, land, and materials that make our work possible.

To learn more, or to contribute your time, talents, or gifts towards the efforts to eliminate substandard housing in our area, contact:

Alachua Habitat

2317 SW 13 Street

Gainesville, FL 32608

352-378-4663

Volunteer sessions are held at 6pm at the above address on the first Wednesday of every month.

Today, thanks to Habitat, Glenda Hamilton looks at the walls of her new home and knows that she will spend the rest of her life here.

Habitat would not exist without community support. We need your help to continue in partnership with people like Glenda Hamilton. We hope that one day all people living in poverty will be able to look back on their struggles from the security of a comfortable home.

APPENDIX C: Team Dance Marathon

Team DM consisted of three communications majors, Natasha Berk, Alexia Davis, and Natalie Day, who were enrolled in an introductory technical writing course. Each of the students lived on or near campus, so a university-connected project was ideal for them. One of the students was already affiliated with a major philanthropic effort taking place on campus, and she served as the initial contact person for the project.

Together the group produced a wide range of marketing materials for this project, the UCF Dance Marathon. They developed several documents for the organization, including a fundraising letter that was distributed to local businesses and a packet of materials for participants in the Dance Marathon that included fundraising tip sheets, donation forms, and directions for participation in an activity called Laundry Lotto. The group also put together a guide to be used by future fundraising coordinators.

In the pages that follow you will find part of the final portfolio of materials that the students submitted to their teacher, Mary Ellen Gomrad. This includes their project proposal, memo of transmittal, and selected fundraising documents.

Proposal

To: Ms. Gomrad
From: Natalie Day, Alexia Davis, and Natasha Berk
Date: 10/09/01
Subject: Collaborative Proposal

Introduction

Children's Miracle Network (CMN) began in 1983 to assist children by actively supporting better health care for kids. There are currently 170 hospitals throughout the United States that make up the Children's Miracle Network and specialize in treating children. One hundred percent of the profits raised locally go to the participating CMN hospital. Greater Orlando's CMN was established in 1997 to coordinate fundraising activities in Orange, Osceola, Seminole, and Brevard counties. Shands Children's Hospital at the University of Florida and Arnold Palmer Hospital for Children & Women share the profits raised through the Orlando CMN. More than 250,000 children go to these two hospitals for treatments for various illnesses each year.

The money raised in fundraisers helps purchase patient care equipment, fund research, and produce patient education materials. With the help of CMN, these hospitals can maintain and upgrade their pediatric programs so they can continue to provide patients with the most advanced medical treatments available. The University of Central Florida community considers this to be an important cause.

Problem Statement

Each year the University of Central Florida holds a Dance Marathon for CMN. It is a non-profit project organized by the UCF students, and all of the proceeds benefit the two central Florida hospitals. Each year a Dance Marathon committee is formed and positions are given to the elected students for tasks crucial to the development and success of the Dance Marathon. For our class project we will be assisting in promotions and fundraising for the 2002 Dance Marathon.

The Dance Marathon is run by UCF students with the help of the CMN advisors, and there is a lot of work to do in a short amount of time. Each task, idea, problem, and solution must go through three people before it is actually done. Help and volunteers are always needed to ensure that the event is a success.

Solution

As students in a technical writing class, we feel that we have the ability to assist the directors of Dance Marathon in the technical writing aspects of the event. Through making promotional material for the event, we will be able to help Dance Marathon reach their audience, which is the dancers who participate in the event. We will also help make promotional material for the fundraising events related to Dance Marathon, which will enable the dancers to raise the money needed to participate in the event.

The bulk of the group project will be to create the fundraising packet that will be distributed to the dancers when they sign up to participate in Dance Marathon. There were some guidelines for the Dance Marathon last year, but since they were not saved on a computer disk, our group has the challenge of creating a whole new one for this year. For the fundraising packet we will create a page of ten ways to raise $100. The packet will also include an informational page informing the reader of exactly what CMN is, a letter from the general director, a calendar created on Excel that will have to be updated, and a letter that the dancers will be able to give to people or companies for donations.

Other work our group project will entail is making flyers, sending out mass emails, and possibly updating the Dance Marathon website. The flyers are needed to let the community know when the organization is putting on a fundraiser. The mass emails will be needed a week in advance to let the dancers know when there are fundraising events such as car washes and cannings. The website would need to be updated because right now there is only a link to the Campus Activities Board (CAB) website but no actual website. This is also a way to reach the audience needed that has never been available to the organization.

Management Overview

As a team, our group has many attributes to give to this project. Natalie Day is the fundraising director for the 2002 Dance Marathon, so she will be the direct link to the Dance Marathon Board. She meets with the board every Thursday and emails her updates on Sunday. Her reports discuss how Dance Marathon fundraising is progressing, so she will know when promotional material is needed. She has also worked on other community service events making promotional material for Mothers Against Drunk Driving and Susan G. Komen events. Her computer skills include Microsoft XP, Word, and Excel.

Alexia Davis's background in public relations and fundraising will add experience and maturity to the group. Her writing experience will help as we create promotional materials essential to recruiting dancers for the event. Her personal goal is to educate others about CMN while promoting the Dance Marathon. Alexia is confident that she will be able to persuade many people to participate. In 1999, she had an internship in the Team Charities Department with the Orlando Magic/RDV Sports organization. Her responsibilities included assisting in the organization of the 8th annual Orlando Magic/Outback Steakhouse Golf Tournament. She organized the golfers' sign-up and payment process and kept track of what corporate fundraising package they purchased. This event is an Orlando Magic tradition and the proceeds go to the Orlando Magic Youth Foundation, similar to the CMN. With her PR and corporate fundraising background, she knows she can help bring about the success of the Dance Marathon.

Natasha Berk will bring her writing skills, her creativity, and her experience in fundraising activities that she has gained through her sorority's philanthropy events. As a member of Zeta Tau Alpha, Natasha was the co-chair for King of the Campus. The goal for the fundraiser was to raise money for Susan G. Komen Breast Cancer Awareness. She met with each fraternity on UCF's campus and had them select one member who would represent their chapter. Natasha then went out into the community looking for sponsorships and donations for the event. The event was a huge success as over 20 businesses in the Orlando area donated either monetarily or physically to the fundraiser. Natasha also has a very extensive writing background. She is currently a writer for the *Central Florida Future.* She works firsthand with meeting deadlines and writing effectively. Due to her marketing minor, Natasha will be able to aid the group in coming up with creative ways to raise money for CMN. Natasha was also a dancer her freshman year for Dance Marathon and knows exactly how important it is to UCF and the community.

Budget and Production

The funds the group will spend on making promotional material will be minimal. We estimate a range from fifteen to thirty dollars. The costs foreseen will be making copies from a copy center, paper, and ink jet cartridges for our computers. We will gladly pay for these items out of our own pockets. Any additional costs will be paid for out of the fundraising money allotted for Dance Marathon by the University of Central Florida's Student Government.

Schedule

The calendar below demonstrates our plan of action on this project for the remainder of the semester. Right now it only includes our planned meeting times and due dates. We will work from this calendar as we add new responsibilities and activities. Each of us has a copy of the calendar on our computers and we will work together to keep it updated and complete.

October 2001						
Sunday	Monday	Tuesday	Wednesday	Thursday	Friday	Saturday
30	1 Group meeting- -library	2	3 Group meeting- -library	4	5	6
7	8 Group meeting- -library	9 Final Proposal Due	10 meeting- -library	11	12	13
14	15 Group meeting- -library	16	17Group meeting- -library	18 Dis-course analysis memo due	19	20
21	22 Group meeting- -library	23	24 Group meeting- -library	25 Prelim. style sheet due	26	27
28	29 Group meeting- -library	30				

November 2001						
Sunday	Monday	Tuesday	Wednesday	Thursday	Friday	Saturday
				1 First draft and progress report due	2	3
4	5	6	7 Group meeting--library	8	9	10
11	12 Group meeting--library	13 Style sheet revision due	14	15 Mand-atory group con-ference	16	17
18	19 Group meeting--library	20	21	22 Field journal and per-formance report due	23	24
25	26 Group meeting--library	27 Class Presenta-tion	28	29 Final Project Due	30	

Conclusion

As technical writers we hope to be successful in the creation of promotional materials. The Dance Marathon is a great organization, and we hope we can do whatever is needed to help the less fortunate children of Central Florida.

Cover Memo

To: Ms. Gomrad
From: Team DM
Date: November 29, 2001
Re: Final Project

For our class project we developed fundraising materials for the 2002 University of Central Florida Dance Marathon. Our packet contains each of our major documents as detailed below.

The first project we worked on was a letter to companies around the area such as Publix, Wal-Mart, Uno Chicago Bar & Grill, and Applebee's. We used a general letter and then adapted it to each one of the companies, specifying what we needed.

The second project we worked on as a team was promotional materials for the Dance Marathon's first fundraiser, Laundry Lotto. With help of a graphic design artist, we made flyers, information sheets, and covers for the envelopes and handed them out to the people who sold raffle tickets. After the Laundry Lotto fundraiser was completed, we created instructions explaining how to "Put on Laundry Lotto" for next year's fundraising director and anyone else who wants to know what is involved with the fundraiser.

Our third project was making the fundraising packet to hand out to the dancers. The packet includes a front page, a letter from the fundraising director, a calendar of events, a page explaining the main fundraising events, a story about one of the miracle children, a page suggesting strategies for raising money, and a letter for the dancers to send to their families to get contributions. The fundraising packet was approved by the Dance Marathon advisor and general director on November 1, 2001. The packet was handed out at the first informational meeting about Dance Marathon on November 2, 2001.

Our final project was a speech that Natasha wrote for the Fundraising Director to give at the first informational meeting about Dance Marathon. The speech went very well.

We've also included a copy of our class presentation on the project. We hope you enjoy this and the entire contents of our project.

Team DM's Final Draft

By: Alexia Davis, Natasha Berk,
and Natalie Day
11/29/01

November 20, 2001

To: Field Marketing Manager
Re: **SUPPORT UCF TO HELP HOSPITALIZED KIDS**

On February 23 and 24, 2002, the University of Central Florida will hold its sixth annual Dance Marathon to benefit Greater Orlando Children's Miracle Network (CMN). We need your help! My name is Natalie Day and I am the fundraising director for this year's Dance Marathon. I am writing to ask your support for this important project.

Proceeds from the UCF Dance Marathon will stay here in Central Florida to provide vital support to Shands Children's Hospital at the University of Florida and Arnold Palmer Hospital for Children & Women in Orlando. Together these hospitals treat more than 250,000 children each year from Central Florida and beyond. Their patients are children suffering from cancer, AIDS, heart disease, and birth defects. They are accident victims and transplant patients. And they are children living with chronic illnesses like diabetes, kidney disease, and cystic fibrosis.

Last year UCF raised nearly $20,000 for CMN. This year we hope to raise more money than ever to help our hospitals help kids. You can help by allowing us to wrap presents for donations, or if you would like to be more involved by becoming a sponsor of Dance Marathon, please contact me at the UCF Campus Activities Board number—(407) 555-6471—or via email at Natucf@email.com.

University of Central Florida students are giving up 18 hours out of their lives to help save 80 years of another's, but they need your help raising the money to dance in Dance Marathon. Your contribution will make a difference in the lives of thousands of children, providing vital support that will help Shands Children's Hospital and Arnold Palmer Hospital continue to give the most advanced level of health care for kids. Thank you in advance for your consideration.

Sincerely,
Natalie Day
2002 Dance Marathon Fundraising Director

Company Letter

How to Put on Laundry Lotto

Laundry Lotto is a raffle targeted at students in dorms and Greek houses. Students will donate 50¢ to put in a raffle for $20 worth of quarters. The prize money comes out of the donations raised.

Supplies

150 envelopes
150 paper clips
150 flyers to glue onto envelopes
150 information sheets to clip on envelopes
500 flyers—both small and large, with and without "see RA/president"
2 rolls of 2000 double tickets—put 20 tickets in each envelope
1 glue stick

Suggestions

- Have public relations or marketing committees or officers make flyers.
- Turn in paper materials to advisor to look at two to three weeks in advance.
- Turn in request for the paper material to be copied two weeks in advance.
- Promote Laundry Lotto a week in advance.
- Call all the contacts.
- Produce material a week in advance to hand out to contacts.
- Possibly make laundry lotto a two-week fundraiser.
- Have enough people to collect envelopes.
- Tell organizations and RAs that it is only $10 worth of tickets and it is not that hard to sell them. Really get the word out.
- Possibly have more than one prize.
- Send thank-you notes after the event.

Contacts

Apollo/Knights Court/Libra Area Coordinator—Jason Jones (407)555-5016
Knight's Krossing Area Coordinator—T.J. Greggs (407)555-5136
Lake Claire Area Coordinator
Presidents of Greek organizations

Results

Made: 125 envelopes
Handed out: 102 envelopes
Received back: 39 envelopes
Costs: $40.76
Raised: $228.89

Laundry Lotto Materials

Laundry Lotto Information Sheet

We would like to thank you for participating in Dance Marathon's first fundraiser of the year. We are so excited and know that with your help, Laundry Lotto will be a success.

The winner of Laundry Lotto will receive $20 in quarters. This will make doing laundry more convenient for those who live in Greek houses, Lake Claire, or the Residence Halls. But this fundraiser is not limited to those who use quarters to do laundry—anyone can play. Quarters can also be used for tolls or vending machines.

In this envelope, we have included raffle tickets, flyers, and a plastic bag. The raffle tickets are 50¢ each. We ask that the person's name and phone number be written on the ticket. The drawing is on Friday, October 12th, and the winner will be notified at 6 p.m. by telephone. When you complete your sales, place the money and the ticket stubs in the plastic bag. I will pick up the money and envelope on the 12th before 5 p.m. Please call me and let me know what time is convenient to pick up the money. If you have any questions or need more raffle tickets, please call Natalie Day at the CAB office—(407)555–6471.

DANCE MARATHON

LAUNDRY LOTTO

Make a 50¢ donation to win $20 in quarters to do your laundry! All donations go to...

Drawing October 12th

Please see your Resident Advisor or club president for tickets!

UCF DANCE MARATHON 2002

Dear Resident Advisors:

On behalf of the Dance Marathon Executive Board, I would like to thank each of you for your dedicated support and assistance with the Laundry Lotto fundraiser. Because of the work you did to promote the project and inform your residents and friends about it, we were able to raise more than $200, which will of course go towards our fundraising efforts for the Children's Miracle Network.

The Dance Marathon committee of the Campus Activities Board works to coordinate this annual Children's Miracle Network event, working hard to promote and educate the entire UCF campus and Orlando community. The Dance Marathon Executive Board and its committee members coordinate fundraising events, catering, and entertaining activities. All the work done helps the committee to raise funds for the Children's Miracle Network while having fun.

Dance Marathon would not happen without the endless support and hard work of these volunteers, including you. Please note that your work and support have not only helped us to raise funds, but have brought smiles to the faces of the many families whom Dance Marathon benefits.

Thanks again for all of your support and assistance with the Laundry Lotto.

For the kids,

Natalie Day
Fundraising Director
2002 Dance Marathon

Hey Everyone!

I am so excited about Dance Marathon this year. It is our goal to raise more money this year than ever before. I know with your help Dance Marathon will be a success! We can't do it without you!

In this packet, I have included a calendar of the fun-raising events planned for this year (fun-raising because we are going to have so much fun while raising money to save kids' lives!). This calendar is subject to change. Throughout the year, you will be informed about extra events that have been planned. Also included in the packet is a sheet explaining the events on the calendar, a story about a child, and a list of fundraising tips. I have also included a form letter to send to family members and businesses. These letters include info about Dance Marathon and ask for assistance in the fun-raising process.

Raising money this year is going to be so easy! A portion of the money raised at each event you participate in will go towards the $100 you have to raise in order to participate in Dance Marathon. So even if it is difficult for you to raise the money on your own, there will be many opportunities for you to raise it with a group effort. If you have any questions or concerns or are interested in working on our committee, please contact me. My email address is Natucf@email.com. This is going to be a wonderful year. We are going to raise a lot of money and have a lot of fun while benefiting children's lives. I am happy that you are participating in Dance Marathon. Thank you for helping to make miracles happen.

For the kids,

Natalie Day
2002 Director of Fun-Raising

Fundraising Packet

FUN RAISING EVENTS

- Laundry Lotto—We will be selling 50¢ raffle tickets for a drawing to win $20 in quarters to do laundry.
- Trick or treat for change—We will go around on Halloween from 6-9pm collecting change from local neighborhoods.
- Jail-n-Bail—We will arrest the presidents of the Greek organizations on campus, and other students, and put them in jail until they can come up with bail money.
- Letter Drop—We will be going to area neighborhoods dropping letters on doorsteps. The letters will explain Dance Marathon and we will attach a return address so donations can be sent to us.
- Links— Links are little strips of paper we are selling to our supporters. People can buy the links and write notes to their friends who are dancing. The links will be hung around the arena, and we will have designated times during the marathon to read them.
- Thanks"Giving"—We will go around to local neighborhoods for change the week before Thanksgiving break.
- Season's Greetings—We will visit local neighborhoods asking for change the 1st week of December.
- DM Invasion—Have a 3-day invasion of the UCF campus with canning in front of the SU, library, and administration the week before Dance Marathon.
- Canning at local stores.

September 2001

Sun	Mon	Tue	Wed	Thur	Fri	Sat
						1
2	3	4	5	6	7	8
9	10	11	12	13	14	15
16	17	18	19	20	21	22
23 DM Retreat 8:30 to 4:00	24	25	26	27	28	29

October 2001

Sun	Mon	Tue	Wed	Thur	Fri	Sat
	1	2	3 1st Recruit-ment	4	5	6 Family Weekend UCF game
7 Laundry Lotto	8 Laundry Lotto	9 Laundry Lotto	10 Laundry Lotto	11	12 Laundry Lotto	13 UCF football game
14	15 Homecoming Week	16	17 Homecoming Week	18	19	20 Home-coming football game
21	22	23	24	25	26	27 Canning
28	29 Flyer Neighbor-hoods	30	31 Halloween Trick or Treat for Change			

November 2001

Sun	Mon	Tue	Wed	Thur	Fri	Sat
				1	2	3 UCF football game
4	5	6	7 Jail-n-Bail	8	9	10 Canning
11	12 Vetrans Day	13	14	15	16	17
18	19	20 Eat at Applebees	21	22 Thanks-giving	23 Thanks-giving Break	24 UCF football game
25	26 Flyer Neighbors	27	28 Sing Holiday Carols for Change	29	30	

December 2001

Sun	Mon	Tue	Wed	Thur	Fri	Sat
						1
2	3	4 Final Exam Week	5	6	7 Final Exam Week	8
9	10	11 Christmas Break Begins	12	13	14	15
16	17	18 Eat at Applebees	19	20	21	22
23	24	25 Christmas	26	27	28	29
30	31 New Year's Eve					

Colton
Sanford, Florida

Colton has the energy of an entire preschool. And while the same may be said of many 4-year-olds, chances are they haven't had to struggle for their lives like Colton has.

When he was one, Colton was diagnosed with a rare life-threatening heart condition that caused his aorta and the arteries that branch from it to narrow, preventing sufficient blood flow to his vital organs. As Colton grew, his arteries stretched even further and his health deteriorated quickly. His life was hanging by a thread. To save him, surgeons at Arnold Palmer Hospital used an artificial patch system in an innovative way, restoring Colton's blood flow and making medical history as they performed the delicate operation for the first time.

If you look for Colton now, you'll probably find this little whirlwind wrestling with his big brother, Austin, and squeezing all the fun he can into every single day.

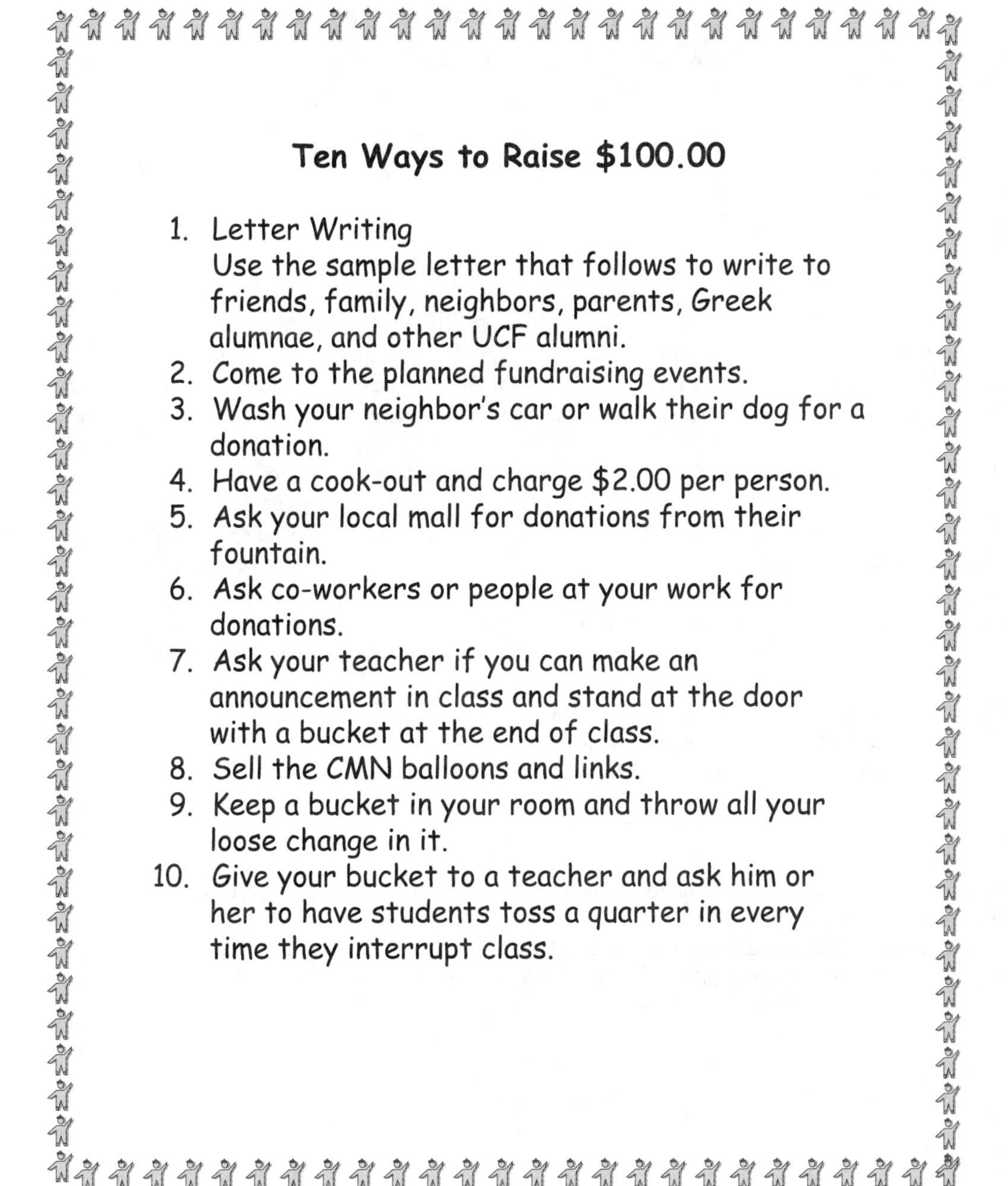

Ten Ways to Raise $100.00

1. Letter Writing
   Use the sample letter that follows to write to friends, family, neighbors, parents, Greek alumnae, and other UCF alumni.
2. Come to the planned fundraising events.
3. Wash your neighbor's car or walk their dog for a donation.
4. Have a cook-out and charge $2.00 per person.
5. Ask your local mall for donations from their fountain.
6. Ask co-workers or people at your work for donations.
7. Ask your teacher if you can make an announcement in class and stand at the door with a bucket at the end of class.
8. Sell the CMN balloons and links.
9. Keep a bucket in your room and throw all your loose change in it.
10. Give your bucket to a teacher and ask him or her to have students toss a quarter in every time they interrupt class.

October 16, 2001

Dear Friends and Family:

On February 23 and 24, 2002, I will be participating in the University of Central Florida Dance Marathon to benefit Greater Orlando Children's Miracle Network. Dance Marathon is UCF's only completely student-organized, student-run, campus-wide philanthropic event. Hundreds of students will raise funds by staying on their feet for 18 hours to benefit critically ill children treated at Shands Children's Hospital at the University of Florida and Arnold Palmer Hospital for Children and Women.

These hospitals treat children who are accident victims and who have all types of other afflictions, including cancer, AIDS, heart and muscular diseases, cerebral palsy, and birth defects. Shands Children's Hospital and Arnold Palmer Hospital are dedicated to treating the whole child, rendering service based on need, not the ability to pay. The funds the hospitals receive will be used to purchase patient care equipment, to sponsor educational programs for patients and the community, and to purchase diversionary items such as red wagons, computer games, and stationary bicycles. In addition, funds will support research that will one day lead to cures for many diseases that afflict children.

It's a worthwhile cause and I need your help! Dancers are asked to collect pledges in advance, and I would appreciate it if you could sponsor me for this event by donating $10, $25, $50, or any amount that you feel comfortable with. Please help me make a difference in the life of a child.

Sincerely,

P.S. If you would like more information about Greater Orlando Children's Miracle Network, visit them on the web at www.gocmn.org.

Mail this form to the following address with a check or money order:

CAB Office	Name______________________________
Dance Marathon, 2002	Address____________________________
P.O. Box 163245	___________________________________
Orlando, FL 32816-3245	Amount attached $____________________
	Dancer/Organization __________________

Please make your check or money order payable to Dance Marathon.

CREDITS

"Other Voices"

- 1.A Michael Moore
- 2.A Bill Wood
- 2.B Thomas P. Miller
- 4.A Andrea Pacini
- 4.B Barbara Heifferon
- 5.A Arthur Padilla
- 5.B Nathan Trenteseaux
- 6.A Summer Smith
- 6.B Sheila Cole
- 7.A Leila Cantara
- 7.B Karen Peters
- 8.A Erica Olmstead and Amber Feldman Goldberg
- 8.B Kristen Kennedy
- 9.A Karla Kitalong
- 9.B Chere Peguesse
- 10.A Maggie Boreman
- 10.B J. Fred Reynolds

Student Material

Chapter 4

Kristianna Fallows, Résumé and Letter of Inquiry

Chapter 5

Maggie Boreman, Proposal

Chapter 6

Kathleen Capozza, Field Journal
Catherine McKeown, Field Journal

Chapter 7

Grace Delorito, Analysis of United Way Discourse Community and Conventions
John Brock, Analysis of United Way Discourse Community and Conventions
Jennifer O'Driscoll, Analysis of United Way Discourse Community and Conventions
Leila Cantara, University of Florida Community Campaign Newsletter Style Sheet
Lisa Pisano, University of Florida Community Campaign Newsletter Style Sheet
Leila Cantara, University of Florida Community Campaign Newsletter. Reprinted by permission of James B. Morgan.
Valerie Hall, Web Style Guide
Rick Hartig, Web Style Guide
Bruce Hickman, Web Style Guide

Chapter 8

Loretta Belfiore, Progress Report
Ryan Pedraza, Progress Report
Katie Roland, Progress Report
Sharon Sandman, Progress Report
Bart Leahy, Progress Report

Chapter 9

Kristi Shook, User Test Report
Katherine Plumlee, Evaluation Report of American Cancer Society Documents
Jason Ryan Law, Evaluation Report of American Cancer Society Documents
Matt Lily, Evaluation Report of American Cancer Society Documents

Chapter 10

Katherine Plumlee, Transmittal Memo
Jason Ryan Law, Transmittal Memo and Résumé
Matt Lily, Transmittal Memo

Appendix A

Loretta Belfiore
Ryan Pedraza
Katie Roland
Sharon Sandman

Appendix B

Sara Hummer
Marisa Lopez
Alexandra Simpson

Appendix C

Natasha Berk
Alexia Davis
Natalie Day

INDEX